# Nonlinear Dispersive Waves

The field of nonlinear dispersive waves has developed enormously since the work of
Stokes, Boussinesq, and Korteweg and de Vries (KdV) in the nineteenth century. In the
1960s researchers developed effective asymptotic methods for deriving nonlinear wave
equations, such as the KdV equation, governing a broad class of physical phenomena.
These equations admit special solutions including those commonly known as solitons.

This book describes the underlying approximation techniques and methods for
finding solutions to these and other equations, such as the nonlinear Schrödinger,
sine–Gordon, Kadomtsev–Petviashvili and Burgers equations. The concepts and
methods covered include wave dispersion, asymptotic analysis, perturbation theory,
the method of multiple scales, deep and shallow water waves, nonlinear optics
including fiber optic communications, mode-locked lasers and dispersion-managed
wave phenomena. Most chapters feature exercise sets, making the book suitable for
advanced courses or for self-directed learning. Graduate students and researchers will
find this an excellent entry to a thriving area at the intersection of applied
mathematics, engineering and physical science.

MARK J. ABLOWITZ is Professor of Applied Mathematics at the University of
Colorado at Boulder.

# Cambridge Texts in Applied Mathematics

All titles listed below can be obtained from good booksellers or from Cambridge University Press. For a complete series listing, visit www.cambridge.org/mathematics

# Nonlinear Dispersive Waves

## Asymptotic Analysis and Solitons

MARK J. ABLOWITZ
*University of Colorado, Boulder*

CAMBRIDGE
UNIVERSITY PRESS

# CAMBRIDGE
## UNIVERSITY PRESS

Shaftesbury Road, Cambridge CB2 8EA, United Kingdom

One Liberty Plaza, 20th Floor, New York, NY 10006, USA

477 Williamstown Road, Port Melbourne, VIC 3207, Australia

314–321, 3rd Floor, Plot 3, Splendor Forum, Jasola District Centre, New Delhi – 110025, India

103 Penang Road, #05–06/07, Visioncrest Commercial, Singapore 238467

Cambridge University Press is part of Cambridge University Press & Assessment, a department of the University of Cambridge.

We share the University's mission to contribute to society through the pursuit of education, learning and research at the highest international levels of excellence.

www.cambridge.org
Information on this title: www.cambridge.org/9781107664104

First published 2011

*A catalogue record for this publication is available from the British Library*

*Library of Congress Cataloging-in-Publication data*
Ablowitz, Mark J.
Nonlinear dispersive waves : asymptotic analysis and solitons / Mark J. Ablowitz.
p. cm. – (Cambridge texts in applied mathematics ; 47)
Includes bibliographical references and index.
ISBN 978-1-107-01254-7 (hardback) – ISBN 978-1-107-66410-4 (pbk.)
1. Wave equation. 2. Nonlinear waves. 3. Solitons. 4. Asymptotic
expansions. I. Title. II. Series.
QC174.26.W28A264 2011
530.15´5355–dc23
2011023918

ISBN    978-1-107-01254-7    Hardback
ISBN    978-1-107-66410-4    Paperback

# Contents

v

# Preface

The field of nonlinear dispersive waves has developed rapidly over the past 50 years. Its roots go back to the work of Stokes in 1847, Boussinesq in the 1870s and Korteweg and de Vries (KdV) in 1895, all of whom studied water wave problems. In the early 1960s researchers developed effective asymptotic methods, such as the method of multiple scales, that allow one to obtain nonlinear wave equations such as the KdV equation and the nonlinear Schrödinger (NLS) equation, as leading-order asymptotic equations governing a broad class of physical phenomena. Indeed, we now know that both the KdV and NLS equations are "universal" models. It can be shown that KdV-type equations arise whenever we have weakly dispersive and weakly nonlinear systems as the governing system. On the other hand, NLS equations arise from quasi-monochromatic and weakly nonlinear systems.

The discovery of solitons associated with the KdV equation in 1965 by Zabusky and Kruskal was a major development. They employed a synergistic approach: computational methods and analytical insight. This was soon followed by a remarkable publication in 1967 by Gardner, Greene, Kruskal and Miura that described the analytical method of solution to the KdV equation, with rapidly decaying initial data. They employed concepts of direct and inverse scattering in the solution of the KdV equation that was perceived by researchers then as nothing short of astonishing. It was the first time such a higher-order nonlinear dispersive wave equation (the KdV equation is third order in space and first order in time) was "solved" or linearized; moreover it was shown how solitons were related to discrete eigenvalues of the time-independent Schrödinger scattering problem. The question of whether this was a single event, i.e., special only to the KdV equation, was answered just a few years later. In 1971 Zakharov and Shabat, using ideas developed by Lax in 1968, obtained the method of solution to the NLS equation with rapidly decaying data. Their solution method also used direct and inverse scattering.

In 1973–1974 Ablowitz, Kaup, Newell and Segur showed that the methods
used to solve the KdV and NLS equations applied to a class of nonlinear wave
equations including physically important equations such as the modified KdV
and sine–Gordon equations. They also showed that the technique was a natural
generalization of the linear method of Fourier transforms. They termed the procedure the inverse scattering transform or IST. Subsequently researchers have
found wide classes of equations, including numerous physically interesting
nonlinear wave equations, solvable by IST, including higher-order PDEs in one
space and one time dimension, multidimensional systems, discrete systems –
i.e., differential–difference and partial difference equations and even singular integral equations. Solutions to the periodic initial value problem, direct
methods to obtain soliton solutions, conservation laws, Hamiltonian structures associated with these equations, and much more, have been obtained.
The development of IST has also motivated researchers to study many of these
and related equations by functional analytic methods in order to establish local,
and whenever feasible, global existence of solutions to the relevant initial value
problems.

   On the other hand, whenever physicists and engineers need to study a specific class of nonlinear wave equations, they invariably consider and frequently
employ direct numerical simulation. This has the advantage of being applicable to a wide class of systems and is often readily carried out. But for complex
multidimensional physical problems it can be extremely difficult or essentially
impossible to carry out direct simulations. For example, researchers in optical
communication rely on asymptotic reductions of Maxwell's equations (with
nonlinear polarization terms) to fundamental NLS models because the scales of
the dynamics differ enormously: indeed by many orders of magnitude ($10^{15}$).
Once an asymptotic model is developed, direct numerical methods are usually
feasible. However, to obtain general information related to specific classes of
solutions, such as solitons or solitary waves, one often finds that an analytically
based approach is highly desirable. Otherwise covering a range of interesting
parameter values becomes a long and arduous chore.

   This book aims to put into perspective concepts and asymptotic methods
that researchers have found useful both for deriving important reduced asymptotic equations from physically significant models as well as for analyzing the
asymptotic equations and solutions under perturbations.

   Part I contains Chapters 1–7; here the fundamental aspects and basic
applications of nonlinear waves and asymptotic analysis are discussed. Also
included is some discussion of linear waves in order to help set ideas and concepts regarding nonlinear waves. Part II consists of Chapters 8 and 9. Here,

the notion of exact solvability or integrability via associated linear compatible systems and the method of the inverse scattering transform (IST) is described. Each of the Chapters 1–9 has exercises that can be used for homework problems or may be considered by the reader as encouraging additional practice and thought. Part III contains applications of nonlinear waves. The material is by and large more recent in nature than Parts I and II. However, the mathematical methods and asymptotic analysis are similar to what has been developed earlier. In most respects the reader will not find the work technically difficult. Indeed the concepts often follow naturally and expand the scope and breadth of our understanding of nonlinear wave phenomena.

A more detailed outline of this book is as follows.

Chapter 1 introduces the Korteweg–de Vries equation and the soliton concept from a historical perspective via the system of anharmonic oscillators originally studied by Fermi, Pasta and Ulam (FPU) in 1955. Kruskal and Zabusky (1965) showed how the KdV equation resulted from the FPU problem and they discussed why the soliton concept of "elastic interaction" explains the recurrence of initial states observed by FPU. In recent years many researchers have adopted the term soliton when they refer to a localized wave, and not necessarily one that maintains its speed/amplitude upon interaction. We will often use the more general notion when discussing physical problems. This chapter also gives additional historical background and examples.

Chapter 2 briefly discusses linear waves, the notion of dispersive and non-dispersive wave systems, the technique of Fourier transforms, the method of characteristics and well-posedness.

Chapter 3 employs asymptotic methods of integrals to analyze the long-time asymptotic solution of linear dispersive wave systems. For the linear KdV equation it is shown that the long-time solution has three regions: exponential decay that matches to an Airy function connection region that in turn matches to a region with decaying oscillations. It is also shown how to extend Fourier analysis to linear differential–difference evolution systems.

Chapter 4 introduces perturbation methods, in particular the method of multiple scales and variants such as the Stokes–Poincaré frequency shift, in the context of ordinary differential equations. Linear and nonlinear equations are investigated including the nonlinear pendulum with slowly varying driving frequency.

In Chapter 5 the equations of water waves are introduced. In the limit of weak nonlinearity and long waves, i.e., shallow water, the KdV equation is derived. The extension to multidimensions of KdV, called the Kadomtsev–Petviashvili (KP) equation, is also discussed.

Nonlinear Schrödinger models are described in Chapter 6. The NLS equation is first derived from a model nonlinear Klein equation. Derivations of NLS equations from water waves in deep water with weak nonlinearity are outlined and some of the properties of NLS equations are described.

Chapter 7 introduces Maxwell's equations with nonlinear polarization terms such as those that arise in the context of nonlinear optics. The derivation of the NLS equation in bulk media is outlined. A brief discussion of how the NLS equation arises in the context of ferromagnetics is also included.

Although the primary focus of this book is directed towards physical problems and methods, the notion of integrable equations and solitons is still extremely useful, especially as a guide. In Chapters 8 and 9 some background information is given about these interesting systems. Chapter 8 shows how the Korteweg–de Vries (KdV), nonlinear Schrödinger (NLS), mKdV, sine–Gordon and other equations can be viewed as a compatibility condition of two linear equations: a linear scattering problem and associated linear time evolution equation under "isospectrality" (constancy of eigenvalues). In Chapter 9 the description of how one can obtain a linearization of these equations is given. It is shown how the solitons are related to eigenvalues of the linear scattering problem. The method is referred to as the inverse scattering transform (IST).

In Chapters 10 and 11 two applications of nonlinear optics are discussed: optical communications and mode-locked lasers. These areas are closely related and NLS equations play a central role.

In communications, NLS equations supplemented with rapidly varying coefficients that take into account damping, gain and dispersion variation is the relevant physically interesting asymptotic system. The latter is associated with the technology of dispersion-management (DM), i.e., the fusing together of optical fibers of substantially different, opposite in sign, dispersion coefficients. Dispersion-management, which is now used in commercial systems, significantly reduces penalties due to noise and multi-pulse interactions in wavelength division multiplexed (WDM) systems. WDM is the technology of the simultaneous transmission of pulses centered in widely separated frequency "windows". The analysis of these NLS systems centrally involves asymptotic analysis, in particular the technique of multiple scales. A key equation associated with DM systems is derived by the multiple-scales method. It is a non-local NLS-type equation that is referred to as the DMNLS equation. For these DM systems special solutions such as dispersion-managed solitons can be obtained and interaction phenomena are discussed.

The study of mode-locked lasers involves the study of NLS equations with saturable gain, filtering and loss terms. In many cases use of dispersion-management is useful. A well-known model, called the master equation,

Taylor-expands the saturable power terms in the loss. It is found that keeping the full saturable loss model leads to mode-locking over wide parameter regimes for constant as well as dispersion-managed models. This equation provides insight to the phenomena that can occur. Localized modes and strings of solitons are found in the anomalous and normal dispersive regimes.

# Acknowledgements

These notes were developed originally for the course on Nonlinear Waves (APPM 7300) at the University of Colorado, taught by M.J. Ablowitz. The notes were written during the 2003–2004 academic year with important contributions and help from Drs C. Ahrens, M. Hoefer, B. Ilan and Y. Zhu. Sincere thanks are also due to Douglas Baldwin who carefully read and helped improve the manuscript. The author is deeply thankful to Dr. T.P. Horikis who improved and enhanced the first draft of the notes during the 2006–2007 and 2007–2008 academic years, and for his major efforts during the spring 2010 semester. Appreciation is due to the National Science Foundation and the Air Force Office of Scientific Research (AFOSR) who have supported the research that forms the underpinnings of this book. Deep thanks are due to Dr. Arje Nachman, Program Director AFOSR, for his vital and continuing support.

# PART I

FUNDAMENTALS AND BASIC
APPLICATIONS

# 1

## Introduction

In 1955 Fermi, Pasta and Ulam (FPU) (Fermi et al., 1955) and Tsingou (see Douxois, 2008) undertook a numerical study of a one-dimensional anharmonic (nonlinear) lattice. They thought that due to the nonlinear coupling, any smooth initial state would eventually lead to an equipartition of energy, i.e., a smooth state would eventually lead to a state whose harmonics would have equal energies. In fact, they did not see this in their calculations. What they found is that the solution nearly recurred and the energy remained in the lower modes.

To quote them (Fermi et al., 1955):

The results of our computations show features which were, from beginning to end, surprising to us. Instead of a gradual, continuous flow of energy from the first mode to the higher modes, ... the energy is exchanged, essentially, among only a few. ... There seems to be little if any tendency toward equipartition of energy among all the degrees of freedom at a given time. In other words, the systems certainly do not show mixing.

Their model consisted of a nonlinear spring–mass system (see Figure 1.1) with the force law: $F(\Delta) = -k(\Delta + \alpha \, \Delta^2)$, where $\Delta$ is the displacement between the masses, $k > 0$ is constant, and $\alpha$ is the nonlinear coefficient. Using Newton's second law and the above nonlinear force law, one obtains the following equation governing the longitudinal displacements:

$$m\ddot{y}_i = k\left[(y_{i+1} - y_i) + \alpha(y_{i+1} - y_i)^2\right] - k\left[(y_i - y_{i-1}) + \alpha(y_i - y_{i-1})^2\right],$$

where $i = 1, \ldots, N - 1$, $y_i$ are the longitudinal displacements of the $i$th mass, and $(\dot{\,}) = d/dt$. Rewriting the right-hand side leads to

$$m\ddot{y}_i = k(y_{i+1} - 2y_i + y_{i-1}) + k\alpha\left[(y_{i+1} - y_i)^2 - (y_i - y_{i-1})^2\right],$$

which can be further rewritten as

$$\frac{m}{k}\ddot{y}_i = \hat{\delta}^2 y_i + \alpha\left[(y_{i+1} - y_i)^2 - (y_i - y_{i-1})^2\right], \qquad (1.1)$$

3

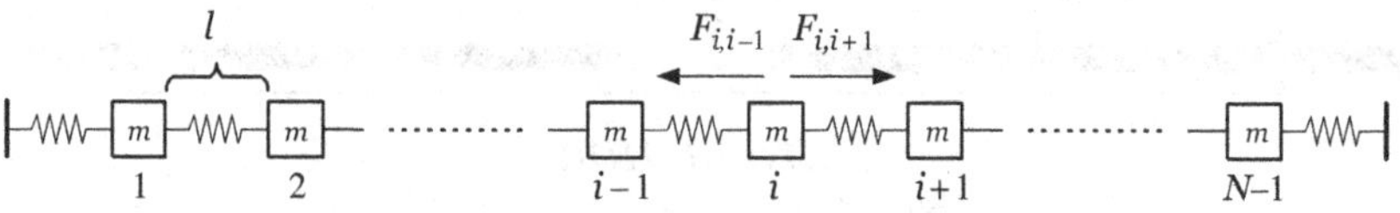

Figure 1.1 Fermi–Pasta–Ulam mass–spring system.

where the operator $\hat{\delta}^2 y_i$ is defined as

$$\hat{\delta}^2 y_i \equiv (y_{i+1} - 2y_i + y_{i-1}).$$

Equation (1.1) is referred to as the FPU equation. Note that if $\alpha = 0$, then (1.1) reduces to the discrete wave equation

$$\frac{m}{k}\ddot{y}_i = \hat{\delta}^2 y_i.$$

The boundary conditions are usually chosen to be either fixed displacements, i.e., $y_0(t) = y_N(t) = 0$; or as periodic ones, $y_0(t) = y_N(t)$ and $\dot{y}_0(t) = \dot{y}_N(t)$; the initial conditions are given for $y_i(t = 0)$ and $\dot{y}_i(t = 0)$. Fermi, Pasta and Ulam chose $N = 65$ and the sinusoidal initial condition

$$y_i(t = 0) = \sin\left(\frac{i\pi}{N}\right), \qquad \dot{y}_i(t = 0) = 0, \qquad i = 1, 2, \ldots, N - 1,$$

with periodic boundary conditions.

The numerical calculations of Fermi, Pasta and Ulam were also pioneering in the sense that they carried out one of the first computer studies of nonlinear wave phenomena. Given the primitive state of computing in the 1950s it was a truly remarkable achievement!

In 1965 Kruskal and Zabusky studied the continuum limit corresponding to the FPU model. To do that, they considered $y$ as approximated by a continuous function of the position and time and expanded $y$ in a Taylor series,

$$y_{i\pm1} = y((i \pm 1)l) = y \pm ly_z + \frac{l^2}{2}y_{zz} \pm \frac{l^3}{3!}y_{zzz} + \frac{l^4}{4!}y_{zzzz} + \cdots,$$

where $z = il$. Setting $h = l/L$, $x = z/L$, $L = Nl$, $t = \tau/(h\omega)$, where $\tau$ is non-dimensional time with $\omega = \sqrt{k/m}$, it follows that

$$\frac{\partial}{\partial t} = h\omega\frac{\partial}{\partial \tau}$$

and using the Taylor series on (1.1) leads to the continuous equation

$$h^2 y_{\tau\tau} = h^2 y_{xx} + \frac{h^4}{12} y_{xxxx} + \alpha \left[ \left( h y_x + \frac{h^2}{2} y_{xx} + \cdots \right)^2 - \left( h y_x - \frac{h^2}{2} y_{xx} + \cdots \right)^2 \right].$$

Hence, to leading order, the continuous limit is given by

$$y_{\tau\tau} = y_{xx} + \frac{h^2}{12} y_{xxxx} + \varepsilon y_x y_{xx} + \cdots, \tag{1.2}$$

where $\varepsilon = 2\alpha h$ and the higher-order terms are neglected. This equation was derived by Boussinesq in the context of shallow-water waves in 1871 and 1872 (Boussinesq, 1871, 1872)!

There are four cases to consider:

(a) When $h^2 \ll 1$ and $|\varepsilon| \ll 1$ (read as $h^2$ and $|\varepsilon|$ are both much less than 1), both the nonlinear term and higher-order derivative term (referred to as the dispersive term) are negligible. Then equation (1.2) reduces to the linear wave equation

$$y_{\tau\tau} = y_{xx}.$$

(b) In the small-amplitude limit where $h^2/12 \gg |\varepsilon|$ (or where $\alpha \to 0$ in the FPU model), the nonlinear term is negligible and the correction to (1.2) is governed by the higher-order linear dispersive wave equation

$$y_{\tau\tau} = y_{xx} + \frac{h^2}{12} y_{xxxx}.$$

(c) If $h^2/12 \ll |\varepsilon|$, then the $y_{xxxx}$ term is negligible and (1.2) yields

$$y_{\tau\tau} = y_{xx} + \varepsilon y_x y_{xx},$$

which has, as can be shown from further analysis or indicated by numerical simulation, breaking or multi-valued solutions in finite time. When breaking occurs one must use (1.2) as a more physical model.

(d) In the case of "maximal balance" where $h^2/12 \approx |\varepsilon| \ll 1$, the wave equation is governed by a different equation.

This case of maximal balance is the most interesting case and we will now analyze it in detail.

Let us look for a solution $y$ of the form[1]

$$y \sim \Phi(X, T; \varepsilon), \qquad X = x - \tau, \qquad T = \frac{\varepsilon\tau}{2}.$$

[1] Later in the book we will see "why".

It follows that

$$\frac{\partial}{\partial \tau} = -\frac{\partial}{\partial X} + \frac{\varepsilon}{2}\frac{\partial}{\partial T},$$

$$\frac{\partial^2}{\partial \tau^2} = \left(\frac{\partial}{\partial \tau}\right)^2 = \frac{\partial^2}{\partial X^2} - \varepsilon\frac{\partial}{\partial X \partial T} + \frac{\varepsilon^2}{4}\frac{\partial^2}{\partial T^2},$$

$$\frac{\partial}{\partial x} = \frac{\partial}{\partial X}.$$

Substituting these relations into the continuum limit, (1.2) yields

$$\left[\frac{\partial^2 \Phi}{\partial X^2} - \varepsilon\frac{\partial \Phi}{\partial X \partial T} + \frac{\varepsilon^2}{4}\frac{\partial^2 \Phi}{\partial T^2}\right] = \frac{\partial^2 \Phi}{\partial X^2} + \frac{h^2}{12}\frac{\partial^4 \Phi}{\partial X^4} + \varepsilon\frac{\partial \Phi}{\partial X}\frac{\partial^2 \Phi}{\partial X^2}.$$

Calling $u = \partial\Phi/\partial X$ and dropping the $O(\varepsilon^2)$ terms, leads to the equation studied by Zabusky and Kruskal (1965) and Kruskal (1965)

$$u_T + uu_X + \delta^2 u_{XXX} = 0, \tag{1.3}$$

where $\delta^2 = h^2/12\varepsilon$ and $u(X,0)$ is the given initial condition. It is important to note that (1.3) is the well-known (nonlinear) Korteweg–de Vries (KdV) equation. It should be remarked that Boussinesq derived (1.3) and other approximate long-wave equations for water waves [e.g., (1.2)] (Boussinesq, 1871, 1872, 1877). Korteweg and de Vries investigated (1.3) in considerable detail and found periodic "cnoidal" wave solutions in the context of long (or shallow) water waves (Korteweg and de Vries, 1895). Before the early 1960s, the KdV equation was primarily of interest only to researchers studying water waves. The KdV equation was not of wide interest to mathematicians during the first half of the twentieth century, since most studies at the time tended to concentrate on linear second-order equations, whereas (1.3) is nonlinear and third order.

Kruskal and Zabusky considered the KdV equation (1.3) with periodic initial values. They initially took $\delta^2$ small with $u(X,0) = \cos(\pi X)$. When $\delta = 0$ one gets the so-called inviscid Burgers equation,

$$u_T + uu_X = 0,$$

which leads to breaking or a multi-valued solution or shock formation in finite time. The inviscid Burgers equation is discussed further in Chapter 2.

When $\delta^2 \ll 1$, a sharp gradient appears at a finite time, which we denote by $t = t_B$, together with "wiggles" (see the dashed line in Figure 1.2). When $t \gg t_B$, the solution develops many oscillations that eventually separate into a train of solitary-type waves. Each solitary wave is localized in space (see the solid line in Figure 1.2). Subsequently, under further propagation, the solitary waves interact and the solution eventually returns to a state that is similar to

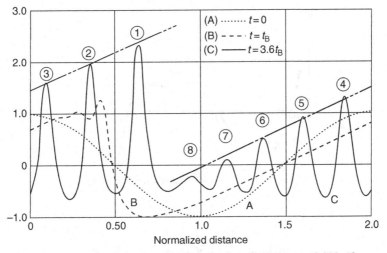

Figure 1.2 Calculations of the KdV equation (1.3), $\delta \approx 0.022$ [from numerical calculations of Zabusky and Kruskal (1965)].

the initial conditions, one which resembles the recurrence phenomenon first observed by FPU in their computations.

An important aspect raised by Kruskal and Zabusky in 1965 was the appearance of the train of solitary waves. To study an individual solitary wave one can look for traveling wave solutions of (1.3); that is, $u = U(\zeta)$, where $\zeta = (X - CT - X_0)$, $C$ is the speed of the traveling wave, and $X_0$ is the phase. Doing so reduces (1.3) to

$$-CU_\zeta + UU_\zeta + \delta^2 U_{\zeta\zeta\zeta} = 0.$$

To look for a solitary wave we take $U \to U_\infty$ as $|\zeta| \to \infty$. First integrate this equation once to find

$$\delta^2 U_{\zeta\zeta} + \frac{U^2}{2} - CU = \frac{E_1}{6},$$

where $E_1$ is a constant of integration. Multiplying by $U_\zeta$ and integrating again leads to

$$\frac{\delta^2}{2}U_\zeta^2 + \frac{U^3}{6} - C\frac{U^2}{2} = \frac{E_1}{6}U + \frac{E_2}{6},$$

where $E_2$ is another constant of integration. Thus, one obtains the equation

$$\frac{\delta^2}{2}U_\zeta^2 = \frac{1}{6}P_3(U)$$

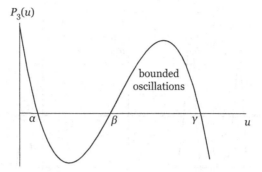

Figure 1.3  Solitons can exist when $\beta < U < \gamma$.

where

$$P_3(U) = -U^3 + 3CU^2 + E_1 U + E_2.$$

We will study the case when the third-order polynomial $P_3(U)$ can be factorized as $P(U) = -(U - \alpha)(U - \beta)(U - \gamma)$, with $\alpha \le \beta \le \gamma$; i.e., three real roots; when there is only one real root, it can be shown that the solution is unbounded. Since $U_\zeta^2$ cannot be negative, one can conclude from the $\left(U_\zeta^2, U\right)$ phase plane diagram (see Figure 1.3) that a real periodic wave can exist only when $U$ is between the roots $\beta$ and $\gamma$, since only in this zone can the solution oscillate. In addition, it is straightforward to derive

$$3C = \alpha + \beta + \gamma, \qquad E_1 = -(\beta\gamma + \alpha\beta + \beta\gamma), \qquad E_2 = \alpha\beta\gamma.$$

Furthermore, the periodic wave solution takes the form

$$U(\zeta) = \beta + (\gamma - \beta)cn^2 \left[ \left( \frac{\gamma - \alpha}{12\delta^2} \right)^{1/2} \zeta; m \right],$$

where $cn(x; m)$ is the cosine elliptic function with modulus $m$ [see Abramowitz and Stegun (1972) or Byrd and Friedman (1971) for more details about elliptic functions] and

$$m = \frac{\gamma - \beta}{\gamma - \alpha}.$$

The above solution is often called a "cnoidal" wave following the terminology of Korteweg and de Vries (1895).

In the special limit $\beta \to \alpha$, i.e., when the factorization has a double root (see Figure 1.4), we can integrate directly; it follows that $m = 1$, $C = (2\alpha + \gamma)/3$, and the solution can be put in the elementary form

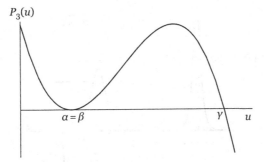

Figure 1.4 The limiting case of a double root ($\alpha = \beta$).

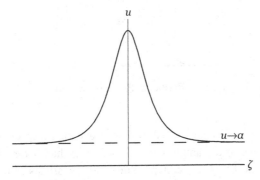

Figure 1.5 Hyperbolic secant solution approaches $\alpha$ as $|\zeta| \to \infty$.

$$U(\zeta) = \alpha + (\gamma - \alpha)\operatorname{sech}^2\left[\left(\frac{\gamma - \alpha}{12\delta^2}\right)^{1/2}\zeta\right].$$

In this case $U \to \alpha$ as $|\zeta| \to \infty$ (see Figure 1.5).

If $\alpha = 0$ then the solution reduces to

$$U(\zeta) = \gamma\operatorname{sech}^2\left[\left(\frac{\gamma}{12\delta^2}\right)^{1/2}\zeta\right] = 3C\operatorname{sech}^2\left(\frac{\sqrt{C}}{2\delta}\zeta\right) = 12\delta^2\kappa^2\operatorname{sech}^2\kappa\zeta,$$

where $\kappa = \sqrt{C}/2\delta$.

We see that such traveling solitary waves propagate with a speed that increases with the amplitude of the waves. In other words, larger-amplitude waves propagate faster than smaller ones. In a truly important discovery, by studying the numerical simulations of the FPU problem, Zabusky and Kruskal (1965) found that these solitary waves had a special property. Namely the solitary waves of the KdV equation collide "elastically"; i.e., they found that after a

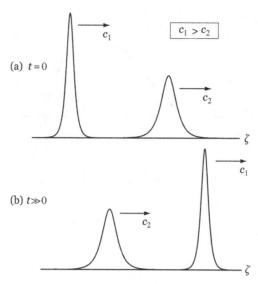

Figure 1.6 "Elastic" collision of two solitons.

large solitary wave overtakes a small solitary wave their respective amplitudes and velocities tend to the amplitude and speed they had before the collision. This suggests that the speeds and amplitudes are invariants of the motion. In fact, the only noticeable change due to the interaction is a phase shift from where the wave would have been if there were no interaction. For example, in Figures 1.6 and 1.7 we see that the smaller soliton is retarded in time whereas the larger one is pushed forward. Zabusky and Kruskal called these elastically interacting waves "solitons". Further, they conjectured that this property of the collisions was the reason for the recurrence phenomenon observed by FPU.[2]

Subsequent research has shown that solitary waves with this elastic interaction property, i.e., solitons, are associated with a much larger class of equations than just the KdV equation. This has to do with the connection of solitons with nonlinear wave equations that are exactly solvable by the technique of the inverse scattering transform (IST). Integrable systems and IST are briefly covered in Chapters 8 and 9. It should also be mentioned that the term soliton has taken on a much wider scope than the original notion of Zabusky and Kruskal: in many branches of physics a soliton represents a solitary or localized type of wave. When we discuss a soliton in the original sense of Zabusky and Kruskal we will relate solitons to the special aspects of the underlying equation and its solutions.

---

[2] The detailed analysis of the recurrence phenomenon is quite intricate and will not be studied here.

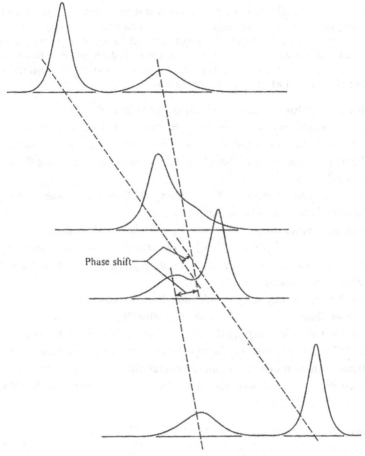

Figure 1.7 A typical interaction of two solitons at succeeding times [from (Ablowitz and Segur, 1981)].

## 1.1 Solitons: Historical remarks

Solitary waves or, as we now know them, *solitons* were first observed by J. Scott Russell in 1834 (Russell, 1844) while riding on horseback beside the narrow Union Canal near Edinburgh, Scotland. He described his observations as follows:

I was observing the motion of a boat which was rapidly drawn along a narrow channel by a pair of horses, when the boat suddenly stopped – not so the mass of water in the channel which it had put in motion; it accumulated round the prow of the vessel in a state of violent agitation, then suddenly leaving it behind, rolled forward with great velocity, assuming the form of a large solitary elevation, a rounded, smooth and well-defined heap of water, which continued its course along the channel apparently without change of form or diminution of speed. I followed it on horseback, and overtook it still rolling on at a rate of some eight or nine miles an hour, preserving its original figure some thirty feet long and a foot to a foot and a half in height. Its height gradually diminished, and after a chase of one or two miles I lost it in the windings of the channel. Such, in the month of August 1834, was my first chance interview with that rare and beautiful phenomenon which I have called the Wave of Translation ...

Subsequently, Russell carried out experiments in a laboratory wave tank to study this phenomenon more carefully. He later called the solitary wave *the Great Primary Wave of Translation*. Russell's work on the wave of translation was the first detailed study of these localized waves. Included among Russell's results are the following:

- he observed solitary waves, which are long, shallow-water waves of permanent form, hence he deduced that they *exist*;
- the speed of propagation, $c$, of a solitary wave in a channel of uniform depth $h$ is given by $c^2 = c_0^2(1 + A/h)$, $c_0^2 = gh$, where $A$ is the maximum amplitude of the wave, $h$ is the mean level above a rigid bottom and $g$ is the gravitational constant.

Russell's results provoked considerable discussion and controversy. Airy, a well-known fluid dynamicist, believed that Russell's wave of translation was a linear phenomenon (Airy, 1845). Subsequent investigations by Boussinesq (1872, 1871) and Rayleigh (1876) confirmed Russell's predictions. From the equations of motion of an inviscid, incompressible fluid, with a free surface, the result $c^2 = g(h + a)$ was derived, and it was also shown that the solitary wave has a profile given by

$$\eta(x,t) = a\,\mathrm{sech}^2[\beta(x - ct - x_0)], \qquad \beta^2 = \frac{3a}{4h^3}, \qquad c = c_0\left(1 + \frac{a}{2h}\right),$$

where $\eta$ is the height of the wave above the mean level $h$ for any $a > 0$, provided that $a \ll h$. Here, $x_0$ is an arbitrary constant phase shift; see Figure 1.8.

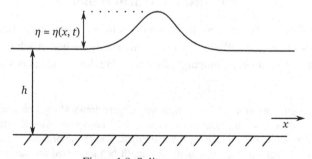

Figure 1.8 Solitary water wave.

Understanding was further advanced by Korteweg and de Vries (1895). They derived a nonlinear evolution equation governing long, one-dimensional, small-amplitude, surface gravity waves propagating in shallow water, now known as the KdV equation (in dimensional form):

$$\frac{1}{c_0}\frac{\partial \eta}{\partial t} + \frac{\partial \eta}{\partial x} + \frac{3}{2h}\eta\frac{\partial \eta}{\partial x} + \frac{h^2}{2}\left(\frac{1}{3} - \hat{T}\right)\frac{\partial^3 \eta}{\partial x^3} = 0, \tag{1.4}$$

where $\eta$ is the surface elevation of the wave, $h$ is the equilibrium level, $g$ is the gravitational constant of acceleration, $c_0 = \sqrt{gh}$, $\hat{T} = T/\rho gh^2$, $T$ is the surface tension and $\rho$ is the density (the terms "long" and "small" are meant in comparison to the depth of the channel, see Chapter 5).

Korteweg and de Vries showed (1.4) has traveling wave solutions, including periodic Jacobian elliptic (cosine) function solutions that they termed "cnoidal" functions, and a special case of a cnoidal function (when the elliptic modulus tends to unity) is a *solitary wave solution*. Equation (1.4) may be brought into non-dimensional form by making the transformation

$$\sigma = \frac{1}{3} - \hat{T}, \qquad t' = \beta t, \qquad x' = \frac{1}{h}(x - c_0 t),$$

$$\beta = \frac{c_0 \sigma}{2h}, \qquad \eta = 2h\sigma u.$$

Hence, we obtain (after dropping the primes)

$$u_t + 6uu_x + u_{xxx} = 0. \tag{1.5}$$

This dimensionless equation is usually referred to as the KdV equation. We note that any constant coefficient may be placed in front of any of the three terms by a suitable scaling of the independent and dependent variables.

Despite this derivation of the KdV equation in 1895, it was not until 1960 that new applications of it were discovered. Gardner and Morikawa (1960) rediscovered the KdV equation in the study of collision-free hydromagnetic waves. Subsequently the KdV equation has arisen in a number of other physical contexts, including stratified internal waves, ion-acoustic waves, plasma physics, lattice dynamics, etc. Actually the KdV equation is "universal" in the sense that it always arises when the governing equation has weak quadratic nonlinearity and weak dispersion. See also Benney and Luke (1964), Benney (1966a,b), Gardner and Su (1969) and Taniuti and Wei (1968).

As mentioned above, it has been known since the work of Korteweg and de Vries that the KdV equation (1.5) possesses the solitary wave solution

$$u(x, t) = 2\kappa^2 \operatorname{sech}^2\left\{\kappa(x - 4\kappa^2 t - \delta_0)\right\}, \tag{1.6}$$

where $\kappa$ and $\delta_0$ are constants. Note that the velocity of this wave, $4\kappa^2$, is proportional to the amplitude, $2\kappa^2$; therefore taller waves travel faster than shorter ones. In dimensional variables the soliton with $\kappa = kh$, $\delta_0 = 0$, takes the form

$$\eta = 2Ah \operatorname{sech}^2\left[k\left(x - c_0\left(1 + \frac{A}{2}\right)t\right)\right], \qquad A = 2\sigma k^2,$$

which agrees with Russell's observations mentioned above.

As discussed earlier, Zabusky and Kruskal (1965) discovered that these solitary wave solutions have the remarkable property that the interaction of two solitary wave solutions is elastic, and called the waves solitons.

Finally, we mention that there are numerous useful texts that discuss nonlinear waves in the context of physically significant problems. The reader is encouraged to consult these references: Phillips (1977); Rabinovich and Trubetskov (1989); Whitham (1974); Lighthill (1978); Infeld and Rowlands (2000); Ostrovsky and Potapov (1986); see also the essay by Miles (1981).

# Exercises

1.1 Following the methods described in this chapter, derive a generalized KdV equation from the FPU problem when the spring force law is given by

$$F(\Delta) = -k(\Delta + \alpha\Delta^3),$$

where $\Delta$ is the displacement between masses and $k$, $\alpha$ are constants.

1.2 Given the modified KdV (mKdV) equation

$$u_t + 6u^2 u_x + u_{xxx} = 0,$$

reduce the problem to an ODE by investigating traveling wave solutions of the form: $u = U(x - ct)$.

(a) Express the bounded periodic solution in terms of Jacobi elliptic functions.

(b) Find all bounded solitary wave solutions.

1.3 Consider the sine–Gordon (SG) equation given by

$$u_{tt} - u_{xx} + \sin u = 0.$$

(a) Use the transformation $u = U(x - ct)$, where $c$ is a constant, to reduce the SG equation to a second-order ODE.

(b) Find a first-order ODE by integrating once.

(c) Make the transformation $U = 2 \tan^{-1} w$ (inverse tan function), and solve the equation for $w$ to find all bounded, real periodic solutions for $w$ and therefore $U$. Express the solution in terms of Jacobi elliptic functions. (Hint: See Chapter 4.)

(d) Use the above to find all bounded wave solutions $U$ that tend to zero at $-\infty$ and $2\pi$ at $+\infty$. These are called kink solutions; they turn out to be solitons.

1.4 Consider the KdV equation

$$u_t + 6uu_x + u_{xxx} = 0.$$

(a) Make the transformation $x \to k^l x, \ t \to k^m t, \ u \to k^n u, \ k \neq 0$, and find $l, m, n$ so that the KdV equation is invariant under the transformation.

(b) Make the transformation $u = t^{-2/3} f(v), \ v = xt^{-1/3}, \ f = 2(\log F)''$. Find an equation for the similarity solution $f$ and an equation for $F$; then obtain a rational solution to the KdV equation. (See Section 3.2 for a discussion of self-similar/similarity solutions.)

1.5 Find the bounded traveling wave solution to the generalized KdV

$$u_t + (n + 1)(n + 2)u^n u_x + u_{xxx} = 0$$

where $n = 1, 2, \ldots$.

1.6 Consider the modified KdV (mKdV) equation

$$u_t - 6u^2 u_x + u_{xxx} = 0.$$

(a) Make the transformation $x \to a^l x, \ t \to a^m t, \ u \to a^n u$ and find $l, m, n$, so that the equation is invariant under the transformation.

(b) Introduce $u(x,t) = (3t)^{-1/3} f(\xi), \ \xi = x(3t)^{-1/3}$ to reduce the mKdV equation to the following ODE

$$f'' = \xi f + 2f^3 + \alpha$$

where $\alpha$ is constant. The equation for $f$ is called the second Painlevé equation, cf. Ablowitz and Segur (1981).

1.7 Show that a solitary wave solution of the Boussinesq equation,

$$u_{tt} - u_{xx} + 3(u^2)_{xx} - u_{xxxx} = 0,$$

is $u(x,t) = a \operatorname{sech}^2[b(x - ct) + d]$, for suitable relations between the constants $a, b, c,$ and $d$. Verify that the Boussinesq solitary wave can propagate in either direction.

1.8 Find the bounded traveling wave solution to the equation

$$u_{tt} = u_{xx} + u_x u_{xx} + u_{xxxx}.$$

Hint: Set $\xi = x - ct$, integrate twice with respect to $\xi$, and use $u_\xi = q(u)$ to solve the resulting equation.

1.9 Consider the sine–Gordon equation

$$u_{xx} - u_{tt} = \sin u.$$

(a) Using the transformation $\chi = \gamma(x - vt)$, $\tau = \gamma(t - vx)$ write the equation in terms of the new coordinates $\chi, \tau$; find $\gamma$ in terms of $v$, $-1 < v < 1$ so that the equation is invariant under the transformation.

(b) Consider the transformation $\xi = (x + t)/2, \eta = (x - t)/2$. Find the equation in terms of the new coordinates $\xi, \eta$. Show that this equation has a self-similar solution of the form $u(\xi, \eta) = f(z), z = \xi\eta$. Then find an equation for $w = \exp(if)$. The equation for $w$ is related to the third Painlevé equation, cf. Ablowitz and Segur (1981). (See Section 3.2 for a discussion of self-similar/similarity solutions.)

1.10 Seek a similarity solution of the Klein–Gordon equation,

$$u_{tt} - u_{xx} = u^3,$$

in the form $u(x, t) = t^m f(xt^n)$ for suitable values of $m$ and $n$. Show that $f(z)$, $z = x/t$ satisfies the equation $(z^2 - 1)f'' + 4zf' + 2f = f^3$. (See Section 3.2 for a discussion of self-similar/similarity solutions.)

# 2

## *Linear and nonlinear wave equations*

In Chapter 1 we saw how the KdV equation can be derived from the FPU problem. We also mentioned that the KdV equation was originally derived for weakly nonlinear water waves in the limit of long or shallow water waves. Researchers have subsequently found that the KdV equation is "universal" in the sense that it arises whenever we have a weakly dispersive and a weakly quadratic nonlinear system. Thus the KdV equation has also been derived from other physical models, such as internal waves, ocean waves, plasma physics, waves in elastic media, etc. In later chapters we will analyze water waves in depth, but first we will discuss some basic aspects of waves.

Broadly speaking, the study of wave propagation is the study of how signals or disturbances or, more generally, information is transmitted (cf. Bleistein, 1984). In this chapter we begin with a study of "dispersive waves" and we will introduce the notion of phase and group velocity. We will then briefly discuss: the linear wave equation, the concept of characteristics, shock waves in scalar first-order partial differential equations (PDEs), traveling waves of the viscous Burgers equation, classification of second-order quasilinear PDEs, and the concept of the well-posedness of PDEs.

## 2.1 Fourier transform method

Consider a PDE in evolution form, first order in time, and in one spatial dimension,

$$u_t = F[u, u_x, u_{xx}, \dots],$$

where $F$ is, say, a polynomial function of its arguments. We will consider the initial value problem on $|x| < \infty$ and assume $u \to 0$ sufficiently rapidly as $|x| \to \infty$ with $u(x,0) = u_0(x)$ given. To begin with, suppose we consider the linear homogeneous case, i.e.,

17

$$u_t = \sum_{j=0}^{N} a_j(x,t) u_{jx},$$

where $u_{jx} \equiv \partial^j u / \partial x^j$ and $a_j(x,t)$ are prescribed coefficients. When the $a_j$ are constants,

$$u_t = \sum_{j=0}^{N} a_j u_{jx}, \tag{2.1}$$

and $u_0(x)$ decays fast enough, then we can use the method of Fourier transforms to solve this equation. Before we do that, however, let us recall some basic facts about Fourier transforms (cf. Ablowitz and Fokas, 2003).

The function $u(x,t)$ can be expressed using the (spatial) Fourier transform as

$$u(x,t) = \frac{1}{2\pi} \int b(k,t) e^{ikx} \, dk, \tag{2.2}$$

where it is assumed that $u$ is smooth and $|u| \to 0$ as $|x| \to \infty$ sufficiently rapidly; also, unless otherwise specified, $\int$ represents an integral from $-\infty$ to $+\infty$ in this chapter. For our purposes it suffices to require that $u \in L_1 \cap L_2$, meaning that $\int |u| \, dx$ and $\int |u|^2 \, dx$ are both finite. Substituting (2.2) into (2.1), assuming the interchange of derivatives and integral, leads to

$$\int e^{ikx} \left\{ b_t - b \sum_{j=0}^{N} (ik)^j a_j \right\} dk = 0.$$

It follows that

$$b_t = b \sum_{j=0}^{N} (ik)^j a_j,$$

or that

$$b_t = -i\omega(k) b,$$

with

$$-i\omega(k) = \sum_{j=0}^{N} (ik)^j a_j. \tag{2.3}$$

We call $\omega(k)$, which we assume is real, the *dispersion relation* corresponding to (2.1). For example, if $N = 3$, then $\omega(k) = ia_0 - ka_1 - ik^2 a_2 + k^3 a_3$ and $\omega(k)$ is real if $a_1, a_3$ are real and $a_0, a_2$ are pure imaginary. We can solve this ODE to get

$$b(k,t) = b_0(k) e^{-i\omega(k)t},$$

where $b_0(k) \equiv b(k, 0)$. From this one obtains

$$u(x, t) = \frac{1}{2\pi} \int b_0(k) e^{i[kx - \omega(k)t]} \, dk.$$

Hence $b_0(k)$ plays the role of a weight function, which depends on the initial conditions according to the inverse Fourier transform:

$$b_0(k) = \int u(x, 0) e^{-ikx} \, dx.$$

Strictly speaking, we now have an "algorithm" in terms of integrals for solving our problem for $u(x, t)$.

Note that, in retrospect, we can also obtain this relation by substituting

$$u(x, t) = \frac{1}{2\pi} \int b_0(k) e^{i[kx - \omega(k)t]} \, dk$$

into the PDE.

There is, in fact, an alternative method for obtaining the dispersion relation. It is based on the observation that in this case one can substitute $u_s = e^{i[kx - \omega(k)t]}$ into the PDE and replace the time and spatial derivatives by

$$\partial_t \to -i\omega, \qquad \partial_x \to ik.$$

Then (2.3) follows from (2.1) directly.

## 2.2 Terminology: Dispersive and non-dispersive equations

Let us define and then explain the terminology that is frequently used in conjunction with these wave problems and Fourier transforms: $k$ is usually called the *wavenumber*, $\omega$ is the *frequency*, $k$ and $\omega$ real, and $\theta \equiv kx - \omega(k)t$ is the *phase* in the exponent or simply the phase.

The temporal *period* (or period for short) is denoted by

$$T \equiv \frac{2\pi}{\omega}.$$

The meaning of the period is that whenever $t \to t + nT$, where $n$ is an integer, then the phase remains the same modulo $2\pi$ and therefore $e^{i\theta}$ remains unchanged. Similarly, we call $\lambda = 2\pi/k$ the *wavelength* and note that whenever $x \to x + n\lambda$ the phase remains the same modulo $2\pi$ and therefore $e^{i\theta}$ remains unchanged. Furthermore, we call

$$c(k) \equiv \frac{\omega(k)}{k}$$

the *phase velocity*, since $\theta = k(x - c(k)t)$. There is also the notion of *group velocity* $v_g(k) \equiv \omega'(k)$; i.e., the speed of a slowly varying group of waves. We will discuss its importance later.

An equation in one space and one time dimension is said to be dispersive when $\omega(k)$ is real-valued and $\omega''(k) \neq 0$. The meaning of dispersion will be further elucidated when we discuss the long time asymptotics of these equations. Consider the first-order linear equation

$$u_t - a_1 u_x = 0, \qquad (2.4)$$

where $a_1 \neq 0$, real and constant. We see that

$$-i\omega = ika_1,$$

which gives the linear dispersion relation

$$\omega(k) = -a_1 k.$$

In this case $\omega''(k) \equiv 0$, which means the PDE is non-dispersive.

Now let us look at the first-order constant coefficient equation (2.1). In this case the dispersion relation (2.3) shows that if all the $a_j$ are real, then $\omega$ is real if and only if all the even powers of $k$ (or even values of the index $j$) vanish; i.e., $a_j = 0$ for $j = 2, 4, 6, \ldots$. In that case it follows from (2.3) that $\omega(k)$ is an odd function of $k$, in which case the dispersion relation takes the general form

$$\omega(k) = -\sum_{j=0}^{N}(-1)^j a_{2j+1} k^{2j+1}.$$

A further example is the linearized KdV equation given by

$$u_t + u_{xxx} = 0, \qquad (2.5)$$

i.e., (1.5) without the nonlinear term. By substituting $u_s = e^{i[kx-\omega(k)t]}$ one obtains its dispersion relation,

$$\omega(k) = -k^3.$$

Thus $\omega$ is real and $\omega'' = -6k \neq 0$, which means that this is a dispersive equation.

We can use Fourier transforms to solve this equation. As indicated above, using the Fourier transform, $u(x, t)$ can be expressed as

$$u(x, t) = \frac{1}{2\pi} \int b(k, t) e^{ikx} \, dk,$$

and for the linearized KdV equation (2.5) one finds that $b_t = ik^3 b$ hence

$$u(x,t) = \frac{1}{2\pi} \int b_0(k) e^{i(kx+k^3 t)} \, dk.$$

However, by itself this is not a particularly insightful solution, since one cannot evaluate this integral explicitly. Often having an integral formula alone does not give useful qualitative information. Similarly, we usually cannot explicitly evaluate the integrals corresponding to most linear dispersive equations. This is exactly the place where asymptotics of integrals plays an extremely important role: it will allow us to approximate the integrals with simple, understandable expressions. This will be a recurring theme in this book: namely, analytical methods can yield solutions that are inconvenient or uninformative and asymptotics can be used to obtain valuable information about these problems. Later we will briefly discuss the long-time asymptotic analysis of the linearized KdV equation.

In so far as we are dealing with dispersive equations, we have seen that requiring $\omega(k)$ to be real implies that $\omega(k)$ is odd. If in addition, $u(x,t)$ is real, this knowledge can be encoded into $b_0(k)$ as follows. We have that

$$u^*(x,t) = \frac{1}{2\pi} \int b_0^*(k) e^{-i[kx-\omega(k)t]} \, dk,$$

where $u^*$ denotes the complex conjugate of $u$. Calling $k' = -k$ yields

$$u^*(x,t) = -\frac{1}{2\pi} \int_{\infty}^{-\infty} b_0^*(-k') e^{-i[-k'x-\omega(-k')t]} \, dk'$$

$$= \frac{1}{2\pi} \int_{-\infty}^{\infty} b_0^*(-k') e^{-i[-k'x-\omega(-k')t]} \, dk'.$$

Then if we require that $u$ is real-valued then $u(x,t) = u^*(x,t)$. In addition, $\omega(-k') = -\omega(k)$. Combined, one obtains that

$$\frac{1}{2\pi} \int b_0(k) e^{i[kx-\omega(k)t]} \, dk = \frac{1}{2\pi} \int b_0^*(-k') e^{i[k'x-\omega(k')t]} \, dk'.$$

This identity is satisfied for all $(x,t)$ if and only if

$$b_0^*(k) = b_0(-k).$$

Note that we cannot apply the Fourier transform method for the nonlinear problem, sometimes called the inviscid Burgers equation,

$$u_t + u u_x = 0,$$

because of the nonlinear product. We study the solution of this equation by using the method of characteristics; this is briefly discussed later in this chapter.

We also note that for some problems, $u_s = \exp(i(kx - \omega(k)t))$ yields an $\omega(k)$ that is not purely real. For example, for the so-called heat or diffusion equation

$$u_t = u_{xx},$$

we find $\omega(k) = -ik^2$. The Fourier transform method works as before giving

$$u(x,t) = \frac{1}{2\pi} \int_{-\infty}^{\infty} b_0(k) e^{ikx - k^2 t} \, dk.$$

We see that the solution decays, i.e., the solution diffuses with increasing $t$.

## 2.3 Parseval's theorem

The $L_2$-norm of a function $f(x)$ is defined by

$$\|f\|_2^2 \equiv \int |f(x)|^2 \, dx.$$

Let $\hat{f}(k)$ be the Fourier transform of $f(x)$. Parseval's theorem [see e.g., Ablowitz and Fokas (2003)] states that

$$\|f(x)\|_2^2 = \frac{1}{2\pi} \|\hat{f}(k)\|_2^2.$$

In many cases the $L_2$-norm of the solutions of PDEs has the meaning of energy (i.e., is proportional to the energy in physical units). The physical meaning associated with Parseval's relation is that the energy in physical space is equal to the energy in frequency (or sometimes called spectral or Fourier) space. Moreover, we know that when $\omega$ is real-valued then

$$\int |u(x,t)|^2 \, dx = \frac{1}{2\pi} \int |b_0(k)|^2 \, dk = \text{constant}.$$

Hence, it follows from Parseval's theorem that energy is conserved in linear dispersive equations. While it is possible to prove this result using direct integration methods, we get it "for free" in linear PDEs using Parseval's theorem.

## 2.4 Conservation laws

We saw above how Parseval's theorem is used to prove energy conservation. However, there may be other conserved quantities. These often play a very useful role in the analysis of problems, as we will see later.

A conservation law (or relation) has the general form

$$\frac{\partial}{\partial t} T(x, t) + \frac{\partial}{\partial x} F(x, t) = 0;$$

we call $T$ the density of the conserved quantity, and $F$ the flux. Let us integrate this relation from $x = -\infty$ to $x = \infty$:

$$\frac{\partial}{\partial t} \int T \, dx + F(x, t) \Big|_{x \to -\infty}^{x \to +\infty} = 0.$$

The second term is zero, since we assume that $F$ decays at infinity, which leads to

$$\frac{\partial}{\partial t} \int T \, dx = 0,$$

i.e., $\int T \, dx = $ constant.

For example, let us study the conservation laws for the linearized KdV equation:

$$u_t + u_{xxx} = 0.$$

This equation is already in the form of a conservation law, with $T_1 = u$ and $F_1 = u_{xx}$; that is, $\int u \, dx$ is conserved. This, however, is only one of many conservation laws corresponding to (2.5). Another example is energy conservation. Using Parseval's theorem, we have already proven that linear dispersive equations with constant coefficients satisfy energy conservation. Since (2.5) is solvable by Fourier transforms we know from Parseval's theorem that energy is conserved. An alternative way of seeing this is by multiplying (2.5) by $u$ and integrating with respect to $x$. It can be checked that this leads to

$$\frac{\partial}{\partial t}\left(\frac{1}{2} u^2\right) + \frac{\partial}{\partial x}\left(u u_{xx} - \frac{1}{2} u_x^2\right) = 0,$$

from which it follows that $\int |u|^2 \, dx = $ constant.

## 2.5 Multidimensional dispersive equations

So far we have focused on dispersive equations in one spatial dimension. The method of Fourier transforms can be generalized to solve constant coefficient multidimensional equations, where a typical solution takes the form:

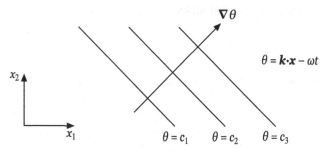

Figure 2.1 Phase contours in two dimensions ($n = 2$).

$$u(x, t) = \frac{1}{2\pi} \int b_0(k)e^{i[k \cdot x - \omega(k)t]}\, dk$$

and $x = (x_1, x_2, \ldots, x_n)$, $k = (k_1, k_2, \ldots, k_n)$, cf. Whitham (1974). As before, $\omega$ is the frequency (assumed real) and $T = 2\pi/\omega$ is the period. The wavenumber becomes the vector $k = \nabla\theta$, where $\theta \equiv k \cdot x - \omega(k)t$, and the wavelength $\lambda = 2\pi\hat{k}/|k|$, where $\hat{k} = k/|k|$ and $|\cdot|$ is the (Euclidean) modulus of the vector. Thus $\theta(x + \lambda) = \theta(x) + 2\pi$. The condition that $\theta$ be constant describes the phase contours (see Figure 2.1). We can also define the phase speed as $c(k) = \omega(k)\hat{k}/|k|$, from which it follows that $c \cdot k = \omega$.

In the multidimensional linear case, an equation is said to be dispersive if $\omega$ is real and (instead of $\omega'' \neq 0$) its Hessian is non-singular, that is

$$\det\left|\frac{\partial^2\omega(k)}{\partial k_i \partial k_j}\right| \not\equiv 0.$$

If we have a nonlinear equation, we call it a *nonlinear dispersive wave equation* if its linear part is dispersive and the equation is energy-preserving.

## 2.6 Characteristics for first-order equations

First let us use the Fourier transform method to solve (2.4), $u_t - a_1 u_x = 0$. As before, we get $\omega(k) = -a_1$ and

$$u(x, t) = \frac{1}{2\pi} \int b_0(k)e^{ik(x+a_1 t)}\, dk.$$

Calling

$$f(x) \equiv u(x, 0) = \frac{1}{2\pi} \int b_0(k)e^{ikx}\, dk,$$

we observe that $u(x, t) = f(x + a_1 t)$. This solution can be checked directly by substitution into the PDE.

We can also solve (2.4) using the so-called method of characteristics. We can think of the left-hand side of (2.4) as a total derivative that is expanded according to the chain rule. That is,

$$\frac{du}{dt} = \frac{\partial u}{\partial t} + \frac{dx}{dt}\frac{\partial u}{\partial x} = 0. \tag{2.6}$$

It follows from (2.6) that $du/dt = 0$ along the curves $dx/dt = -a_1$, or, said differently, that $u = c_2$, a constant, along the curves $x + a_1 t = c_1$. Thus for the initial value problem $u(x, t = 0) = f(x)$, at any point $x = \xi$ we can identify the constant $c_2$ with $f(\xi)$ and $c_1$ with $\xi$. Hence the solution to the initial value problem is given by

$$u = f(x + a_1 t).$$

Alternatively, from (2.4) and (2.6) (solving for $du$), we can write the above in the succinct form

$$\frac{dt}{1} = \frac{dx}{-a_1} = \frac{du}{0}.$$

These equations imply that

$$t = \frac{-1}{a_1}x + \frac{c_1}{a_1}, \qquad u = c_2,$$

where $c_1$ and $c_2$ are constants. Hence

$$x + a_1 t = c_1, \qquad u = c_2.$$

For first-order PDEs the solution has one arbitrary function degree of freedom and the constants are related by this function. In this case $c_2 = g(c_1)$, where $g$ is an arbitrary function, leads to

$$u = g(x + a_1 t).$$

As above, suppose we specify the initial condition $u(x, 0) = f(x)$. Then if we take $c_1 = \xi$, so that $t = 0$ corresponds to the point $x = \xi$ on the initial data, this then implies that $u(\xi, t) = g(\xi) = f(\xi)$; now in general, along $\xi = x + a_1 t$, that agrees with the solution obtained by Fourier transforms. The meaning of $\xi$ is that of a characteristic curve (a line in this case) in the $(x, t)$-plane, along which the solution is non-unique; in other words, along a characteristic $\xi$, the solution can be specified arbitrarily. Moreover as we move from one characteristic, say $\xi_1$, to another, say $\xi_2$, the solution can change abruptly.

The method of characteristics also applies to quasilinear first-order equations. For example, let us consider the inviscid Burgers equation mentioned earlier:

$$u_t + uu_x = 0. \tag{2.7}$$

From (2.7) we have that $du/dt = 0$ along the curves $dx/dt = u$, or, said differently, that $u = c_2$, a constant, along the curves $x - ut = c_1$. Hence an implicit solution is given by

$$u = f(x - ut).$$

Alternatively, if we specify the initial condition $u(x, 0) = f(x)$, with $f(x)$ a smooth function of $x$, and we take $c_1 = \xi$ so that corresponding to $t = 0$ is a point $x = \xi$ on the initial data, this then means that $u(x, t) = f(\xi)$ along $x = \xi + f(\xi)t$. The latter equation implies that $\xi$ is a function of time, i.e., $\xi = \xi(x, t)$, which in turn leads to the solution $u = u(x, t)$. If we have a "hump-like" initial condition such as $u(x, 0) = \text{sech}^2 x$ then either points at the top of the curve, e.g., the maximum, "move" faster than the points of lower amplitude and eventually break, or a multi-valued solution occurs. The breaking time follows from $x = \xi + f(\xi)t$. Taking the derivative of this equation yields $\partial \xi / \partial x = \xi_x$ and $\xi_t$:

$$\xi_x = \frac{1}{f'(\xi)t + 1}, \quad \xi_t = -\frac{f(\xi)}{f'(\xi)t + 1} \tag{2.8}$$

and the breaking time is given by $t = t_B = 1/\max(-f')$. This is the breaking time, depicted in Figure 1.2, that is associated with the KdV equation (dashed line) in Chapter 1.

Thus the solution to (2.7) can be written in the form

$$u = u(\xi(x, t)).$$

Prior to the breaking time $t = t_B$ the solution is single valued. So we can verify that the solution (2.8) satisfies the equation:

$$u_t = u'(\xi)\xi_t$$
$$u_x = u'(\xi)\xi_x$$

and using (2.7) and (2.8)

$$u_t + uu_x = -\frac{f(\xi)}{f'(\xi)t + 1} + \frac{f(\xi)}{f'(\xi)t + 1} = 0.$$

In the exercises, other first-order equations are studied by the method of characteristics.

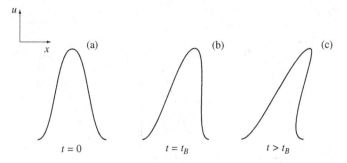

Figure 2.2 The solution of (2.7) at different times.

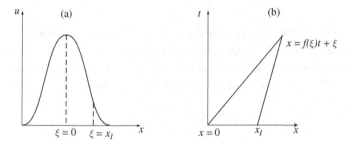

Figure 2.3 (a) A typical function $u(\xi, 0) = f(\xi)$. (b) Characteristics associated with points $\xi = 0$, $\xi = x_I$ that intersect at $t = t_B$. The point $x_I$ is the inflection point of $f(\xi)$.

In Figure 2.2 is shown a typical case and we can formally find the solution for values $t > t_B$ ignoring the singularities and triple-valued solution. The value of time $t$, when characteristics first intersect, is denoted by $t_B$ – see Figure 2.3.

## 2.7 Shock waves and the Rankine–Hugoniot conditions

Often we wish to take the solution further in time, beyond $t = t_B$, but do not want the multi-valued solution for physical or mathematical reasons. In many important cases the solution has a nearly discontinuous structure. This is shown schematically in Figure 2.4. Such a situation occurs in the case of the so-called viscous Burgers equation

$$u_t + u u_x = \nu u_{xx} \tag{2.9}$$

where $\nu$ is a constant; in this case there is a rapidly changing solution that can be viewed as an approximation to a discontinuous solution for small

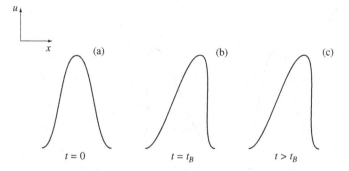

Figure 2.4 Evolution of the solution to (2.7) with a "shock wave".

$v$ ($0 < v \ll 1$). One way of introducing discontinuities, i.e., shocks, without resorting to adding a "viscous" term (i.e., we keep $v = 0$) is to consider (2.7) as coming from a conservation law and its corresponding integral form. In other words, (2.7) can be written in conservation law form as

$$\frac{\partial}{\partial t}u + \frac{\partial}{\partial x}\left(\frac{u^2}{2}\right) = 0$$

which in turn can be derived from the *integral* form

$$\frac{d}{dt}\int_x^{x+\Delta x} u(x,t)\,dx + \frac{1}{2}\left(u^2(x+\Delta x,t) - u^2(x,t)\right) = 0, \qquad (2.10)$$

in the limit $\Delta x \to 0$. Equation (2.10) can support a shock wave since it is an integral relation. Suppose between two points $x_1$ and $x_2$ we have a discontinuity that can change in time $x = x_s(t)$ – see Figure 2.5. Then (2.10) reads

$$\frac{d}{dt}\left(\int_{x_1}^{x_s(t)} u(x,t)\,dx + \int_{x_s(t)}^{x_2} u(x,t)\,dx\right) + \frac{1}{2}\left(u^2(x_2,t) - u^2(x_1,t)\right) = 0.$$

Letting $x_2 = x_s(t) + \epsilon$, $x_1 = x_s(t) - \epsilon$, and $\epsilon > 0$, then as $\epsilon \to 0$, we have $x_2 \to x_+$ and $x_1 \to x_-$ and so

$$(u(x_-,t) - u(x_+,t))\frac{dx_s}{dt} = -\frac{1}{2}\left(u^2(x_+,t) - u^2(x_-,t)\right)$$

and find

$$v_s = \frac{dx_s}{dt} = \frac{1}{2}(u(x_+,t) - u(x_-,t)) = \frac{u_+ + u_-}{2}, \qquad (2.11)$$

where $u_\pm = u(x_\pm,t)$. The last equation, (2.11), describes the speed of the shock, $v_s = dx_s/dt$, in terms of the jump discontinuities.

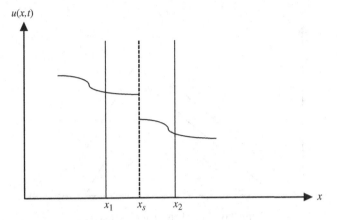

Figure 2.5 A discontinuity in $u(x, t)$ at $x = x_s(t)$.

We note that if (2.7) is generalized to

$$u_t + c(u)u_x = 0,$$

the corresponding conservation law is

$$\frac{\partial}{\partial t}u + \frac{\partial}{\partial x}(q(u)) = 0, \tag{2.12}$$

where $q'(u) = c(u)$, and its integral form is

$$\frac{d}{dt}\int_x^{x+\Delta x} u(x, t)\,dx + q(u(x + \Delta x)) - q(u(x, t)) = 0.$$

Then following the same procedure as above, we find that the shock speed is

$$v_s = \frac{dx_s}{dt} = \frac{q(x_+, t) - q(x_-, t)}{u_+ - u_-}. \tag{2.13}$$

We note that in the limit $x_+ \to x_-$ we have $v_s \to q'(x_+)$.

Equations (2.11) and (2.13) are sometimes referred to as the *Rankine–Hugoniot* (RH) relations that were originally derived for the Euler equations of fluid dynamics, cf. Lax (1987) and LeVeque (2002).

The RH relations, sometimes referred to as shock conditions, are used to avoid multi-valued behavior in the solution, which would otherwise occur after characteristics cross – see Figure 2.6. In order to make the problem well-posed one needs to add admissibility conditions, sometimes called entropy-satisfying conditions, to the relation. In their simplest form these conditions indicate characteristics should be going into a shock as time increases, rather than emanating, as that would be unstable, see Lax (1987) and LeVeque (2002).

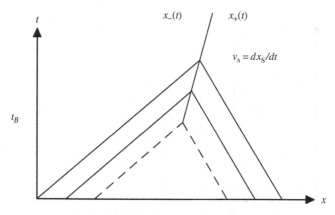

Figure 2.6 Typical characteristic diagram.

Another viewpoint is that an admissibility relation should be the limit of, for example, a viscous solution to Burgers' equation, see (2.9), where the admissible solution $u_A(x, t) = \lim_{\nu \to 0} u(x, t; \nu)$. We discuss Burgers' equation in detail later.

Let us consider the problem of fitting the discontinuities satisfying the shock conditions into the smooth part of the solution, e.g., for $x \to x_+(t)$, $x \to x_-(t)$.

For example, consider (2.7) with the following initial condition:

$$u(x, 0) = \begin{cases} u_+, & x > 0 \\ u_-, & x < 0. \end{cases}$$

where $u_\pm$ are constants. In the case where $u_+ < u_-$, we see that the characteristics cross immediately; i.e., they satisfy $dx/dt = u$, so

$$x = u_+ t + \xi, \quad \xi > 0,$$
$$x = u_- t + \xi, \quad \xi < 0,$$

as indicated in Figure 2.7. The speed of the shock is given by

$$v_s = \frac{dx}{dt} = \frac{u_+ + u_-}{2}.$$

If $u_+ = -u_-$ the shock is stationary.

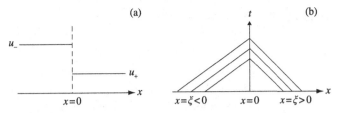

Figure 2.7  (a) Shock wave $u_+ > u_-$. (b) Characteristics cross.

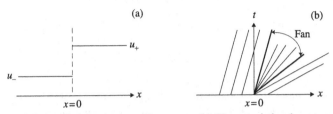

Figure 2.8  (a) "Rarefaction wave" $u_+ < u_-$. (b) Characteristics do not cross. There is a "fan" at $x = 0$.

On the other hand, if $u_+ < u_-$, as in Figure 2.8, then the discontinuities do not cross. In this case

$$x = u_+t + \xi, \quad \xi > 0,$$
$$x = u_-t + \xi, \quad \xi < 0,$$
$$x = ut, \quad \xi = 0,$$

and the solution is termed a *rarefaction wave* with a *fan* at $\xi = 0$, $u = x/t$ where $u_- < x/t < u_+$.

Next we return briefly to discuss the viscous Burgers equation (2.9). If we look for a traveling solution $u = u(\zeta)$, $\zeta = x - Vt - x_0$, where $V, x_0$ are constants, and $V$ the velocity of the (viscous shock) traveling wave, then

$$-Vu_\zeta + uu_\zeta = \nu u_{\zeta\zeta}.$$

Integrating yields

$$-Vu + \frac{1}{2}u^2 + A = \nu u_\zeta, \tag{2.14}$$

where $A$ is the constant of integration. Further, if $u \to u_\pm$ as $x \to \pm\infty$, where $u_\pm$ are constants, then

$$A = -\left(\frac{1}{2}u_\pm^2 - Vu_\pm\right)$$

and by eliminating the constant $A$, we find

$$V = \frac{1}{2}\frac{u_+^2 - u_-^2}{u_+ - u_-} = \frac{1}{2}(u_+ + u_-),$$

which is the same as the shock wave speed found earlier for the inviscid Burgers equation. Indeed, if we replace Burgers' equation by the generalization

$$u_t + (q(u))_x = \nu u_{xx},$$

then the corresponding traveling wave $u = u(\zeta)$ would have its speed satisfy

$$V = \frac{q(u_+) - q(u_-)}{u_+ - u_-}$$

[by the same method as for (2.7)], which agrees with the shock wave velocity (2.13) associated with (2.12). For the Burgers equation we can give an explicit expression for the solution $u$ by carrying out the integration of (2.14)

$$d\zeta = \frac{\nu du}{u^2/2 - Vu + A}$$

with $A = -((1/2)u_+^2 - (1/2)(u_+ + u_-)u_+) = (1/2)u_+u_-$. This results in

$$\begin{aligned} u &= \frac{u_+ + u_- e^{(u_+ - u_-)\zeta/(2\nu)}}{1 + e^{(u_+ - u_-)\zeta/(2\nu)}} \\ &= \frac{u_+ + u_-}{2} + \frac{u_- - u_+}{2} \tanh\left(\frac{u_+ - u_-}{4\nu}\zeta\right), \end{aligned}$$

like in Figure 2.9. The *width* of the viscous shock wave is proportional to $w = 4\nu/(u_+ - u_-)$.

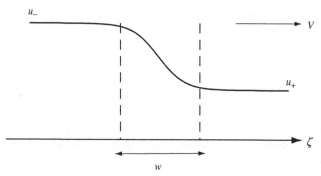

Figure 2.9 Traveling viscous shock wave with velocity $V$ and width $w = 4\nu/(u_+ - u_-)$.

## 2.8 Second-order equations: Vibrating string equation

Let us turn our attention to deriving an approximate equation governing the transverse vibration of a long string that is initially plucked and left to vibrate (see Figure 2.10).

Let $\rho$ be the mass density of the string: we assume it to be constant with $\rho = m/L$, where $m$ is the total mass of the string and $L$ is its length. The string is in equilibrium (i.e., at rest) when it is undisturbed along a straight line, which we will choose to be the $x$-axis. When the string is plucked we can approximately describe its vertical displacement from equilibrium at each point $x$ as a function $y(x, t)$. Consider an infinitesimally small segment of the string, i.e., a segment of length $|\Delta s| \ll 1$, where $\Delta s^2 = \Delta x^2 + \Delta y^2$ (see Figure 2.11). We will assume no external forces and that horizontal acceleration is negligible. It follows from Newton's second law that

- difference in vertical tensions:

$$(T \sin \theta)|_{x+\Delta x} - (T \sin \theta)|_x = \Delta m\, y_{tt},$$

- difference in horizontal tensions:

$$(T \cos \theta)|_{x+\Delta x} - (T \cos \theta)|_x = 0.$$

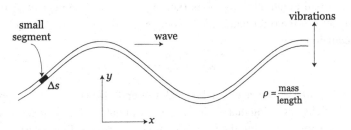

Figure 2.10 Vibrations of a long string.

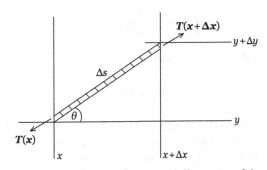

Figure 2.11 Tension forces acting on a small segment of the string.

In the limit $|\Delta x| \to 0$ we assume that $y = y(x,t)$; then one has (in that limit)

$$\sin\theta = \frac{\Delta y}{\Delta s} = \frac{\Delta y}{\sqrt{(\Delta x)^2 + (\Delta y)^2}} \to \frac{y_x}{\sqrt{1 + y_x^2}},$$

$$\cos\theta = \frac{\Delta x}{\Delta s} \to \frac{1}{\sqrt{1 + y_x^2}}.$$

Hence with these assumptions, in this limit, and using $dm = \rho\,ds$, Newton's equations become

$$\left(\rho\frac{ds}{dx}\right)y_{tt} - \frac{\partial}{\partial x}\left(T\frac{y_x}{\sqrt{1 + y_x^2}}\right) = 0,$$

$$\frac{\partial}{\partial x}\left(T\frac{1}{\sqrt{1 + y_x^2}}\right) = 0.$$

From the second equation we get that $T = T_0 \sqrt{1 + y_x^2}$, $T_0$ constant and, using $ds/dx = \sqrt{1 + y_x^2}$, the first equation yields

$$y_{tt} = \frac{T_0}{\rho \sqrt{1 + y_x^2}} y_{xx}. \tag{2.15}$$

This is a nonlinear equation governing the vibrations of the string. If we consider small-amplitude vibrations (i.e., $|y_x| \ll 1$) and so neglect $y_x^2$ in the denominator of the right-hand side we obtain, as an approximation, the linear wave equation,

$$y_{tt} - c^2 y_{xx} = 0,$$

where $c^2 = T_0/\rho$ is the wave speed (speed of sound). However, we can approximate (2.15) by assuming that $|y_x|$ is small but without neglecting it completely. That can be done by taking the first correction term from a Taylor series of the denominator:

$$\frac{1}{\sqrt{1 + y_x^2}} \approx \frac{1}{1 + y_x^2/2} \approx 1 - \frac{1}{2}y_x^2.$$

Doing so leads to the nonlinear equation

$$y_{tt} = c^2 y_{xx}\left(1 - \frac{1}{2}y_x^2\right),$$

which is more amenable to analysis than (2.15), but still retains some of its nonlinearity. It is interesting that this equation is a cubic nonlinear version of (1.2) when $\varepsilon \gg h^2/12$. Note: if $y_x$ becomes large (e.g., a "shock" is formed), then we need to go back to (2.15) and reconsider our assumptions.

## 2.9 Linear wave equation

Consider the linear wave equation

$$u_{tt} - c^2 u_{xx} = 0,$$

where $c = $ constant with the initial conditions $u(x,0) = f(x)$ and $u_t(x,0) = g(x)$. Below we derive the solution of this equation given by D'Alembert.

It is helpful to change to the coordinate system

$$\xi = x - ct, \qquad \eta = x + ct.$$

Thus, from the chain rule we have that

$$\partial_x = \frac{1}{2}(\partial_\xi + \partial_\eta), \qquad \partial_t = \frac{1}{2c}(-\partial_\xi + \partial_\eta)$$

and the equation transforms into

$$\left[(\partial_\xi + \partial_\eta)^2 - (-\partial_\xi + \partial_\eta)^2\right]u = 0.$$

Simplifying leads to

$$4u_{\xi\eta} = 0.$$

The latter equation can be integrated with respect to $\xi$ to give

$$u_\eta = \tilde{F}(\eta),$$

where $\tilde{F}$ is an arbitrary function. Integrating once more gives

$$u(\xi, \eta) = F(\eta) + G(\xi) = F(x + ct) + G(x - ct).$$

This is the general solution of the wave equation. It remains to incorporate the initial data. We have that

$$u(x,0) = F(x) + G(x) = f(x),$$
$$u_t(x,0) = cF'(x) - cG'(x) = g(x).$$

Differentiating the first equation leads to

$$f'(x) = F'(x) + G'(x),$$
$$g(x) = cF'(x) - cG'(x).$$

By addition and subtraction of these equations one gets that

$$F' = \frac{1}{2}\left(f' + \frac{g}{c}\right), \qquad G' = \frac{1}{2}\left(f' - \frac{g}{c}\right).$$

It follows that

$$F(x) = \frac{1}{2}f(x) + \frac{1}{2c}\int_0^x g(\zeta)\,d\zeta + c_1,$$

$$G(x) = \frac{1}{2}f(x) - \frac{1}{2c}\int_0^x g(\zeta)\,d\zeta + c_2,$$

where $c_1, c_2$ are constants. Hence,

$$u(x,t) = \frac{1}{2}\Big[f(x+ct) + f(x-ct)\Big] + \frac{1}{2c}\int_{x-ct}^{x+ct} g(\zeta)\,d\zeta + c_3, c_3 = c_1 + c_2.$$

Since $u(x,0) = f(x)$ it follows that $c_3 = 0$ and that

$$u(x,t) = \frac{1}{2}\Big[f(x+ct) + f(x-ct)\Big] + \frac{1}{2c}\int_{x-ct}^{x+ct} g(\zeta)\,d\zeta.$$

This is the well-known solution of D'Alembert to the wave equation. From this solution we can see that the wave (or disturbance) or a discontinuity propagates along the lines $\xi = x + ct =$ constant and $\eta = x - ct =$ constant, called the characteristics. More precisely, when looking forward in time we see that a disturbance at $t = 0$ located in the region $a < x < b$ propagates within the "range of influence" that is bounded by the lines $x - ct = \xi_R = b$ and $x + ct = \eta_L = a$ (see Figure 2.12). Similarly, looking back in time we see that the solution at time $t$ was generated from a region in space that is bounded by the "domain of dependence" bounded by the characteristics $x - ct = a$ and $x + ct = b$. Thus no "information" originating inside $a < x < b$ can affect the region outside the range of influence; $c$ can be thought of as the "speed of light".

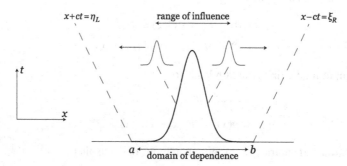

Figure 2.12 The characteristics of the wave equation; domain of dependence and range of influence.

## 2.10 Characteristics of second-order equations

We already encountered the method of characteristics in Section 2.6 for the case of first-order equations (in the evolution variable $t$). The notion of characteristics, however, can be extended to higher-order equations, as we just saw in D'Alembert's solution to the wave equation. Loosely speaking, characteristics are those curves along which discontinuities can propagate. A more formal definition for characteristics is that they are those curves along which the Cauchy problem does not have a unique solution. For second-order quasilinear PDEs the Cauchy problem is given by

$$Au_{xx} + Bu_{xy} + Cu_{yy} = D, \qquad (2.16)$$

where we assume that $A$, $B$, $C$, and $D$ are real functions of $u$, $u_x$, $u_y$, $x$, and $y$ and the boundary values are given on a curve $C$ in the $(x, y)$-plane in terms of $u$ or $\partial u/\partial n$ (the latter being the normal derivative of $u$ with respect to the curve $C$).

We will not go into the formal derivation of the equations for the characteristics, which can be found in many PDE textbooks (see, e.g., Garabedian, 1984). A quick method for deriving these equations can be accomplished by keeping in mind the basic property of a characteristic – a curve along which discontinuities can propagate, cf. Whitham (1974). One assumes that $u$ has a small discontinuous perturbation of the form

$$u = U_s + \varepsilon\Theta(v(x, y))V_s, \qquad (2.17)$$

where $U_s, V_s$ are smooth functions, $\varepsilon$ is small, and

$$\Theta(v) = \begin{cases} 0, & v < 0 \\ 1, & v > 0 \end{cases}$$

is the Heaviside function (sometimes denoted by $H(v)$). Here $v(x, y) = $ constant are the characteristic curves to be found. We recall that $\Theta'(v) = \delta(v)$, i.e., the derivative of a Heaviside function is the Dirac delta function (cf. Lighthill 1958). Roughly speaking, $\delta$ is less smooth than $\Theta$ and near $v = 0$ is taken to be "much larger" than $\Theta$, which is less smooth than $U_s$, etc. In turn, $\delta'$ is more singular (much larger) than $\delta$ near $v = 0$, etc. When we substitute (2.17) into (2.16) we keep the highest-order terms, which are those terms that multiply $\delta'$, and neglect the less singular terms that multiply $\delta$, $\Theta$, and $U_s$. This gives us

$$\left(Av_x^2 + Bv_xv_y + Cv_y^2\right)\varepsilon\delta'V_s + \text{l.o.t.} = 0,$$

(here l.o.t., lower-order terms, means terms that are less singular) or

$$Av_x^2 + Bv_xv_y + Cv_y^2 = 0. \tag{2.18}$$

Since $v(x, y) =$ constant are the characteristics, we have that $dv = v_xdx + v_ydy = 0$ and therefore

$$\frac{dy}{dx} = -\frac{v_x}{v_y}.$$

Combined with (2.18), we obtain

$$A\frac{dy}{dx}^2 - B\frac{dy}{dx} + C = 0 \tag{2.19}$$

as the equation for the characteristic $y = y(x)$.

## 2.11 Classification and well-posedness of PDEs

Based on the types of solutions to (2.19), it is possible to classify quasilinear PDEs in two independent variables. Letting $\lambda \equiv dy/dx$, we have the following classification (recall $A, B, C$ are assumed real):

- Hyperbolic: Two, real roots $\lambda_1$ and $\lambda_2$;
- Parabolic: Two, real repeated roots $\lambda_1 = \lambda_2$; and
- Elliptic: Two, complex conjugate roots $\lambda_1$ and $\lambda_2 = \lambda_1^*$.

The terminology comes from an analogy between the quadratic form (2.18) and the quadratic form for a plane conic section (Courant and Hilbert, 1989).

Some prototypical examples are:

(a) The wave equation: $u_{tt} - c^2u_{xx} = 0$ (note the interchange of variables $x \to t$, $y \to x$). Here $A = 1$, $B = 0$, $C = -c^2$. Thus (2.19) gives

$$\left(\frac{dx}{dt}\right)^2 - c^2 = 0,$$

which implies the characteristics are given by $x \pm ct =$ constant. There are two, *real* solutions. Therefore, the wave equation is hyperbolic.

(b) The heat equation: $u_t - u_{xx} = 0$ (in this example, $y \to t$). Here $A = -1$, $B = 0$, $C = 0$. Thus (2.19) gives

$$\left(\frac{dt}{dx}\right)^2 = 0,$$

which implies the characteristics are given by $t =$ constant. There are two, *repeated* real solutions. Therefore, the heat equation is parabolic.

(c) The potential equation (often called Laplace's equation): $u_{xx} + u_{yy} = 0$. Here $A = 1$, $B = 0$, $C = 1$. Thus (2.19) gives

$$\left(\frac{dy}{dx}\right)^2 + 1 = 0,$$

which implies the characteristics are given by $y = \pm ix +$ constant. There are two, *complex* solutions. Therefore, the potential equation is elliptic.

That the potential equation is elliptic suggests that we can specify "initial" data on any real curve and obtain a unique solution. However, we will see in doing this that something goes seriously wrong. This leads us to the concept of *well-posedness*. To this end, consider the problem

$$u_{yy} + u_{xx} = 0,$$
$$u(x, 0) = f(x),$$
$$u_y(x, 0) = g(x),$$

and look for solutions of the form $u_s = \exp\left[i(kx - \omega(k)y)\right]$. This leads to the dispersion relation $\omega = \pm ik$. We then have the formal solution

$$u(x, y) = \frac{1}{2\pi} \int b_+(k, 0) \exp(ikx) \exp(ky)\, dk$$
$$+ \frac{1}{2\pi} \int b_-(k, 0) \exp(ikx) \exp(-ky)\, dk,$$

where $b_+$ and $b_-$ are determined from the initial data. In general, these integrals do not converge. To make them converge we need to require that $b_\pm(k)$ decays faster than any exponential, which generally speaking is not physically reasonable. The "problem" is that the dispersion relation growth rate, $\omega(k) = \pm ik$, is unbounded and imaginary.

Now suppose we consider the following initial data on $y = 0$: $f(x) = 0$ and $g(x) = g_k(x) = \sin(kx)/k$, where $k$ is a positive integer. Using separation of variables, we find, for each value of $k$, the solution is

$$u_k(x, y) = [A_0 \cosh(ky) + B_0 \sinh(ky)] \sin(kx),$$

where $A_0$ and $B_0$ are constants to be determined. Using $u_k(x, 0) = 0$, we find that $A_0 = 0$. And

$$\frac{\partial u_k}{\partial y}(x, 0) = \frac{\sin(kx)}{k} = B_0 k \sin(kx)$$

implies $B_0 = 1/k^2$. Thus, the solution is

$$u_k(x,y) = \frac{\sinh(ky)\sin(kx)}{k^2}.$$

Now from a physical point of view, we expect that if the initial data becomes "small", so should the solution. However, for any non-zero fixed value of $y$, we see that for each $x$, $u_k(x,y) \to \infty$ as $k \to \infty$, even though $g_k \to 0$. The solution does not depend continuously on the data. We say for a given initial/boundary value problem:

**Definition 2.1** The problem is well-posed if there exists a unique solution and given a sequence of data $g_k$ converging to $g$ as $k \to \infty$, there exists a solution $u_k$ and $u_k \to u_*$ as $k \to \infty$, where $u_*$ is the solution corresponding to the data $g$.

In other words, for the equation to be well-posed it must have a unique solution that depends continuously on the data. Hence the potential equation (i.e., Laplace's equation) is not well-posed. See also Figure 2.13.

As a final example, we will look at the Klein–Gordon equation:

$$u_{tt} - c^2 u_{xx} + m^2 u(x,t) = 0, \tag{2.20}$$

$$u(x,0) = f(x), \qquad u_t(x,0) = g(x).$$

This equation describes elastic vibrations with a restoring force proportional to the displacement and also occurs in the description of the so-called weak interaction, which is somewhat analogous to electromagnetic interactions. In the latter case, the constant $m$ is proportional to the boson (analogous to the photon) mass and $c$ is the speed of light. Remembering the formal correspondence $\partial_t \leftrightarrow -i\omega$ and $\partial_x \leftrightarrow ik$, we find the dispersion relationship $\omega_\pm = \pm\sqrt{c^2 k^2 + m^2} = \pm\omega(k)$. Then

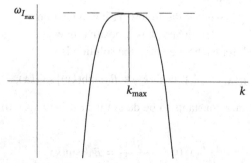

Figure 2.13 $\omega = \omega(k) = (\omega_R + i\omega_I)(k)$: ill-posed when $\omega_I \to \infty$ as $k \to \infty$; well-posed when $\omega$ is bounded as in the above figure.

$$u(x,t) = \frac{1}{2\pi} \int b_+(k,0) \exp\left[i\left(kx + ck\sqrt{1 + \left(\frac{m}{ck}\right)^2}\,t\right)\right] dk$$

$$+ \frac{1}{2\pi} \int b_-(k,0) \exp\left[i\left(kx - ck\sqrt{1 + \left(\frac{m}{ck}\right)^2}\,t\right)\right] dk, \quad (2.21)$$

where $b_+$ and $b_-$ will be determined from the initial data. We have

$$f(x) = \frac{1}{2\pi}\left[\int b_+(k)\exp(ikx)dk + \int b_-(k)\exp(ikx)\,dk\right] \quad (2.22a)$$

$$g(x) = \frac{i}{2\pi}\left[\int b_+(k)\omega(k)\exp(ikx)\,dk - \int b_-(k)\omega(k)\exp(ikx)\,dk\right]. \quad (2.22b)$$

Let

$$\widehat{f}(k) \equiv \int f(x)\exp(-ikx)\,dx,$$

$$\widehat{g}(k) \equiv \int g(x)\exp(-ikx)\,dx;$$

then (2.22) implies that

$$\widehat{f} = b_+ + b_-,$$

$$\frac{-i\widehat{g}}{\omega} = b_+ - b_-.$$

Solving this system and substituting back into (2.21), we find that

$$u(x,t) = \frac{1}{4\pi} \int \left(\widehat{f}(k) - \frac{i\widehat{g}(k)}{\omega(k)}\right) \exp\left[i(kx + \omega(k)t\right]\,dk$$

$$+ \frac{1}{4\pi} \int \left(\widehat{f}(k) + \frac{i\widehat{g}(k)}{\omega(k)}\right) \exp\left[i(kx - \omega(k)t\right]\,dk.$$

As an example, for the specific initial data $u(x,0) = \delta(x)$, where $\delta(x)$ is the Dirac delta function, we have $u_t(x,0) = 0$, and so $\widehat{f}(k) = 1$ and $\widehat{g}(k) = 0$. Thus,

$$u(x,t) = \frac{1}{2\pi} \int e^{ikx} \cos\left[\omega(k)t\right]\,dk,$$

which due to $\sin(kx)$ being an odd function simplifies to

$$u(x,t) = \frac{1}{2\pi} \int \cos(kx)\cos\left[\omega(k)t\right]\,dk.$$

Finally, note that the solution method based on Fourier transforms may be summarized schematically:

$$u(x,0) \xrightarrow{\text{Direct transformation}} \hat{u}(k,0)$$

$$\left\downarrow \text{Evolution}\right.$$

$$u(x,t) \xleftarrow{\text{Inverse transformation}} \hat{u}(k,t)$$

where $\hat{u}(k,t)$ is the Fourier transform of $u(x,t)$. In Chapters 8 and 9, this will be generalized to certain types of nonlinear PDEs.

## Exercises

2.1   Solve the following equations using Fourier transforms (all functions are assumed to be Fourier transformable):

     (a)   $u_t + u_{5x} = 0, u(x,0) = f(x)$.

     (b)   $u_t + \int K(x - \xi)u(\xi,t)d\xi = 0, u(x,0) = f(x), \hat{K}(k) = e^{-k^2}$. Recall the convolution theorem for Fourier transforms (cf. Ablowitz and Fokas, 2003: $\mathcal{F} \int f(x - \xi)g(\xi,t)d\xi = \hat{F}(k)\hat{G}(k)$ where $\mathcal{F}$ represents the Fourier transform).

     (c)   $u_{tt} + u_{4x} = 0, u(x,0) = f(x), u_t(x,0) = g(x)$.

     (d)   $u_{tt} - c^2 u_{xx} - m^2 u(x,t) = 0, u(x,0) = f(x), u_t(x,0) = g(x)$. Contrast this solution with that of the standard Klien–Gordon equation, see (2.20).

2.2   Analyze the following equations using the method of characteristics:

     (a)   $u_t + c(u)u_x = 0$, with $u(x,0) = f(x)$ where $c(u)$ and $f(x)$ are smooth functions of $u$ and $x$, respectively.

     (b)   $u_t + uu_x = -\alpha u$, where $\alpha$ is a constant and $u(x,0) = f(x)$, with $f(x)$ a smooth function of $x$.

2.3   Find two conservation laws associated with

     (a)   $u_{tt} - u_{xx} + m^2 u = 0$, where $m \in \mathbb{R}$.

     (b)   $u_{tt} + u_{xxxx} = 0$.

2.4   Show that the center of mass $\bar{x} = \int xu(x,t)\,dx$ satisfies the following relations:

     (a)   For $u_t + u_{xxx} = 0, \dfrac{d\bar{x}}{dt} = 0$.

     (b)   For $u_t + uu_x + u_{xxx} = 0, \dfrac{d\bar{x}}{dt} = c$, where $c$ is proportional to $\int u^2\,dx$.

2.5   Use D'Alembert's solution to solve the initial value problem of the wave
      equation on the semi-infinite line

$$u_{tt} - c^2 u_{xx} = 0,$$

$c$ constant, where $u(x, 0) = f(x), u_t(x, 0) = g(x), x > 0$ and either
(a) $u(0, t) = h(t)$ or
(b) $u_t(0, t) = h(t)$.

2.6   Consider the PDE

$$u_{tt} - u_{xx} + \sigma u_{xxxx} = 0,$$

where $\sigma = \pm 1$. Determine for which values of $\sigma$ that the equation is well-posed and solve the Cauchy problem for the case where the equation is well-posed.

2.7   Consider the equation

$$u_t + u u_x = 0,$$

with initial condition $u(0, x) = \cos(\pi x)$.
(a) Find the (implicit) solution of the equation.
(b) Show that $u(x, t)$ has a point $t = t_B$ where $u_x$ is infinite. Find $t_B$.
(c) Describe the solution beyond $t = 1/\pi$.

2.8   (a) Solve the Cauchy problem

$$u_x + u u_y = 1, \qquad u(0, y) = ay,$$

where $a$ is a non-zero constant.
(b) Find the solution of the equation in part (a) with the data

$$x = 2t, \qquad y = t^2, \qquad u(0, t^2) = t.$$

Hint: The following forms are equivalent:

$$\frac{dx}{1} = \frac{dy}{u} = \frac{du}{1}$$

and

$$\frac{dx}{ds} = 1, \qquad \frac{dy}{ds} = u, \qquad \frac{du}{ds} = 1.$$

2.9   Consider the equation

$$u_t + u u_x = 1, \qquad -\infty < x < \infty, \qquad t > 0.$$

(a) Find the general solution.
(b) Discuss the solution corresponding to: $u = \frac{1}{2} t$ when $t^2 = 4x$.
(c) Discuss the solution corresponding to: $u = t$ when $t^2 = 2x$.

2.10  Solve the equation

$$u_t + uu_x = 0, \qquad -\infty < x < \infty, t > 0,$$

subject to the initial data

$$u(x,0) = \begin{cases} 0, & x \leq 0 \\ x/a, & 0 < x < a \\ 1, & x \geq a. \end{cases}$$

Examine the solution as $a \to 0$.

2.11  Find the the solution of the equation

$$yu_x - xu_y = 0,$$

corresponding to the data $u(x,0) = f(x)$. Explain what happens if we give $u(x(s), y(s)) = f(s)$ along the curve defined by $\{s : x^2(s) + y^2(s) = a^2\}$.

2.12  Find the solution to the initial value problem

$$u_t + u^3 u_x = 0,$$
$$u(x,0) = f(x).$$

Find the solution when $f(x) = x^{1/3}$ and discuss its behavior as $t \to \infty$.

2.13  Use the traveling wave coordinate $\xi = x - ct$ to reduce the PDE

$$u_t - u_{xx} = \begin{cases} u, & 0 \leq u \leq 1/2 \\ 0, & 1/2 \leq u \leq 1. \end{cases}$$

to an ODE. Solve the ODE subject to the boundary conditions $u(-\infty) = 1$ and $u(+\infty) = 0$. Discuss the solution for $c = 2$ and $c \neq 2$.

# 3

## Asymptotic analysis of wave equations: Properties and analysis of Fourier-type integrals

In the previous chapter we constructed the solution of equations using Fourier transforms. This yields integrals that depend on parameters describing space and time variation. When used with asymptotic analysis the Fourier method becomes extremely powerful. Long-time asymptotic evaluation of integrals yields the notion of group velocity and the structure of the solution. In this chapter, we apply these techniques to the linear Schrödinger equation, the linear KdV equation, discrete equations, and to the Burgers equation, which is reduced to the heat equation via the Hopf–Cole transformation.

Consider the asymptotic behavior of Fourier integrals resulting from the use of Fourier transforms (see Chapter 2). These integrals take the form

$$u(x,t) = \frac{1}{2\pi} \int \widehat{u}(k) \exp\left[i(kx - \omega(k)t)\right] dk \qquad (3.1)$$

in the limit as $t \to \infty$ with $\chi \equiv x/t$ held fixed, i.e., we will study approximations to (3.1) for large times and for large distances from the origin. We can rewrite this integral in a slightly more general form by defining $\phi(k) \equiv k\chi - \omega$ (for simplicity of notation here we suppress the dependence on $\chi$ in $\phi$) and considering integrals of the form

$$I(t) = \int_a^b \widehat{u}(k) \exp\left[i\phi(k)t\right] dk, \qquad (3.2)$$

where we assume $\phi(k)$ is smooth on $-\infty \leq a < b \leq \infty$. First, by the Riemann–Lebesgue lemma (cf. Ablowitz and Fokas, 2003), provided $\widehat{u}(k) \in L^1(a,b)$, we have

$$\lim_{t \to \infty} I(t) = 0.$$

But this does not provide any additional qualitative information. Insight can be obtained if we assume that both $\hat{u}(k)$ and $\phi(k)$ are sufficiently smooth. If $\phi'(k) \neq 0$ in $[a,b]$, then we can rewrite (3.2) as

45

$$\int_a^b \frac{\widehat{u}(k)}{i\phi'(k)t} \frac{d}{dk} e^{i\phi(k)t} dk;$$

then integrating by parts gives

$$I(t) = e^{i\phi(k)t} \frac{\widehat{u}(k)}{i\phi'(k)t} \Bigg|_a^b - \int_a^b \exp\left[i\phi(k)t\right] \frac{d}{dk}\left(\frac{\widehat{u}(k)}{i\phi'(k)t}\right) dk.$$

Repeating the process we can show that the second term is of order $1/t^2$.[1] Thus, we have

$$I(t) = i\left[\frac{\widehat{u}(a)e^{i\phi(a)t}}{\phi'(a)} - \frac{\widehat{u}(b)e^{i\phi(b)t}}{\phi'(b)}\right] \frac{1}{t} + O(1/t^2).$$

If $\widehat{u}(k)$ and $\phi'(k)$ are $C^\infty$ (i.e., infinitely differentiable) functions, then we may continue to integrate by parts to get the asymptotic expansion

$$I(t) \sim \sum_{n=1}^\infty \frac{c_n(t)}{t^n},$$

where the $c_n$ are bounded functions. Notice that if $\widehat{u} \in L^1(-\infty, \infty)$, then

$$\lim_{a \to -\infty} \widehat{u}(a) = \lim_{b \to \infty} \widehat{u}(b) = 0.$$

Therefore, if $\phi'(k) \neq 0$ for all $k$ and if both $\widehat{u}(k), \phi(k) \in C^\infty$, then we have

$$u(x, t) = o(t^{-n}), \qquad n = 1, 2, 3, \ldots, \tag{3.3}$$

i.e., $u$ goes to zero faster than any power of $t$.

It is more common, however, that there exists a point or points where $\phi'(c) = 0$. The technique used to analyze this case is called the *method of stationary phase*, which was originally developed by Kelvin (cf. Jeffreys and Jeffreys, 1956; Ablowitz and Fokas, 2003). We will explore this technique now.

## 3.1 Method of stationary phase

When considering the Fourier integral (3.2), we wish to determine the long-time asymptotic behavior of $I(t)$. As our previous analysis shows, intuitively, for large values of $t$, there is very rapid oscillation, leading to cancellation. But, the oscillation is less rapid near points where $\phi'(k) = 0$ and the cancellation will not be as complete. We will call a point, $k$, where $\phi'(k) = 0$, a *stationary point*.

---

[1] We say that $u = O(t^{-n})$ if $\lim_{t \to \infty} |u|t^n = $ constant and $u = o(t^{-n})$ if $\lim_{t \to \infty} |u|t^n = 0$. See also Section 3.2 for the definitions of big "O", little "o" and asymptotic expansions.

Suppose there is a so-called stationary point $c$, $a < c < b$, such that $\phi'(c) = 0$, $\phi''(c) \neq 0$ and $\widehat{u}(c)$ exists and is non-zero. In addition, suppose $\widehat{u}(k)$ and $\phi(k)$ are sufficiently smooth. Then we expect that the dominant contribution to (3.2) should come from a neighborhood of $c$, i.e., the leading asymptotic term is

$$I(t) \sim \int_{c-\epsilon}^{c+\epsilon} \widehat{u}(c) \exp\left[i\left(\phi(c) + \frac{\phi''(c)}{2}(k-c)^2\right)t\right] dk, \qquad (3.4)$$

$$= \widehat{u}(c) \exp(i\phi(c)t) \int_{c-\epsilon}^{c+\epsilon} \exp\left[i\frac{\phi''(c)}{2}(k-c)^2 t\right] dk,$$

where $\epsilon$ is small. Let

$$\xi^2 = \frac{|\phi''(c)|}{2}(k-c)^2 t, \qquad \mu \equiv \text{sgn}(\phi''(c)).$$

Then (3.4) becomes

$$I(t) \sim \sqrt{\frac{2}{t|\phi''(c)|}} \widehat{u}(c) \exp(i\phi(c)t) \int_{-\epsilon\sqrt{\frac{t|\phi''(c)|}{2}}}^{\epsilon\sqrt{\frac{t|\phi''(c)|}{2}}} \exp(i\mu\xi^2) d\xi.$$

For $|\phi''(c)| = O(1)$, extending the limits of integration from $-\infty$ to $\infty$, i.e., we take $\epsilon\sqrt{t} \to \infty$, yields the dominant term. The integral

$$\int_{-\infty}^{\infty} \exp(i\mu\xi^2) d\xi = \sqrt{\pi} \exp\left(i\mu\frac{\pi}{4}\right),$$

is well known, hence

$$I(t) \sim \widehat{u}(c) \sqrt{\frac{2\pi}{t|\phi''(c)|}} \exp\left(i\phi(c)t + i\mu\frac{\pi}{4}\right). \qquad (3.5)$$

If we are considering Fourier transforms, we have $\phi(k) = k\chi - \omega(k)$, for which $\phi'(k) = 0$ implies

$$\chi = \frac{x}{t} = \omega'(k_0) \qquad (3.6)$$

for a stationary point $k_0$. Or, assuming $\omega'$ has an inverse, $k_0 = (\omega')^{-1}(x/t) \equiv k_0(x/t)$. In addition, if we are working with a dispersive PDE, then by definition, except at special points, $\phi''(k) \neq 0$. Substituting this into (3.5),

$$u(x,t) \sim \frac{\widehat{u}(k_0)}{\sqrt{2\pi t|\phi''(k_0)|}} \exp\left(i\phi(k_0)t + i\mu\frac{\pi}{4}\right), \qquad (3.7)$$

which can be rewritten as

$$u(x,t) \sim \frac{A(x/t)}{\sqrt{t}} \exp(i\phi(k_0)t), \qquad (3.8)$$

where $A(x/t)/\sqrt{t}$ is the decaying in time, slowly varying, complex amplitude. This decay rate, $O(1/\sqrt{t})$, is slow, especially when compared to those cases when the solution decays exponentially or as $O(1/t^n)$ for $n$ large. Note that dimensionally, $\omega'$ corresponds to a speed. Stationary phase has shown that the leading-order contribution comes from a region moving with speed $\omega'$, which is termed the group velocity. This is the velocity of a slowly varying packet of waves. We will see later that this is also the speed at which energy propagates.

**Example 3.1**  Linear KdV equation: $u_t + u_{xxx} = 0$. The dispersion relation is $\omega = -k^3$ and so $\phi(k) = kx/t + k^3$. Thus, the stationary points, satisfying (3.6), occur at $k_0 = \pm\sqrt{-x/(3t)}$ so long as $x/t < 0$. Note that we might expect something interesting to happen as $x/t \to 0$, since $\phi'(k)$ vanishes there and implies a higher-order stationary point. We also expect different behavior when $x/t > 0$ since $k_0$ becomes imaginary there.

**Example 3.2**  Free-particle Schrödinger equation: $i\psi_t + \psi_{xx} = 0$, $|x| < \infty$, $\psi(x,0) = f(x)$, and $\psi \to 0$ as $|x| \to \infty$. This is a fundamental equation in quantum mechanics. The wavefunction, $\psi$, has the interpretation that $|\psi(x,t)|^2 dx$ is the probability of finding a particle in $dx$, a small region about $x$ at time $t$. The dispersion relation $\omega = k^2$ shows that the stationary point satisfying (3.6) occurs at $k_0 = x/(2t)$.

## 3.2  Linear free Schrödinger equation

Now we will consider a specific example of the method of stationary phase as applied to the linear free Schrödinger equation

$$iu_t + u_{xx} = 0. \tag{3.9}$$

First assume a wave solution of the form $u_s = e^{i(kx-\omega(k)t)}$, substitute it into the equation and derive the dispersion relation

$$\omega(k) = k^2.$$

Notice that $\omega$ is real and $\omega''(k) = 2 \neq 0$ and hence solutions to (3.9) are in the dispersive wave regime. Thus the solution, via Fourier transforms, is

$$u(x,t) = \frac{1}{2\pi} \int \hat{u}_0(k) e^{i(kx-k^2t)} \, dk,$$

where $\hat{u}_0(k)$ is the Fourier transform of the initial condition $u(x,0) = u_0(x)$. As before, to apply the method of stationary phase, we rewrite the exponential in the Fourier integral the following way

$$u(x,t) = \frac{1}{2\pi} \int \hat{u}_0(k) e^{i\phi(k)t}\, dk,$$

where $\phi(k) = k\chi - k^2$ with $\chi = x/t$. The dominant contribution to the integral occurs at a stationary point $k_0$ when $\phi'(k_0) = 0$; $k_0 = x/(2t)$ in this case. Noting that $\operatorname{sgn}(\phi''(k_0)) = -1$, the asymptotic estimate for $u(x,t)$ using (3.7) is

$$u(x,t) = \frac{\hat{u}_0\left(\frac{x}{2t}\right)}{\sqrt{4\pi t}} \exp\left(i\left(\frac{x}{2t}\right)^2 t - i\frac{\pi}{4}\right) + o\left(\frac{1}{\sqrt{t}}\right). \tag{3.10}$$

A more detailed analysis, using the higher-order terms arising from the Taylor series of $\hat{u}(k)$ and $\phi(k)$ near $k = k_0$, shows that if $\hat{u}_0(k) \in C^\infty$ then we have the more general result

$$u(x,t) \sim \frac{1}{\sqrt{4\pi t}} \exp\left(i\left(\frac{x}{2t}\right)^2 t - i\frac{\pi}{4}\right)\left[\hat{u}_0(\tfrac{x}{2t}) + \sum_{n=1}^{\infty} \frac{g_n(\frac{x}{2t})}{(4it)^n}\right],$$

where $g_n(y)$ are related to the higher derivatives of $\hat{u}_0$ (see Ablowitz and Segur, 1981).

**Definition 3.3** The function $I(t)$ has an *asymptotic expansion* around $t = t_0$ (Ablowitz and Fokas, 2003) in terms of an ordered *asymptotic sequence* $\{\delta_j(t)\}$ if
(a) $\delta_{j+1}(t) \ll \delta_j(t)$, as $t \to t_0$, $j = 1, 2, \ldots, n$, and
(b) $I(t) = \sum_{j=1}^n a_j \delta_j(t) + R_n(t)$ where the remainder $R_n(t) = o(\delta_n(t))$ or

$$\lim_{t \to t_0} \left|\frac{R_n(t)}{\delta_n(t)}\right| = 0.$$

The notation $\delta_{j+1}(t) \ll \delta_j(t)$, means

$$\lim_{t \to t_0} \left|\frac{\delta_{j+1}(t)}{\delta_j(t)}\right| = 0.$$

We also write

$$I(t) \sim \sum_{j=1}^n a_j \delta_j(t),$$

which says $I(t)$ has the asymptotic expansion given on the right-hand side of the symbol "$\sim$". Notice that if $n = \infty$, then we mean $R_n(t) = o(\delta_n(t))$ for all $n$. Asymptotic series expansions may not (usually do not) converge.

The asymptotic solution (3.10) can be viewed as a "slowly varying" similarity or self-similar solution. In general, a similarity solution takes the form

$$u(x,t) = \frac{1}{t^p} f\left(\frac{x}{t^q}\right), \tag{3.11}$$

for suitable constants $p$ and $q$. This important concept describes phenomena in terms of a reduced number of variables, when time and space are suitably rescaled. Similarity solutions find important applications in many areas of science. One well-known example is that of G.I. Taylor who in the 1940s found a self-similar solution to the problem of determining the shock-wave radius and speed after an (nuclear) explosion (Taylor, 1950). He found that the front of an approximately spherical blast wave is described self-similarly with the independent variables scaled according to a two-fifths law, $r/t^{2/5}$, where $r$ is the radial distance from the origin.

We determine the similarity solution to the linear Schrödinger equation by assuming the form (3.11) and inserting it into (3.9); i.e., we assume the ansatz $u_{\text{sim}}(x,t) = \frac{1}{t^p} f(\eta)$ where $\eta = \frac{x}{t^q}$. Then we find the relation

$$\frac{f''}{t^{p+2q}} - i\left(\frac{pf}{t^{p+1}} + \frac{q\eta f'}{t^{p+1}}\right) = 0.$$

In order to keep all the powers of $t$ the same, we require $2q = 1$ but $p$ can be arbitrary. In general, linear equations do not determine both $p$ and $q$ but nonlinear equations usually fix both. For linear equations, $p$ is fixed by additional information; e.g., initial values or side conditions. For the linear free Schrödinger equation we take $p = 1/2$, motivated by our previous analysis with the stationary phase technique. Then our equation takes the form

$$f'' - \frac{i}{2}(\eta f' + f) = 0,$$

which we can integrate directly to obtain

$$f' - \frac{i}{2}\eta f = c_1, c_1 \text{ const.}$$

Introducing the integrating factor $e^{-i\eta^2/4}$, we integrate this equation to find

$$f(\eta) = c_1 \int e^{i(\eta^2 - (\eta')^2)/4} d\eta' + c_2 e^{i\eta^2/4}, c_2 \text{ const.}$$

If $c_1 = 0$, we have the similarity solution

$$u_{\text{sim}}(x,t) = \frac{c_2}{\sqrt{t}} \exp\left(\frac{i}{4}\left(\frac{x}{t}\right)^2 t\right) = \frac{\tilde{c}_2}{\sqrt{4\pi t}} \exp\left(i\tilde{\eta}^2 - i\pi/4\right),$$

where $\tilde{\eta} = \frac{x}{2\sqrt{t}}$, $c_2 = \frac{\tilde{c}_2}{\sqrt{4\pi}} e^{-i\pi/4}$.

By identifying the "constant of integration" $\tilde{c}_2$, with the slowly varying function $\hat{u}_0(x/t)$, we can write the asymptotic solution, obtained earlier via the method of stationary phase, (3.10), as

$$u(x,t) \sim \hat{u}_0(x/(2t))u_{\text{sim}}(x,t) = \frac{\hat{u}_0(x/(2t))}{\sqrt{4\pi t}}e^{i(x/(2t))^2 t}e^{-i\pi/4}. \qquad (3.12)$$

In the argument of the first term $\hat{u}_0$, the spatial coordinate, $x$, is scaled by $1/2t$ but in the second term $(e^{ix^2/(4t)} = e^{i(x/(2\sqrt{t}))^2})$, $x$ is scaled by $1/\sqrt{t}$. For large values of $t$, the first term varies in $x$ much more slowly than the second term. We refer to this as a slowly varying similarity solution. In a sense $u_{\text{sim}}(x,t)$ is an "attractor" for this linear problem. Slowly varying similarity solutions arise frequently – even for certain nonlinear problems such as the nonlinear Schrödinger equation

$$iu_t + u_{xx} \pm |u|^2 u = 0,$$

e.g., see Ablowitz and Segur (1981).

## 3.3 Group velocity

Recall that the "group velocity", defined to be $\omega'(k)$, enters into the method of stationary phase through the solution of the equation $\phi'(k) = 0$; i.e., through the stationary points. We have seen that, for linear problems, the dominant terms in the solution come from the region near the stationary points. The energy can be shown to be transported by the group velocity. Next we describe this using a different asymptotic procedure.

Frequently we look directly for slowly varying wave groups in order to approximate the solution; see Whitham (1974). Suppose we consider, based on our stationary phase result, that the solution to a linear dispersive wave problem [e.g., the linear Schrödinger equation (3.9)] is of the form

$$u(x,t) \sim A(x,t)e^{i\theta(x,t)}, \qquad t \gg 1, \qquad (3.13)$$

$$A(x,t) = \frac{1}{\sqrt{t}}g(x/t),$$

$$\theta(x,t) = (k_0(x/t)x/t - \omega(k_0(x/t)))t = \phi(x/t)t,$$

where $g$ and $\phi$ are slowly varying. That is,

$$\frac{\partial g}{\partial x} = \frac{1}{t}g'\left(\frac{x}{t}\right) \ll 1 \quad \text{and} \quad \frac{\partial g}{\partial t} = -\frac{x}{t^2}g'\left(\frac{x}{t}\right) \ll 1;$$

also $\dfrac{\partial \phi}{\partial x} \ll 1$ and $\dfrac{\partial \phi}{\partial t} \ll 1$. Notice that $\theta$ is rapidly varying. To generalize the notion of wavenumber and frequency as defined in Section 2.2, we take

$$\frac{\partial \theta}{\partial x} = \phi'(x/t) = k_0 \tag{3.14}$$

$$\frac{\partial \theta}{\partial t} = \phi(x/t) - (x/t)\phi'(x/t) = k_0 x/t - \omega(k_0) - k_0 x/t = -\omega(k_0), \tag{3.15}$$

where we used (3.13). So we can define the slowly varying wavenumber and frequency by the above relations. The ansatz (3.13) with a rapidly varying phase (or the real part in cases where the solution $u$ is real) for wave-like problems is sometimes called the WKB approximation (after Wentzel–Kramer–Brillouin) for wave problems. Many linear problems we see naturally have solutions of the WKB type. In fact there exists a wide range of nonlinear problems that admit generalizations of such rapidly varying phase solutions. In Chapter 4 we will study ODEs where such rapidly varying phase methods (i.e., WKB methods) are useful.

From relations (3.14) and (3.15), we find that (dropping the subscript on $k$)

$$k_t = \frac{\partial^2 \theta}{\partial t \partial x} = \frac{\partial^2 \theta}{\partial x \partial t} = -\omega_x \tag{3.16a}$$

$$k_t + \omega_x = 0 \tag{3.16b}$$

$$k_t + \omega'(k)k_x = 0. \tag{3.16c}$$

Equation (3.16b), which follows from (3.16a), is the so-called conservation of wave equation. Assuming $\omega = \omega(k)$, equation (3.16c) is a first-order hyperbolic PDE that, as we have seen, can be solved via the method of characteristics. It follows (see Chapter 2) that $k(x, t)$ is constant along group lines (curves) in the $(x, t)$-plane defined by $dx/dt = \omega'(k(x, t))$ for any slowly varying wave packet.

Now we will show that the energy in the asymptotic solution to (3.13) propagates with the group velocity. Recall that the energy in the solution of a PDE is proportional to the $L^2$-norm in space. We will limit ourselves to the interval $[x_1(t), x_2(t)]$ as the region of integration to find the energy in a wave packet. For this approximation, we have

$$Q(t) = \int_{x_1(t)}^{x_2(t)} |A|^2 \, dx.$$

Differentiating this quantity with respect to $t$ we find

$$\frac{dQ}{dt} = \int_{x_1(t)}^{x_2(t)} \frac{\partial}{\partial t}|A|^2\, dx + |A|^2(x_2)\frac{dx_2}{dt} - |A|^2(x_1)\frac{dx_1}{dt}$$

$$= \int_{x_1(t)}^{x_2(t)} \left\{ \frac{\partial}{\partial t}|A|^2 + \frac{\partial}{\partial x}(|A|^2\omega'(k)) \right\}\, dx, \qquad (3.17)$$

where we used the characteristic curves $\dfrac{dx_2}{dt} = \omega'(k_2)$ and $\dfrac{dx_1}{dt} = \omega'(k_1)$ with $k$ remaining constant along them.

Notice in our notation that $A = g(x/t)/\sqrt{t}$ is associated with the long-time asymptotic result (3.7)–(3.8). So let us substitute this into the expression for $Q(t)$ to get

$$Q(t) = \int_{x_1(t)}^{x_2(t)} \frac{1}{t}\left|g\left(\tfrac{x}{t}\right)\right|^2\, dx.$$

Now make the change of variable $u = x/t$ and, along with the relations from the characteristic curves, $\dfrac{dx_2}{dt} = \omega'(k_2)$ and $\dfrac{dx_1}{dt} = \omega'(k_1)$, we have

$$Q(t) = \int_{\omega'(k_1)+c_1/t}^{\omega'(k_2)+c_2/t} |g(u)|^2\, du,$$

where $c_1, c_2$ are constants, and that for $t \gg 1$ is asymptotically constant in time. Thus to leading order the quantity $Q(t)$ is conserved, i.e., $dQ/dt = 0$. Thus returning to (3.17), since $x_1, x_2$ are arbitrary, we find the conservation equation

$$\frac{\partial}{\partial t}|A|^2 + \frac{\partial}{\partial x}(\omega'(k)|A|^2) = 0. \qquad (3.18)$$

The energy density, $|A|^2$, shows that the group velocity $\omega'(k)$, associated with the energy flux term $\omega'(k)|A|^2$, is the velocity with which the energy is transported. We also note that this equation is of the same form as the conservation of mass equation from fluid dynamics

$$\frac{\partial\rho}{\partial t} + \frac{\partial}{\partial x}(u\rho) = 0,$$

which is discussed later in this book. Here, $\rho$ is the fluid density and $u$ is the velocity of the fluid.

Another way to derive (3.18) is directly from the PDE itself. For the linear free Schrödinger equation, using

$$u^*(iu_t + u_{xx}) = 0,$$
$$u(-iu_t^* + u_{xx}^*) = 0,$$

where $u^*$ is the complex conjugate, and then subtracting these two equations, we find

$$i\left(u_t u^* + u u_t^*\right) + u^* u_{xx} - u u_{xx}^* = 0,$$

$$i\frac{\partial}{\partial t}|u|^2 + \frac{\partial}{\partial x}\left(u^* u_x - u u_x^*\right) = 0,$$

$$\frac{\partial}{\partial t}T + \frac{\partial}{\partial x}F = 0,$$

which we have seen is a conservation law with density $T$ and flux $F$. Assuming $u = Ae^{i\theta}$, we have $T = |A|^2$, as the conserved energy, and the term $F$ is

$$F = -i\left(u^* u_x - u u_x^*\right),$$

$$= -i\left(A^* e^{-i\theta}(iA\theta_x + A_x)e^{i\theta} - Ae^{i\theta}\left(-iA^*\theta_x + A_x^*\right)e^{-i\theta}\right).$$

Recall that $A$ is slowly varying so $|A_x| = \left|A_x^*\right| \ll 1$; dropping these terms and substituting the generalized wavenumber relation (3.14), $\theta_x \sim k$ and $\omega' \sim 2k$, the flux turns out to be, at leading order for $t \gg 1$,

$$F = 2k|A|^2 = \omega'(k)|A|^2.$$

Thus we have again found (3.18) for large time.

## 3.4 Linear KdV equation

The linear Korteweg–de Vries (KdV) equation (which models the unidirectional propagation of small-amplitude long water waves or shallow-water waves) is given by

$$u_t + u_{xxx} = 0. \tag{3.19}$$

In this section we will use multiple levels of asymptotics, the method of stationary phase as discussed before, the method of steepest descent, and self-similarity, to find the long-time behavior of the solution. There are three asymptotic regions to consider:
- $x/t < 0$;
- $x/t > 0$;
- $x/t \to 0$.

The asymptotic solutions found in each of these regions will be "connected" via the Airy function, which is the relevant self-similar solution for this problem.

The Fourier solution to (3.19) is given by

$$u(x, t) = \frac{1}{2\pi} \int \hat{u}_0(k) e^{i(kx - \omega(k)t)} \, dk.$$

Substituting this into (3.19), the dispersion relation is found to be $\omega(k) = -k^3$, which implies

$$u(x, t) = \frac{1}{2\pi} \int \widehat{u}_0(k) e^{i(kx + k^3 t)} \, dk.$$

By the convolution theorem, see Ablowitz and Fokas (2003) this can also be written in the form

$$u(x, t) = \frac{1}{(3t)^{1/3}} \int \mathrm{Ai}\left(\frac{x - x'}{(3t)^{1/3}}\right) u_0(x') \, dx',$$

where Ai denotes the the Airy function; see (3.22)–(3.23).

### 3.4.1 Stationary phase, $x/t < 0$

The method of stationary phase uses the Fourier integral in the form

$$u(x, t) = \frac{1}{2\pi} \int \hat{u}_0(k) e^{i\phi(k)t} \, dk,$$

where $\phi = k\chi + k^3$ and $\chi = x/t$. The important quantities are $\phi' = \chi + 3k^2 = 0$, $k_{0\pm} = \pm\sqrt{\frac{-x}{3t}}$, and $\phi'' = 6k$. Stationary phase is applied only for real stationary points. In this case, when $x/t < 0$, $k_{0\pm} = \pm\sqrt{|x/(3t)|}$ are real.

Since we have two stationary points, we add the contributions from both of them to form our asymptotic estimate. Recalling that the general form of the stationary phase result is

$$u(x, t) \sim \sum_j \frac{\hat{u}_0(k_{0j})}{\sqrt{2\pi t |\phi''(k_{0j})|}} e^{i\phi(k_{0j})t + i\mu_j \pi/4},$$

where the sum is over all stationary points, we have, in our particular case,

$$\phi(k_{0\pm}) = \mp 2 \left|\frac{x}{3t}\right|^{3/2},$$

$$\phi''(k_{0\pm}) = \pm 2\sqrt{3}\sqrt{\left|\frac{x}{t}\right|},$$

$$\mu_\pm = \mathrm{sgn}(\phi''(k_{0\pm})) = \pm 1,$$

$$u(x, t) \sim \frac{\left|\hat{u}_0\left(\sqrt{\left|\frac{x}{3t}\right|}\right)\right|}{\sqrt{\pi t}\left|\frac{3x}{t}\right|^{1/4}} \cos\left[2\left|\frac{x}{3t}\right|^{3/2} t - \frac{\pi}{4} - \widehat{\psi}_0\left(\sqrt{\left|\frac{x}{3t}\right|}\right)\right], \tag{3.20}$$

where we define $\hat{u}_0(k) \equiv |\hat{u}_0(k)|e^{i\widehat{\psi_0}(k)}$ and use the relation $\hat{u}_0(k) = \hat{u}_0^*(-k)$. Equation (3.20) is the leading asymptotic estimate for $t \gg 1$, when $x/t < 0$.

### 3.4.2 Steepest descent, $x/t > 0$

In this section we introduce the asymptotic method of steepest descent by working through an example with the linear KdV equation (3.19). The steepest descent technique is similar to that of stationary phase except we define our Fourier integral with complex-valued $\phi$ as follows

$$I(t) = \int_C f(z)e^{t\phi(z)} \, dz,$$

where $C$ is, in general, a specified contour, which for Fourier integrals is $(-\infty, \infty)$. In contrast, the stationary phase method uses the exponential in the form $e^{i\phi(k)t}$ where $\phi(k) \in \mathbb{R}$ and the contour $C$ is the interval $(-\infty, \infty)$. We are interested in the long-time asymptotic behavior, $t \gg 1$. We can find the dominant contribution to $I(t)$ whenever we can deform (by Cauchy's theorem if deformations are allowed) on to a new contour $C'$ called the steepest descent contour, defined to be the contour where $\phi(z)$ has a constant imaginary part: $\text{Im } \phi(\chi) = $ constant. The theory of complex variables then shows that $\text{Re}\{\phi\}$ has its most rapid change along the steepest descent path (Ablowitz and Fokas, 2003; Erdelyi, 1956; Copson, 1965; Bleistein and Handelsman, 1986). Hence, this asymptotic scheme will incorporate contributions from points along the steepest descent contour with the most rapid change, namely at saddle points, $\phi'(z_0) = 0$ corresponding to a local maxima. The method takes the following general steps:

(a) Find the saddle points $\phi'(z_0) = 0$; they can be complex.
(b) Use Cauchy's theorem to deform the contour $C$ to $C'$, the steepest descent contour through a saddle point $z_0$. The steepest descent contour is defined by $\text{Im}\{\phi(z)\} = \text{Im}\{\phi(z_0)\}$. Use Taylor series to expand about the saddle point and keep only the low-order terms when finding the dominant contribution.

Let us now study the asymptotic solution for the KdV equation (3.19) with $x/t > 0$. The Fourier solution takes the form

$$u(x, t) = \frac{1}{2\pi} \int_C \hat{u}_0(k)e^{t\phi(k)} \, dk$$

where $\phi(k) = i(k \, x \, / \, t + k^3)$ and $C$ is the real axis: $-\infty < k < \infty$. In this regime, $x/t > 0$, $\phi'(k) \neq 0$ for all $k \in C$, and we can perform integration by parts, as in (3.3), to prove that $u(x, t) \sim o(t^{-n})$ for all $n$. But this does not give us the exact rate of decay, though we expect it is exponential. Steepest descent gives the exponential rate of decay under suitable assumptions. First we find the saddle points: $\phi'(k) = i(x/t + 3k^2) = 0$ implies that we have two saddle points, $k_{\pm} = \pm i\sqrt{\frac{x}{3t}}$.

Next, we determine the steepest descent contour: we wish to find the curves defined by the relations $\mathrm{Im}\{\phi(k) - \phi(k_{\pm})\} = 0$ and choose one that "makes sense", namely that the integral will converge along our choice of contour. In practice, one can expand $\phi(k)$ in a Taylor series near a saddle point, keep the second-order terms, and use them to determine the steepest descent contour. In our case, this gives

$$\mathrm{Im}\{\phi(k) - \phi(k_{\pm})\} = \mathrm{Im}\left\{\frac{\phi''(k_{\pm})}{2}(k - k_{\pm})^2\right\}$$

$$= \mathrm{Im}\left\{\mp 3\left(\tfrac{x}{3t}\right)^{1/2} r_{\pm}^2 e^{2i\theta_{\pm}}\right\} = 0$$

where we have defined $k - k_{\pm} = r_{\pm}e^{i\theta_{\pm}}$.

Let us first look at the case $k_+$: We wish to make $\mathrm{Im}\left\{-3\left(\tfrac{x}{3t}\right)^{1/2} r^2 e^{2i\theta}\right\} = 0$ for all $x/t > 0$. This requires that $\mathrm{Im}\{e^{2i\theta}\} = 0$ or $\sin(2\theta) = 0$. This is true when $\theta = 0, \pi/2, \pi, 3\pi/2, 2\pi, \ldots$ In order to choose a specific contour, we must get a convergent result when integrating along this contour. In particular, we need $\mathrm{Re}\left\{-3(x/3t)^{1/2}r^2 e^{2i\theta}\right\} < 0$ or $-\cos(2\theta) < 0$ so that we have a decaying exponential. Therefore, we choose $\theta = 0, \pi$.

Now, for the $k_-$ case, the choices for $\theta$ are the same as before, namely $\mathrm{Im}\{\phi(k) - \phi(k_-)\} = 0$ when $\theta = 0, \pi/2, \pi, 3\pi/2, 2\pi, \ldots$ But this time, the convergent integral relation for the real part requires that $\cos(2\theta) < 0$, so $\theta$ must be $\pi/2$ or $3\pi/2$. We can rule out these choices by considering the contour that passes through the point $k_- = -i\sqrt{\frac{x}{3t}}$. Starting at $-\infty$, it is impossible to pass through the point $k_-$ with local angle $\pi/2$ or $3\pi/2$ and continue on to $+\infty$. There is a "turn" at $k_-$ that sends the contour down to $-i\infty$ and we cannot reach $+\infty$ as required.

Our contour locally (near $k_+$) follows the curve $C' = \left\{re^{i\theta} \in \mathbb{C} \,\middle|\, re^{i\theta} = k - i\sqrt{\frac{x}{3t}}, k \in (-\infty, \infty)\right\}$. Of the two possible choices for $\theta$, we take $\theta = 0$ so that we have a positively oriented contour. In order to satisfy Cauchy's theorem, we must smoothly transition from $-\infty$ up to the curve defined near $k_+$ and then back down to $+\infty$. This is not a problem for us as we are only interested in what happens on the contour near $k_+$.

Now that we have found our steepest descent contour, we evaluate the Fourier integral along this contour using Cauchy's theorem. Assuming no poles between the contours $C$ and $C'$, we deform the contour $C$ to $C'$ and find the asymptotic estimate, with the dominant contribution from the saddle point $k_+$, to be

$$
\begin{aligned}
u(x,t) &\sim \frac{1}{2\pi} \int_{C'} \hat{u}_0(k_+) e^{t\phi(k_+)} e^{\frac{t}{2}\phi''(k_+)(z-k_+)^2}\, dz, \\
&= \frac{1}{2\pi} \hat{u}_0\left(i\sqrt{\frac{x}{3t}}\right) e^{-2\left(\frac{x}{3t}\right)^{3/2}t} \int_{-\infty}^{\infty} e^{-3\sqrt{\frac{x}{3t}}r^2 t}\, dr, \\
&= \frac{\hat{u}_0\left(i\sqrt{\frac{x}{3t}}\right)}{2\pi\sqrt{t}\left(\frac{3x}{t}\right)^{1/4}} e^{-2\left(\frac{x}{3t}\right)^{3/2}t} \int_{-\infty}^{\infty} e^{-s^2}\, ds, \\
&= \frac{\hat{u}_0\left(i\sqrt{\frac{x}{3t}}\right)}{2\sqrt{\pi t}\left(\frac{3x}{t}\right)^{1/4}} e^{-2\left(\frac{x}{3t}\right)^{3/2}t}.
\end{aligned}
\tag{3.21}
$$

In going from the first line to the second line, we made the change of variable $z - k_+ = r$. From the second line to the third, the substitution $r = s/\sqrt{3t\left(\frac{x}{3t}\right)^{1/2}}$ was made. The last line results from the evaluation of the Gaussian integral $\int e^{-s^2}\, ds = \sqrt{\pi}$. We also assumed that $\hat{u}_0(k)$ is analytically extendable in the upper half-plane. If $\hat{u}_0(k)$ had poles between $C$ and $C'$, then these contributions would need to be included; indeed they would be larger than the steepest descent contribution.

### 3.4.3 Similarity solution, $x/t \to 0$

Motivated by the formulas (3.20) and (3.21), let us look for a similarity solution of the linear KdV equation (3.19) with the form

$$
u_{\text{sim}}(x,t) = \frac{1}{(3t)^p} f\left(\frac{x}{(3t)^q}\right).
$$

Substituting this ansatz into (3.19) and calling $\eta = x/(3t)^q$, we obtain the equation

$$
\frac{f'''}{(3t)^{p+3q}} - \frac{1}{(3t)^{p+1}}(3pf + 3q\eta f') = 0.
$$

To obtain an ODE independent of $t$ requires $q = 1/3$; for a linear problem; $p$ is free and is fixed depending on additional (e.g., initial) conditions. The above asymptotic analysis to the IVP implies (based on asymptotic matching of the solution – see below) that $p = 1/3$ as well. Integrating the above equation once gives

$$
f'' - \eta f = c_1 = 0,
\tag{3.22}
$$

where we take $f \to 0$ as $\eta \to \infty$. This equation is known as Airy's equation. The bounded solution to this ODE is a well-known special function of mathematical physics and is denoted by $\mathrm{Ai}(\eta)$ (up to a multiplicative constant), which has the integral representation (Ablowitz and Fokas, 2003)

$$\mathrm{Ai}(\eta) = \frac{1}{2\pi} \int_{-\infty}^{\infty} e^{i(s\eta + s^3/3)} \, ds. \tag{3.23}$$

As with the linear Schrödinger equation (3.12), we can now write the self-similar asymptotic solution to the linear KdV equation as

$$u(x,t) \sim \hat{u}_0(0) u_{\mathrm{sim}}(x,t) = \frac{\hat{u}_0(0)}{(3t)^{1/3}} \, \mathrm{Ai}\left(\tfrac{x}{(3t)^{1/3}}\right).$$

An alternative, and in this case improved, way of doing this is starting with the Fourier integral solution of this equation (as we have done before)

$$u(x,t) = \frac{1}{2\pi} \int \widehat{u}_0(k) e^{i(kx + k^3 t)} \, dk$$

and changing the integration variable to $s$ defined by $k = s/(3t)^{1/3}$. This gives

$$u(x,t) = \frac{1}{2\pi(3t)^{1/3}} \int \widehat{u}_0\left(\frac{s}{(3t)^{1/3}}\right) \exp\left\{ i\left( \frac{sx}{(3t)^{1/3}} + \frac{s^3}{3} \right) \right\} \, ds.$$

Note that when $x/(3t)^{1/3} = O(1)$ (recall $x/t$ is small) the exponent is not rapidly varying. We assume $\hat{u}(k)$ is smooth. Therefore, as $t \to \infty$ we can expand $u_0$ inside the integral in a Taylor series, which gives

$$u(x,t) = \frac{\widehat{u}_0(0)}{2\pi(3t)^{1/3}} \int e^{i(s\eta + s^3/3)} \, ds + \frac{\widehat{u}_0'(0)}{2\pi(3t)^{2/3}} \int s e^{i(s\eta + s^3/3)} \, ds + \cdots,$$

where $\eta = x/(3t)^{1/3}$. Using

$$\mathrm{Ai}'(\eta) = \frac{1}{2\pi} \int_{-\infty}^{\infty} i s e^{i(s\eta + s^3/3)} ds,$$

we arrive at the asymptotic expansion for $x/(3t)^{1/3} = O(1)$,

$$u(x,t) \sim \frac{\widehat{u}_0(0)\,\mathrm{Ai}(\eta)}{(3t)^{1/3}} - \frac{i\widehat{u}_0'(0)\,\mathrm{Ai}'(\eta)}{(3t)^{2/3}} + \cdots. \tag{3.24}$$

### 3.4.4 Matched asymptotic expansion

We have succeeded in determining the asymptotic behavior of the solution to the linear KdV equation (3.19) in three different regions: $x/t < 0$, $x/t \to 0$, and $x/t > 0$. Each region required a different asymptotic technique. Now we will see how the Airy function matches the seemingly disparate solutions together.

Either applying the method of steepest descent directly on the integral solution (3.23) of the Airy equation (Ablowitz and Fokas, 2003) or consulting a reference such as the *Handbook of Mathematical Functions* (Abramowitz and Stegun, 1972) reveals that

$$\text{Ai}(\eta) \sim \frac{e^{-\frac{2}{3}\eta^{3/2}}}{2\sqrt{\pi}\eta^{1/4}} \qquad \text{as} \qquad \eta \to +\infty,$$

$$\text{Ai}(\eta) \sim \frac{\cos\left(\frac{2}{3}\eta^{3/2} - \pi/4\right)}{\sqrt{\pi}|\eta|^{1/4}} \qquad \text{as} \qquad \eta \to -\infty.$$

Then we rewrite (3.20) and (3.21) in the following form:

$$u(x,t) \sim \frac{\left|\hat{u}_0\left(i\left(\frac{x}{3t}\right)^{1/2}\right)\right|}{2\sqrt{\pi}} \frac{e^{-\frac{2}{3}\left(\frac{x}{(3t)^{1/3}}\right)^{3/2}}}{(3t)^{1/3}\left(\frac{x}{(3t)^{1/3}}\right)^{1/4}}, \qquad \frac{x}{t} > 0$$

$$u(x,t) \sim \frac{\left|\hat{u}_0\left(\left|\frac{x}{3t}\right|^{1/2}\right)\right|}{\sqrt{\pi}} \frac{\cos\left(\frac{2}{3}\left|\frac{x}{(3t)^{1/3}}\right|^{3/2} - \frac{\pi}{4} - \widehat{\psi}_0\left(\left|\frac{x}{3t}\right|^{1/2}\right)\right)}{(3t)^{1/3}\left|\frac{x}{(3t)^{1/3}}\right|^{1/4}}, \qquad \frac{x}{t} < 0.$$

Comparing the previous two sets of equations, we will see that this can be interpreted as a slowly varying similarity solution involving the Airy function that matches the asymptotic behavior in all three regions. Calling $\eta = x/(3t)^{1/3}$, we see that the regime $x/t < 0$, $\eta \to -\infty$, exhibits oscillatory behavior and decays like $|\eta|^{-1/4}$ just like the Airy function. When $x/t > 0$, $\eta \to \infty$, we have exponential decay that behaves like $e^{-\frac{2}{3}\eta^{3/2}}/\eta^{1/4}$. We further notice that the formulas (3.20) and (3.21) are valid for $x/t = O(1)$ and $|\eta| \gg 1$. As $x/t \to 0$, they match to the solution (3.24), which is valid for $\eta = O(1)$. We also note that further analysis shows that there is a uniform asymptotic expansion that is valid for all three regions of $\left(x/(3t)^{1/3}\right)$:

$$u(x,t) \sim \frac{\hat{u}_0(k_0) + \hat{u}_0(-k_0)}{2(3t)^{1/3}} \text{Ai}(\eta) - \frac{\hat{u}_0(k_0) - \widehat{u}_0(-k_0)}{2ik_0(3t)^{2/3}} \text{Ai}'(\eta) + \dots,$$

where $k_0 = \sqrt{-x/(3t)}$ (see Figure 3.1). There is, in fact, a systematic method of finding uniform asymptotic expansions (Chester et al., 1957).

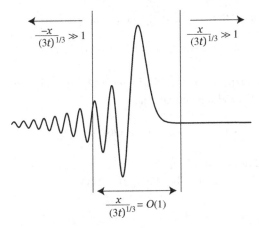

Figure 3.1 Asymptotic expansion for the linear KdV equation (3.19).

## 3.5 Discrete equations

It is well known that discrete equations arise in the study of numerical methods for differential equations. However, discrete equations also arise in important physical applications, cf. Ablowitz and Musslimani (2003) and Christodoulides and Joseph (1998). For example, consider the semidiscrete linear free Schrödinger equation for the function $u_n(t) = u(nh, t)$

$$i\frac{\partial u_n}{\partial t} + \frac{u_{n+1} - 2u_n + u_{n-1}}{h^2} = 0, \tag{3.25}$$

with $h > 0$ constant and given initial values $u_n(t = 0) = f_n$, with $f_n \to 0$ sufficiently rapidly as $|n| \to \infty$. This equation can be used to solve the continuous free Schrödinger equation numerically, in which case the parameter $h$ is taken to be sufficiently small. However, as described in the references, this semidiscrete equation also arises in the study of waveguide arrays in optics, in which case $h$ can be $O(1)$. In either case, we can study this equation using integral asymptotics (Ablowitz and Segur, 1979, 1981).

### 3.5.1 Continuous limit

Before deriving the exact solutions of (3.25) and their long-time asymptotics, it is important to understand the "continuous limit" of this equation as $h \to 0$. To do that we imagine that $x = nh$ is a continuous variable with $n$ large and $h = \Delta x$ small; that is, $n \to \infty$, $h \to 0$, and $nh \to x$. We can then formally expand the solution in a Taylor series around $x$ in the following way:

$$u_{n\pm1}(t) = u(nh \pm h, t) = u(x \pm h, t),$$

$$\approx u(x,t) \pm hu_x(x,t) + \frac{h^2}{2}u_{xx}(x,t) + \cdots.$$

Substituting this expansion into (3.25) and keeping $O(1)$ terms leads to the continuous free Schrödinger equation,

$$iu_t + u_{xx} = O(h^2).$$

The initial conditions can be approximated in a similar manner, i.e., if $u_n(t = 0) = f_n$ then in the continuous limit $u(x, 0) = f(x)$.

### 3.5.2 Discrete dispersion relation

In analogy to the continuous case, we can find wave solutions to (3.25) that serve as the basic building blocks for the solution. Similar to the constant coefficient continuous case, we can look for wave solutions (or special solutions) in the form

$$u_{ns}(t) = Z^n e^{-i\omega t},$$

where $Z = e^{ikh}$. Note that in the continuous limit i.e., as $nh \to x$, we have $Z^n = e^{iknh} \to e^{ikx}$, so this solution becomes $e^{i(kx-\omega t)}$, i.e., the continuous wave solution. Substituting $u_{ns}$ into (3.25) leads to the discrete dispersion relation

$$\omega + \frac{Z - 2 + 1/Z}{h^2} = 0.$$

It is convenient to use Euler's formula $(Z + 1/Z)/2 = \cos(kh)$ to rewrite the dispersion relation as

$$\omega(k) = \frac{2[1 - \cos(kh)]}{h^2}. \tag{3.26}$$

This is the dispersion relation for (3.25). Taking the continuous limit $h \to 0$, $k = O(1)$ fixed, and using a Taylor expansion leads to

$$\omega(k) = k^2,$$

which is exactly the dispersion relation for the continuous Schrödinger equation. But it is important to note that the dispersion relations for the discrete and continuous problems are markedly different for large values of $k$.

### 3.5.3 *Z* transforms and discrete Fourier transforms

The next step in the analysis is to represent the general solution as an integral of the special (wave) solutions, in analogy to the Fourier transform. This kind of representation is sometimes referred to as the Z transform. To understand it better, let us recall some properties of the Fourier transform. In the continuous case, recall we have the pair

$$u(x) = \frac{1}{2\pi} \int_{-\infty}^{\infty} \widehat{u}(k) e^{ikx} \, dk,$$

$$\widehat{u}(k) = \int_{-\infty}^{\infty} u(x) e^{-ikx} \, dx.$$

Substituting the second into the first and interchanging the integrals yields

$$
\begin{aligned}
u(x) &= \frac{1}{2\pi} \int_{-\infty}^{\infty} \left[ \int_{-\infty}^{\infty} u(x') e^{-ikx'} \, dx' \right] e^{ikx} \, dk, \\
&= \frac{1}{2\pi} \int_{-\infty}^{\infty} u(x') \left( \int_{-\infty}^{\infty} e^{ik(x-x')} \, dk \right) dx', \\
&= \int_{-\infty}^{\infty} u(x') \delta(x - x') \, dx', \\
&= u(x),
\end{aligned}
\tag{3.27}
$$

where we have used that the delta function satisfies

$$\delta(x) = \frac{1}{2\pi} \int_{-\infty}^{\infty} e^{ikx} \, dk;$$

see Ablowitz and Fokas (2003) for more details.

Now, the Z transform (ZT) and its inverse (IZT) are given by

$$u_n = \frac{1}{2\pi i} \oint_C \tilde{u}(z) z^{n-1} \, dz, \tag{3.28}$$

$$\tilde{u}(z) = \sum_{m=-\infty}^{\infty} u_m z^{-m}, \tag{3.29}$$

where the contour $C$ is the unit circle in the complex domain. In analogy with the Fourier transform (FT) pair, we substitute (3.29) into (3.28) and use complex analysis to get that

$$u_n = \frac{1}{2\pi i} \oint_C \left[ \sum_{m=-\infty}^{\infty} u_m z^{-m} \right] z^{n-1} \, dz$$

$$= \sum_{m=-\infty}^{\infty} u_m \frac{1}{2\pi i} \oint_C z^{n-m-1} \, dz$$

$$= \sum_{m=-\infty}^{\infty} u_m \delta_{n,m} = u_n, \qquad (3.30)$$

where the Kronecker delta $\delta_{n,m}$ is 1 when $n = m$ and zero otherwise. Here we have relied on the fact that $\dfrac{1}{2\pi i} \oint z^{\alpha} \, dz$, $\alpha$ integer, is equal to 1 only for a simple pole, i.e., $\alpha = -1$, and is otherwise equal to zero. Equation (3.30) using complex analysis is the analog to (3.27) for the ZT.

In discrete problems it is convenient to put the ZT in a "Fourier dress". We can do that by substituting $z$ with $k$ according to $z = e^{ikh}$. Since the contour of integration is the (counterclockwise) unit circle, $k$ ranges from $-\pi/h$ to $\pi/h$ and $z$ goes from $-1$ to 1. Thus, the ZT is transformed into the discrete Fourier transform (DFT)

$$u_n = \frac{1}{2\pi i} \int_{-\pi/h}^{\pi/h} \tilde{u}(z) e^{ik(nh-h)} i h e^{ikh} \, dk = \frac{h}{2\pi} \int_{-\pi/h}^{\pi/h} \tilde{u}(z) e^{iknh} \, dk, \qquad (3.31)$$

using $z(k) = e^{ikh}$. In the continuous limit as $h \to 0$, we have $h\tilde{u}(z) = h\tilde{u}(e^{ikh}) \to \widehat{u}(k)$, $nh \to x$, $u_n = u(nh) \to u(x)$ and we get the continuous FT

$$u(x) = \frac{1}{2\pi} \int_{-\infty}^{\infty} \widehat{u}(k) e^{ikx} \, dk.$$

For the IZT there is a corresponding inverse DFT (IDFT), which can also be obtained using $z(k) = e^{ikh}$:

$$h\tilde{u}(z) = h \sum_{m=-\infty}^{\infty} u_m z^{-m} = \sum_{m=-\infty}^{\infty} u(mh) e^{-i(mh)k} h,$$

or, with $mh \to x$, and noting that the Riemann sum tends to an integral with $h = \Delta x$:

$$\widehat{u}(k) = \int_{-\infty}^{\infty} u(x) e^{-ixk} \, dx.$$

Thus, in the continuous limit one obtains the continuous IFT.

### 3.5.4 Asymptotics of the discrete Schrödinger equation

Using the DFT (3.31) replacing $\tilde{u}(z)$ by $\tilde{u}(z, t) = \tilde{u}_0(z)e^{-i\omega t}$ on the semidiscrete linear free Schrödinger equation (3.25) leads to the solution represented as

$$u_n(t) = \frac{h}{2\pi} \int_{-\pi/h}^{\pi/h} \tilde{u}_0(e^{ikh})e^{iknh}e^{-i\omega(k)t}\, dk,$$

where from (3.26) we have that $\omega(k) = 2[1 - \cos(kh)]/h^2$ and the initial condition is $\tilde{u}_0(e^{ikh}) = \sum_{m=-\infty}^{\infty} u_m(0)e^{-ikmh}$. Therefore, we get that

$$u_n(t) = \frac{h}{2\pi} \int_{-\pi/h}^{\pi/h} \tilde{u}_0(e^{ikh})e^{i\phi(k)t}\, dk,$$

where

$$\phi(k) = k\chi_n - \frac{2[1 - \cos(kh)]}{h^2}$$

and $\chi_n = nh/t$. Using the method of stationary phase we have that

$$\phi'(k) = \frac{nh}{t} - \frac{2\sin(kh)}{h} = 0,$$

from which the stationary points, $k_0$, satisfy

$$\sin(k_0 h) = \frac{nh^2}{2t}. \tag{3.32}$$

Note that the group velocity for the discrete equation is

$$v_g = \omega'(k) = 2\sin(kh)/h \to 2k$$

as $h \to 0$ ($k$ fixed), which is the group velocity corresponding to the continuous Schrödinger equation. In contrast to the continuous case, where $v_g(k)$ is unbounded, in the discrete case we now have that the group velocity is bounded

$$|v_g| = \left| \frac{2\sin(kh)}{h} \right| \leq \frac{2}{h}, \tag{3.33}$$

where equality holds only when $kh = \pm\pi/2$ (the degenerate case). This means that in contrast to the continuous Schrödinger equation, where information can travel with an arbitrarily large speed, in the semidiscrete Schrödinger equation information can travel at most with a finite speed. This is typical in discrete equations. Only in the continuous limit $h \to 0$, $k$ fixed, does $v_g$ become unbounded.

We can further obtain that

$$-\phi''(k) = 2\cos(kh).$$

Using (3.32) gives that at the stationary points

$$-\phi''(k_0) = 2\cos(k_0 h),$$

$$= \pm 2\sqrt{1 - \sin^2(k_0 h)},$$

$$= \pm 2\sqrt{1 - \left(-\frac{nh^2}{2t}\right)^2},$$

$$= \pm \frac{\sqrt{(2t)^2 - (nh^2)^2}}{t}.$$

Thus, using the stationary phase method, we obtain that

$$u_n(t) \sim \sum_{k_0} \frac{\widehat{hu_0}(k_0)e^{i\phi(k_0)t + i\mu\pi/4}}{\sqrt{2\pi t|\phi''(k_0)|}},$$

where $\mu = \text{sgn}(\phi''(k_0))$. Note also that from (3.32), $k_0$ is a function of $n$ and from (3.32) there are generically two values of $k_0$ for each $n$. Note that in the degenerate case $kh = \pm\pi/2$, or $(nh)^2 = (2t)^2$, one gets that $\phi'' = 0$, which means that we need to take higher-order terms in the stationary phase method.

There are three regions that, in principle, can be studied: $|nh^2/2t| < 1$, $|nh^2/2t| > 1$, and near $nh^2/2t = \pm 1$. The first, which is the most important region in this problem, is obtained by the stationary phase method given above. It gives rise to oscillations decaying like $O(1/\sqrt{t})$. The "outside" region has exponential decay (similar to the KdV equation as $x/t \to 0$ and $x/t > 0$). The matching region has the somewhat slower $O(t^{-1/3})$ decay rate (this corresponds to $kh \approx \pi/2$).

This discrete problem has behavior that is similar to the Klein–Gordon equation studied in Chapter 2,

$$u_{tt} - u_{xx} + m^2 u = 0,$$

where the dispersion relation is given by $\omega(k) = \sqrt{k^2 + m^2}$ and the group velocity is bounded,

$$|v_g| = |\omega'(k)| = \left|\frac{k}{\sqrt{k^2 + m^2}}\right| \le 1,$$

cf. equation (3.33).

In a similar way one can also study the asymptotic analysis of the solution to the doubly discrete Schrödinger equation,

$$i\frac{u_n^{m+1} - u_n^{m-1}}{\Delta t} + \frac{u_{n+1}^m - 2u_n^m + u_{n-1}^m}{h^2} = 0,$$

but we leave this as an exercise to the interested reader (see Exercise 3.5).

### 3.5.5 Fully discrete wave equation and the CFL condition

Consider the wave equation

$$u_{tt} = c^2 u_{xx},$$

where $c$ is a constant. Suppose we want to solve this equation using a standard second-order explicit centered finite-difference scheme, i.e.,

$$\frac{u_n^{m+1} - 2u_n^m + u_n^{m-1}}{\Delta t^2} = c^2 \left( \frac{u_{n+1}^m - 2u_n^m + u_{n-1}^m}{\Delta x^2} \right), \tag{3.34}$$

where $\Delta x > 0, \Delta t > 0$ are constant and $u_n^m = u(n\Delta x, m\Delta t)$. When using the scheme (3.34), we are faced with the problem of numerical stability/instability. We can use the ZT in order to understand the origin of this issue as follows. Let us define the special (wave) solutions

$$u_{n\,s}^m = Z^n \Omega^m,$$

where $Z = e^{ikh}$ and $\Omega = e^{-i\omega t}$. Note that as in the continuous limit, substituting $u_{n\,s}^m$ into the scheme above leads to the "dispersion relation"

$$\frac{\Omega - 2 + \dfrac{1}{\Omega}}{\Delta t^2} = c^2 \frac{Z - 2 + \dfrac{1}{Z}}{\Delta x^2}.$$

It is convenient to define

$$p = \frac{\Delta t}{\Delta x} c$$

and rewrite the dispersion relation as a quadratic polynomial in $\Omega$ (or in $Z$) as

$$\Omega^2 - \left[ 2 + p^2 \left( Z - 2 + \frac{1}{Z} \right) \right] \Omega + 1 = 0. \tag{3.35}$$

In this way it is easier to spot the features of the dispersion relation. Indeed, since this relation is quadratic in $\Omega$ it has two roots $\Omega_{1,2}$ and can be written as

$$(\Omega - \Omega_1)(\Omega - \Omega_2) = \Omega^2 - (\Omega_1 + \Omega_2)\Omega + \Omega_1 \Omega_2 = 0.$$

Thus the sum of the two roots must be equal to the square brackets in (3.35) and the product of the two roots must be equal to 1. Hence, there are two cases to consider: either one root is greater than 1 and the other smaller, i.e., $|\Omega_1| < 1$ and $|\Omega_2| > 1$ (the distinct-real root case), or both roots are equal to 1 in magnitude. Defining $b = 1 + p^2(Z - 2 + 1/Z)/2 = 1 + p^2[\cos(kh) - 1]$, where $Z = e^{ikh}$, we see that the discriminant, $\Delta$, of (3.35) is given by $\Delta/4 = b^2 - 1$. Hence when $|b| > 1$ there are two distinct real-valued solutions and when $|b| < 1$ there are two complex roots. The double-root case corresponds to $b = 1$ or $kh = \pm \pi/2$.

When $|b| > 1$ the solutions are unstable, because one of the roots is larger than 1. From a numerical perspective this results in massively unstable growth, or "numerical ill-posedness"; note, the growth rate becomes arbitrarily large as $\Delta t \to 0, m \to \infty$ for fixed $m\Delta t$. The condition for stability $|b| \leq 1$ corresponds to $|1 + p^2[\cos(k\Delta x) - 1]| \leq 1$, which is satisfied by $-2 \leq p^2[\cos(k\Delta x) - 1] \leq 0$, or we can guarantee this inequality for all $k$ if $p^2 \leq 1$, i.e.,

$$0 \leq \frac{\Delta t}{\Delta x} c \leq 1 \quad \text{or} \quad 0 \leq \frac{\Delta t}{\Delta x} \leq \frac{1}{c}.$$

This condition was first discovered and named after Courant, Friedrichs, and Lewy (Courant et al., 1928) and is often referred to as the CFL condition.

## 3.6 Burgers' equation and its solution: Cole–Hopf transformation

In this section we study the solution of the viscous Burgers equation

$$u_t + uu_x = \nu u_{xx}, \tag{3.36}$$

where $\nu$ is a constant, with the initial condition

$$u(x, 0) = f(x),$$

where $f(x)$ is a sufficiently decaying function as $|x| \to \infty$ and $t > 0$. It is convenient to transform (3.36) using $u = \partial\psi/\partial x$ to get

$$\psi_t + \frac{1}{2}\psi_x^2 = \nu\psi_{xx},$$

where we assume $\psi \to$ constant as $|x| \to \infty$. Now let

$$\psi(x, t) = -2\nu \log \phi(x, t), \tag{3.37}$$

so

$$u(x, t) = -2\nu\frac{\phi_x}{\phi}(x, t), \tag{3.38}$$

which is called the *Cole–Hopf* transformation (Cole, 1951; Hopf, 1950). This yields

$$(-2\nu)\frac{\phi_t}{\phi} + \frac{1}{2}(-2\nu)^2\left(\frac{\phi_x}{\phi}\right)^2 = \nu(-2\nu)\left[\frac{\phi_{xx}}{\phi} - \left(\frac{\phi_x}{\phi}\right)^2\right],$$

which simplifies to

$$\phi_t = \nu\phi_{xx}, \tag{3.39}$$

the linear heat equation!

Later in this book we will discuss the linearization and complete solution of some other important equations, such as the KdV equation, by the method of the *inverse scattering transform* (IST).

For Burgers' equation, the solution is obtained by solving (3.39) and transforming back to $u(x, t)$ via (3.38). To solve (3.39) we need the initial data $\phi(x, 0)$, which are found from

$$-2v\frac{\phi_x}{\phi}(x, 0) = f(x)$$

or

$$\phi(x, 0) = h(x) = C \exp\left(-\frac{1}{2v}\int_0^x f(x')\,dx'\right). \tag{3.40}$$

The constant $C$ does not appear in (3.37) so without loss of generality we take $C = 1$.

The solution to the heat equation with $\phi(x, 0) = h(x)$ can be obtained from the fundamental solution or Green's function $G(x, t)$:

$$\phi(x, t) = \int_{-\infty}^{\infty} h(x')G(x - x', t)\,dx',$$

where it is well known that

$$G(x, t) = \frac{1}{\sqrt{4\pi v}}e^{-x^2/4vt}. \tag{3.41}$$

We note that $G(x, t)$ satisfies

$$G_t = vG_{xx},$$
$$G(x, 0) = \delta(x).$$

Further, $G(x, t)$ can be obtained by Fourier transforms

$$G(x, t) = \frac{1}{2\pi}\int_{-\infty}^{\infty} \hat{G}(k, t)e^{ikx}\,dk,$$
$$\hat{G}(k, t) = \int_{-\infty}^{\infty} G(x, t)e^{-ikx}\,dx.$$

Then

$$\hat{G}_t = -vk^2\hat{G}$$

and with $\hat{G}(k, 0) = 1$ we have

$$G(x, t) = \frac{1}{2\pi}\int_{-\infty}^{\infty} e^{ikx-k^2vt}\,dk. \tag{3.42}$$

The integral equation (3.42) can be evaluated by completing the square

$$G(x, t) = e^{-x^2/4vt}\frac{1}{2\pi}\int_{-\infty}^{\infty} e^{-\left(k-\frac{x}{2vt}\right)^2}\,dk = \frac{e^{-x^2/4vt}}{2\sqrt{\pi vt}},$$

which yields (3.41). Thus the solution to the heat equation is

$$\phi(x,t) = \frac{1}{\sqrt{4\pi vt}} \int_{-\infty}^{\infty} h(x')e^{-(x-x')^2/4vt}\, dx',$$

with $\phi(x,0) = h(x)$ given by (3.40). Then the solution of the Burgers equation takes the form

$$u(x,t) = -2v\frac{\phi_x}{\phi} = \frac{\int_{-\infty}^{\infty} \frac{x-x'}{t} h(x')e^{-(x-x')^2/4vt}\, dx'}{\int_{-\infty}^{\infty} h(x')e^{-(x-x')^2/4vt}\, dx'}$$

or using (3.40)

$$u(x,t) = \frac{\int_{-\infty}^{\infty} \frac{x-x'}{t} e^{-\Gamma(x,t,x')/2v}\, dx'}{\int_{-\infty}^{\infty} e^{-\Gamma(x,t,x')/2v}\, dx'}, \qquad (3.43)$$

with

$$\Gamma(x,t,x') = \frac{(x-x')^2}{2t} + \int_0^{x'} f(\zeta)\, d\zeta.$$

In principle we can determine the solution at any time $t$ by carrying out the quadrature in (3.43). But for general integrable functions $f(x)$ the task is still formidable and little qualitative information is obtained from (3.43) by itself. However, in the limit $v \to 0$ we can recover valuable information. When $v \to 0$ we can use Laplace's method (see Ablowitz and Fokas, 2003) to determine the leading-order contributions of the integral. We note that the minimum (in $x'$) of $\Gamma(x,t,x')/2v$, determined from

$$\frac{\partial \Gamma}{\partial x'} = \Gamma'(x') = f(x') - \frac{x-x'}{t} = 0,$$

gives the dominant contribution. An integral of the form

$$I(v) = \int_{-\infty}^{\infty} g(x')e^{-\Gamma(x,t,x')/2v}\, dx',$$

can be evaluated as $v \to 0$ by expanding $\Gamma$ near its minimum $x' = \xi$,

$$\Gamma(x,t,x') = \Gamma(x,t,\xi) + \frac{1}{2}(x'-\xi)^2\Gamma''(x,t,\xi) + \dots,$$

and

$$I(v) \underset{v\to 0}{\sim} \sqrt{\frac{4\pi v}{\Gamma''(x,t,x')}}\, e^{-\Gamma(x,t,\xi)/2v}.$$

Thus the solution $u(x,t)$ from (3.43) is given, as $v \to 0$, by

$$u(x,t) \sim \frac{\frac{x-\xi}{t} e^{-\Gamma(x,t,\xi)/2v} \sqrt{\frac{4\pi v}{\Gamma''(x,t,\xi)}}}{e^{-\Gamma(x,t,\xi)/2v} \sqrt{\frac{4\pi v}{\Gamma''(x,t,\xi)}}}$$

or

$$u(x,t) \sim \frac{x-\xi}{t} = f(\xi).$$

We recall that this is the exact solution to the inviscid Burgers equation

$$u_t + uu_x = 0,$$

with $u(x,0) = f(x)$ in Section 2.6; we found $u(x,t) = f(\xi)$ along $x = \xi + f(\xi)t$ in regions where the solution is smooth. We remark that shocks can occur when there are multiple minima of equal value to $\Gamma(x,t,\nu)$, but we will not go into further details of that solution here.

## 3.7 Burgers' equation on the semi-infinite interval

In this section we will discuss Burgers' equation, (3.36), on the semi-infinite interval $[0, \infty)$.[2] We assume $u(x,t) \to 0$ rapidly as $x \to \infty$ and at $x = 0$ the mixed boundary condition

$$\alpha(t)u(0,t) + \beta(t)u_x(0,t) = \gamma(t), \qquad (3.44)$$

where $\alpha, \beta, \gamma$ are prescribed functions of $t$ and we take for convenience $u(x,0) = 0$. It will be helpful to employ the Dirichlet–Neumann (DN) relationship (often termed the DN map) associated with the heat equation (3.39)

$$\begin{aligned}
\phi_x(0,t) &= -\frac{1}{\sqrt{\nu\pi}} \int_0^t \frac{\phi_\tau(0,\tau)}{\sqrt{t-\tau}}, \\
&= -\frac{1}{\sqrt{\nu}} D_t^{1/2} \phi_0(t), \qquad (3.45)
\end{aligned}$$

where $\phi_0(t) = \phi(0,t)$ and the "half-derivative" is defined as

$$D_t^{1/2}\phi_0(t) = \frac{1}{\sqrt{\pi}} \int_0^t \frac{\phi_\tau(0,\tau)}{\sqrt{t-\tau}}.$$

Thus if we know $\phi(0,t)$ associated with (3.39) then (3.45) determines $\phi_x(0,t)$. Equation (3.45) can be found in many ways, e.g., via Laplace transforms (Ablowitz and Fokas, 2003): define the Laplace transform of $\phi(x,t)$ as

$$\hat{\Phi}(x,s) = \mathcal{L}\{\phi\} = \int_0^\infty \phi(x,t)e^{-st}\,dt;$$

then (3.39) implies

$$\nu\hat{\Phi}_{xx} - s\hat{\Phi} + 1 = 0,$$

[2] The analysis here follows the work of M. Hoefer developed for the "Nonlinear Waves" course taught by M.J. Ablowitz in 2003.

where we take $\phi(x, 0) = 1$ with no loss of generality. Thus

$$\Phi(x, s) = A(s)e^{-\sqrt{\frac{s}{\nu}}x} + \frac{1}{s},$$

where we omitted the exponentially growing part. Then using (3.39) at $x = 0$,

$$\hat{\Phi}_{xx}(0, s) = \mathcal{L}\left\{\frac{1}{\nu}\phi_t(0, t)\right\},$$

implies

$$A(s) = \frac{1}{\sqrt{\nu s}}\mathcal{L}\{\phi_t(0, t)\}.$$

Hence

$$\hat{\Phi}_x(0, s) = -\frac{1}{\sqrt{\nu s}}\mathcal{L}\{\phi_t(0, t)\}$$

or by taking the inverse Laplace transform, using the convolution theorem, and $\mathcal{L}\{1/\sqrt{s}\} = 1/\sqrt{\pi t}$, we find (3.45).

The boundary condition (3.44) is converted to

$$\alpha(t)\frac{\phi_x}{\phi}(0, t) + \beta(t)\left[\underbrace{\frac{\phi_{xx}}{\phi}}_{\frac{1}{\nu}\frac{\phi_t}{\phi}} - \left(\frac{\phi_x}{\phi}\right)^2\right](0, t) + \frac{\gamma(t)}{2\nu} = 0$$

via the Hopf–Cole transformation (3.38). Or calling $\phi_0(t) = \phi(0, t)$ and using (3.45) we have

$$\alpha(t)\frac{D_t^{1/2}\phi_0}{\phi_0} - \frac{\beta(t)}{\sqrt{\nu}}\left[\frac{\phi_{0t}}{\phi_0} - \left(\frac{D_t^{1/2}\phi_0}{\phi_0}\right)^2\right](0, t) - \frac{\gamma(t)}{2\sqrt{\nu}} = 0. \qquad (3.46)$$

We see that, in general, (3.46) is a nonlinear equation. However, in the case of Dirichlet boundary conditions for Burgers' equation, $\beta(t) = 0$, we find from (3.46) a linear integral equation of the Abel type for $\phi_0$:

$$D_t^{1/2}\phi_0 = \frac{g(t)}{2\sqrt{\nu}}\phi_0,$$

where $g(t) = \gamma(t)/\alpha(t)$. Using $D_t^{1/2}D_t^{1/2}\phi_0(t) = \phi_0(t) - 1$, i.e.,

$$\frac{1}{\sqrt{\pi}}\int_0^t \frac{d\tau_1}{\sqrt{t - \tau_1}} \cdot \frac{1}{\sqrt{\pi}}\int_0^t \frac{\phi_{0,\tau_2}d\tau_2}{\sqrt{t - \tau_2}}$$

$$= \frac{1}{\pi}\int_0^t d\tau_2\,\phi_{0,\tau_2}\underbrace{\int_{\tau_2}^t \frac{d\tau_1}{\sqrt{t - \tau_1}\sqrt{\tau_1 - \tau_2}}}_{=\pi} = \phi_0(t) - 1$$

(recall $\phi_0(0) = 1$), we have

$$\phi_0(t) = 1 + \frac{1}{2\sqrt{\nu\pi}} \int_0^t \frac{g(t')\phi_0(t')}{\sqrt{t-t'}}\, dt', \qquad (3.47)$$

which is an inhomogeneous weakly singular *linear* integral equation. This equation was obtained by a different procedure by Calogero and DeLillo (1989). Note, solving (3.47) gives $\phi(0, t)$, which combined with (3.40) allows Burgers' equation to be linearized via (3.39) with the Cole–Hopf transformation (3.38). However, in general, the Hopf–Cole transformation does not linearize the equation. Rather, we see that one has a nonlinear integral equation (3.46) to study; but the number of variables has been reduced, from $(x, t)$ to $t$.

# Exercises

3.1 Analyze the long-time asymptotic solution of
   (a) $iu_t + u_{4x} = 0$,
   (b) $u_t + u_{5x} = 0$, and
   (c) $u_{tt} - u_{xx} + m^2 u = 0, m > 0$ constant.
3.2 Consider the equation

$$u_t + cu_x = 0,$$

   with $c > 0$ constant and $u(x, 0) = f(x)$. Show that the solution is $u = f(x - ct)$ and that this does not decay as $t \to \infty$. Reconcile this with the Riemann–Lesbegue lemma.
3.3 Consider the linear KdV equation

$$u_t + u_{xxx} = 0.$$

   Use the asymptotic analysis discussed in this chapter to analyze the behavior of the energy inside each of the three asymptotic regions as $t \to \infty$. At what speed does each propagate?
3.4 For the differential–difference, free Schrödinger equation,

$$i\frac{\partial u_n}{\partial t} + \frac{u_{n+1} - 2u_n + u_{n-1}}{h^2} = 0,$$

   with $h > 0$ constant, find the long-time solution in the regions $|nh^2/2t| > 1$ and near $nh^2/2t = \pm 1$.
3.5 Consider the partial-difference equation,

$$i\frac{u_n^{m+1} - u_n^{m-1}}{\Delta t} + \frac{u_{n+1}^m - 2u_n^m + u_{n-1}^m}{\Delta x^2} = 0,$$

with $\Delta x > 0, \Delta t > 0$ constant, assuming $u_n^m = u(n, m)$ and $u(n, 0)$ and $u(n, 1)$ are given as initial conditions. Investigate the "long-time", i.e., $m \gg 1$, asymptotic solution.

3.6   Find a nonlinear PDE that is third order in space, first order in time that is linearized by the Cole–Hopf transformation.

3.7   Use the Fourier transform method to solve the boundary value problem

$$u_{xx} + u_{yy} = -x\exp(-x^2),$$

$$u(x, 0) = 0,$$

$$-\infty < x < \infty, \quad 0 < y < \infty,$$

where $u$ and its derivative vanish as $y \to \infty$. Hint: It is convenient to use the inverse Fourier transform result: $\mathcal{F}^{-1}\left\{\frac{1}{k}\hat{f}(k)\,dk\right\} = \int f(x)\,dx$.

3.8   Solve the inhomogeneous partial differential equation

$$u_{xt} = -\omega\sin(\omega t), \qquad t > 0,$$

$$u(x, 0) = x, \qquad u(0, t) = 0.$$

3.9   Find the solution to Burgers' equation (3.36) on the semi-infinite interval $0 < x < \infty$ with $u(x, 0) = f(x)$ where $f(x)$ decays rapidly as $|x| \to \infty$ and $u(0, t) = g(t)$.

3.10  Solve the linear one-dimensional linear Schrödinger equation with quadratic potential (the "simple harmonic oscillator")

$$iu_t = u_{xx} - V_0 x^2 u,$$

with $V_0 > 0$ constant and $u(x, 0) = f(x)$ where $f(x)$ decays rapidly as $|x| \to \infty$. In what sense is the "ground state" (i.e., the lowest eigenvalue) the most important solution in the long-time limit? Hint: Use separation of variables.

# 4

## *Perturbation analysis*

In terms of the methods of asymptotic analysis, so far we have studied integral asymptotics associated with Fourier integrals that represent solutions of linear PDEs. Now, suppose we want to study physical problems like the propagation of waves in the ocean, or the propagation of light in optical fibers; the general equations obtained from first principles in these cases are the Euler or Navier–Stokes equations governing fluid motion on a free surface and Maxwell's electromagnetic (optical wave) equations with nonlinear induced polarization terms. These equations are too difficult to handle using linear methods or, in most situations, by direct numerical simulation. Loosely speaking, these physical equations describe "too much".

Mathematical complications often arise when one has widely separated scales in the problem, e.g., the wavelength of a typical ocean wave is small compared to the ocean's depth and the wavelength of light in a fiber is much smaller than the fiber's length or transmission distance. For example, the typical wavelength of light in an optical fiber is of the order of $10^{-6}$ m, whereas the length (distance) of an undersea telecommunications fiber is of the order of 10,000 km or $10^7$ m, i.e., 13 orders of magnitude larger than the wavelength! Therefore, if we were to try solving the original equations numerically – and resolve both the smallest scales as well as keep the largest ones – we would require vast amounts of computer time and memory.

Yet it is imperative to retain some of the features from all of these scales, otherwise only limited information can be obtained. Therefore, before we can study the solutions of the governing equations we must first obtain useful and manageable equations. For that we need to simplify the general equations, while retaining the essential phenomena we want to study. This is the role of perturbation analysis and the reason why perturbation analysis often plays a decisive role in the physical sciences.

In this chapter we will introduce some of the perturbation methods that will be used later in physical problems. The reader can find numerous references on perturbation techniques, cf. Bender and Orszag (1999), Cole (1968), and Kevorkian and Cole (1981). An early paper with many insightful principles of asymptotic analysis ("asymptotology") is Kruskal (1963). We will begin with simpler "model" problems (ODEs) before discussing more complex physical problems.

## 4.1 Failure of regular perturbation analysis

Consider the ODE

$$\frac{d^2y}{dt^2} + y = \varepsilon y, \tag{4.1}$$

where $\varepsilon$ is a small (constant) parameter, i.e., $|\varepsilon| \ll 1$. Suppose we try to expand the solution as

$$y = y_0 + \varepsilon y_1 + \varepsilon^2 y_2 + \cdots,$$

where the $y_j$, $j = 0, 1, 2, \ldots$, are assumed to be $O(1)$ functions that are to be found. This is usually called regular perturbation analysis, because it is the simplest and nothing out of the ordinary is used. Substituting this expansion into (4.1) leads to an infinite number of equations, i.e., a perturbation series of equations. We group terms according to their power of $\varepsilon$. To $O(1)$, i.e., for those terms that have no $\varepsilon$ before them, we get

$$O(1): \qquad y_{0,tt} + y_0 = 0.$$

This is called the leading-order equation. Its solution, assumed to be real, called the leading-order solution, is conveniently given by

$$y_0(t) = A_0 e^{it} + A_0^* e^{-it} = A_0 e^{it} + \text{c.c.},$$

where $A_0$ is a complex constant and c.c. denotes the complex conjugate of the terms to its left. Clearly, this solution does not have any $\varepsilon$ in it, i.e., it completely disregards the $\varepsilon y$ term in (4.1). However, it may still be relatively close to the exact solution, because $|\varepsilon| \ll 1$. Two questions naturally arise:
(a) How can we improve the approximation?
(b) Is the solution we obtained a good approximation of the exact solution?

We can try to improve the approximation by going to the next order of the perturbation series, i.e., equating the terms that multiply $\varepsilon^1$. This leads to

$$y_{1,tt} + y_1 = y_0.$$

Since we already found $y_0$, this is an equation for $y_1$, the next order in the perturbation series (called the $O(\varepsilon)$ correction). Substituting in the $y_0$ above gives

$$O(\varepsilon): \qquad y_{1,tt} + y_1 = A_0 e^{it} + \text{c.c.} \qquad (4.2)$$

Ignoring unimportant homogeneous solutions, the above equation has a solution of the form

$$y_1(t) = A_1 t e^{it} + \text{c.c.},$$

where $A_1$ is a constant, and substituting it into (4.2) leads to the condition

$$2i\,A_1 = A_0.$$

Therefore, we get that

$$y_1(t) = -\frac{it}{2} A_0 e^{it} + \text{c.c.},$$

where we will omit additional terms due to the homogeneous solution. We can continue in a similar manner to obtain $y_2$: the equation for $y_2$ is found to be

$$y_{2,tt} + y_2 = y_1;$$

we can guess a solution of the form $y_2 = A_2 t^2 e^{it} + A_3 t e^{it} + \text{c.c.}$ and we obtain

$$8A_2 = -A_0, \, 8A_3 = -iA_0.$$

Using the first condition, $2iA_1 = A_0$, we arrive at

$$y_2 = -\left(\frac{t^2}{8} + i\frac{t}{8}\right) A_0 e^{it} + \text{c.c.}$$

Let us inspect the solution we have so far obtained:

$$y = y_0 + \varepsilon y_1 + \varepsilon^2 y_2,$$

$$= (A_0 e^{it} + \text{c.c.}) - \varepsilon\left(\frac{it}{2} A_0 e^{it} + \text{c.c.}\right) - \varepsilon^2\left(\left(\frac{t^2}{8} + i\frac{t}{8}\right) A_0 e^{it} + \text{c.c.}\right), \qquad (4.3)$$

$$= A_0\left(1 - \varepsilon\frac{it}{2} - \varepsilon^2\left(\frac{t^2}{8} + \frac{it}{8}\right)\right) e^{it} + \text{c.c.} \qquad (4.4)$$

It is not difficult to convince ourselves that if we continue in the same fashion to include higher-order terms we will be adding higher-order monomials of $t$ inside the parentheses. So long as $t$ is $O(1)$, our approximate solution should be quite close to the exact one. However, when $t$ becomes large a problem arises with our solution. Indeed, when $t = O(1/\varepsilon)$ the terms $(\varepsilon t)^n$ inside the brackets are of $O(1)$, hence our formal solution is not asymptotic (i.e., the remainder of any finite number of terms is not smaller than the previous terms). It must be realized at this stage that this problem is not a feature of the exact solution itself. Indeed, it is straightforward to obtain the exact solution of the linear equation (4.1):

$$y_{\text{exact}}(t) = Ae^{i\sqrt{1-\varepsilon}\,t} + \text{c.c.}, \qquad (4.5)$$

where $A$ is an arbitrary constant. This solution is finite and infinitely differentiable for any value of $t$. Moreover, from the binomial expansion $\sqrt{1-\varepsilon} = 1 - \varepsilon/2 - \varepsilon^2/8 + \cdots$ and by further expanding the exponential we find the approximation above; i.e., (4.4).

The failure is therefore on the part of the method we used. Inspecting the solution (4.4) we see that the problem arises when $t = O(1/\varepsilon)$ because the terms inside the parentheses become of the same order when $t = O(1/\varepsilon)$. Hence the perturbation method breaks down because the higher-order terms are not small, violating our assumption. The violation is due to the terms that *grow arbitrarily large with t*. Such terms are referred to as "secular" or "resonant". Thus $t^n e^{it}$ is a secular term for any $n > 1$. If we trace our steps back we see that the secular terms arise because the right-hand side of the non-homogeneous equations we obtain are in the kernels of the left-hand sides, e.g., in (4.2) the left-hand side $Ae^{it}$ is a solution of the homogeneous equation for $y_1$. This problem is inevitable when using the regular perturbation method as outlined above. So how can we resolve it? This is where so-called "singular" perturbation methods enter the picture.

## 4.2 Stokes–Poincaré frequency-shift method

We would like to find an approximate solution to (4.1) that is valid for large values of $t$, e.g., up to $t = O(1/\varepsilon)$. One perturbative approach is what we will call the Stokes–Poincaré or frequency-shift method; the essential ideas go back to Stokes (see Chapter 5) and later Poincaré. Although it does not work in every case, it is often the most efficient method at hand. The idea of the frequency-shift method is explained below using (4.1) as an example. We define a new time variable $\tau = \omega t$, where $\omega$ is called the "frequency". Since $d/dt = \omega d/d\tau$, the equation takes the form

$$\omega^2 \frac{d^2 y}{d\tau^2} + y = \varepsilon y.$$

We now expand the solution as before

$$y = y_0 + \varepsilon y_1 + \varepsilon^2 y_2 + \cdots,$$

*but* also expand $\omega$ perturbatively as

$$\omega = 1 + \varepsilon\omega_1 + \varepsilon^2\omega_2 + \cdots.$$

Substituting these expansions into the ODE and collecting the same powers of $\varepsilon$ gives the leading-order equation

$$O(1): \qquad y_{0,\tau\tau} + y_0 = 0,$$

which is the same as we got using regular perturbation analysis. The leading-order solution is therefore

$$y_0(\tau) = Ae^{i\tau} + \text{c.c.}$$

The next-order equation is given by

$$O(\varepsilon): \qquad y_{1,\tau\tau} + y_1 = -2\omega_1 y_{0,\tau\tau} + y_0.$$

This equation has a term containing $\omega_1$, which is a modification of the regular perturbation analysis. Substituting the leading-order solution into the right-hand side leads to

$$y_{1,\tau\tau} + y_1 = (1 + 2\omega_1)Ae^{i\tau} + \text{c.c.}$$

By choosing $\omega_1 = -1/2$, we eliminate the right-hand side and therefore avoid the secular term in the solution. Without loss of generality, we choose the trivial homogeneous solution, i.e., $y_1 = 0$ (since we can "incorporate them" in the leading-order solution). We can continue in this fashion to obtain $\omega_2$; the next-order equation is given by

$$y_{2,\tau\tau} + y_2 = -\left(2\omega_2 + \omega_1^2\right)y_{0,\tau\tau} + y_1 - 2\omega_1 y_1.$$

Using $y_1 = 0$ gives

$$y_{2,\tau\tau} + y_2 = -\left(2\omega_2 + \omega_1^2\right)y_{0,\tau\tau},$$

and choosing $\omega_2 = -\omega_1^2/2 = -1/8$ eliminates the secular terms; finally, we choose $y_2 = 0$.

Let us inspect our approximate solution: the only non-zero part of $y$ is $y_0$, but it contains $\omega$ to order $O(\varepsilon^2)$:

$$y(t) = Ae^{i\tau} + \text{c.c.} = Ae^{i\omega t} + \text{c.c.}$$

$$= Ae^{i(1+\varepsilon\omega_1+\varepsilon^2\omega_2+\cdots)t} + \text{c.c.} = Ae^{i\left[1-\frac{\varepsilon}{2}-\frac{\varepsilon^2}{8}+\cdots\right]t} + \text{c.c.}$$

Hence we expect the solution to be valid for $t = O(1/\varepsilon^2)$. To verify this, let us expand the square-root in the exact solution (4.5) in powers of $\varepsilon$:

$$y_{\text{exact}}(t) = Ae^{i\sqrt{1-\varepsilon}\,t} + \text{c.c.} = Ae^{i\left[1-\frac{\varepsilon}{2}-\frac{\varepsilon^2}{8}+O(\varepsilon^3)\right]t} + \text{c.c.},$$

so our exact and approximate solution agree to the first three terms in the exponent. Note that the remainder term in the exponent is $O(\varepsilon^3 t)$, which is small for $t = o(1/\varepsilon^3)$. Furthermore, the approximate solution can be made valid for larger values of $t$ by finding a sufficient number of terms in the frequency expansion. Note that the higher-order homogeneous solutions are chosen as zero (i.e, $y_k = 0$ for $k \geq 1$) and the higher non-homogeneities only change the amplitude, not the frequency, of the leading-order solution.

This example demonstrates the ease and efficiency of the frequency-shift method. Unfortunately this method does not always work. For example, let us consider the equation with a damping term

$$\frac{d^2y}{dt^2} + y = -\varepsilon\frac{dy}{dt}$$

and repeat the analysis using the frequency-shift method. Defining $\tau = \omega t$ as before leads to the equation

$$\omega^2\frac{d^2y}{d\tau^2} + y = -\varepsilon\omega\frac{dy}{d\tau}.$$

Expanding the solution and $\omega$ in powers of $\varepsilon$ we get

$$(1 + \varepsilon\omega_1 + \varepsilon^2\omega_2 + \cdots)^2\frac{d^2y}{d\tau^2} + y = -\varepsilon(1 + \varepsilon\omega_1 + \varepsilon^2\omega_2 + \cdots)\frac{dy}{d\tau}.$$

The leading-order solution is found to be

$$y_0(\tau) = Ae^{i\tau} + \text{c.c.}$$

The next-order equation is then

$$O(\varepsilon): \qquad y_{1,\tau\tau} + y_1 = -2\omega_1 y_{0,\tau\tau} - y_{0,\tau}.$$

Substituting the leading-order solution into the right-hand side gives that

$$y_{1,\tau\tau} + y_1 = 2\omega_1(Ae^{i\tau} + \text{c.c.}) - (iAe^{i\tau} + \text{c.c.})$$
$$= (2\omega_1 - i)Ae^{i\tau} + (2\omega_1 + i)A^*e^{-i\tau}.$$

It may appear at first glance that we can avoid the secularity by choosing $\omega_1 = i/2$; however, this is not so, because that only eliminates the terms that multiply $e^{i\tau}$. Indeed, to eliminate the terms that multiply $e^{-i\tau}$ we would need to choose $\omega_1 = -i/2$. In other words, we cannot choose $\omega_1$ consistently to eliminate all the secular terms. Without eliminating the secular terms we are back to the original problem, i.e., the solution will not be valid for $t \sim O(1/\varepsilon)$. Another method is required.

The frequency-shift method usually works for conservative systems. The addition of the damping term results in a non-conservative (non-energy conserving) problem.

## 4.3 Method of multiple scales: Linear example

Since the frequency-shift method works in some cases but not in others, we introduce the method of multiple scales, cf. Bender and Orszag (1999), Cole (1968) and Kevorkian and Cole (1981). The method of multiple scales has its roots in the method of averaging, which have "rapid phases" and the other terms are "slowly varying", cf. Krylov and Bogoliubov (1949), Bogoliubov and Mitropolsky (1961); see also Sanders et al. (2009a).

The method of multiple scales is much more robust than the frequency-shift method. It works for a wide range of problems and, in particular, for the problems we will tackle here and for many problems that arise in nonlinear waves, as we will see later in this book. The idea is to introduce *fast* and *slow* variables into the equation. Let us consider the damped example above, where the frequency-shift method failed,

$$\frac{d^2y}{dt^2} + y = -\varepsilon\frac{dy}{dt}. \qquad (4.6)$$

We can introduce the slow variable $T = \varepsilon t$ and consider $y$ to be a function of *two variables*: the "slow" variable $T$ and the "fast" one $t$. We can then denote $y$ as

$$y = y(t, T; \varepsilon),$$

where the $\varepsilon$ is added here in order to stress that we will expand the solutions in powers of $\varepsilon$. To be precise, we introduce two variables $\tilde{t} = t$ and $T = \varepsilon t$, however, we will omit the "~" from $\tilde{t}$ hereafter for notational convenience.

Using the chain rule for differentiation gives

$$\frac{dy}{dt} \rightarrow \frac{\partial y}{\partial t} + \varepsilon\frac{\partial y}{\partial T}$$

or in operator form

$$\frac{d}{dt} \rightarrow \frac{\partial}{\partial t} + \varepsilon\frac{\partial}{\partial T}$$

and the equation is transformed into

$$\left(\frac{\partial}{\partial t} + \varepsilon\frac{\partial}{\partial T}\right)^2 y + y = -\varepsilon\left(\frac{\partial}{\partial t} + \varepsilon\frac{\partial}{\partial T}\right)y,$$

or

$$y_{tt} + 2\varepsilon y_{tT} + \varepsilon^2 y_{TT} + y = -\varepsilon(y_t + \varepsilon y_T). \qquad (4.7)$$

It is important to note here that we have transformed an ODE into a PDE! Thus, formally, we have complicated the problem by this transformation, but this complication is valuable, as we will now see. We now expand the solution as

$$y = y_0 + \varepsilon y_1 + \varepsilon^2 y_2 + \cdots,$$

substitute into the PDE (4.7), and collect like powers of $\varepsilon$. This leads to the leading-order equation

$$O(1): \qquad y_{0,tt} + y_0 = 0.$$

Strictly speaking, this is a PDE since $y$ depends on $T$ as well. So the general solution is given by

$$y_0(t, T) = A(T)e^{it} + \text{c.c.},$$

where $A(T)$ is an arbitrary function of $T$. The next-order equation is found to be

$$O(\varepsilon): \qquad y_{1,tt} + y_1 = -2y_{0,tT} - y_{0t}.$$

Substituting the leading-order solution into the right-hand side gives that

$$y_{1,tt} + y_1 = -i(2A_T + A)e^{it} + \text{c.c.} \qquad (4.8)$$

As in the frequency-shift method, we would like to eliminate the secular terms. So we require that

$$2A_T + A = 0. \qquad (4.9)$$

This ODE for $A(T)$ gives

$$A(T) = A_0 e^{-T/2}, \qquad (4.10)$$

where $A_0$ is constant.

Now we can choose the homogeneous solution $y_1 = 0$, as in the frequency-shift method. So far, our approximate solution is

$$y(t, T) \sim A(T)e^{it} + \text{c.c.} = A_0 e^{it-T/2} + \text{c.c.}$$

In terms of the original variable $t$, this reads

$$y(t) \sim A_0 e^{it-\varepsilon t/2} + \text{c.c.} = A_0 e^{(i-\frac{\varepsilon}{2})t} + \text{c.c.} \qquad (4.11)$$

We can continue to obtain higher orders by allowing $A(T)$ to vary perturbatively, in such a way as to avoid secular terms in the expansion of $y_k$ for $k \geq 1$. Our method is to expand the equation for $A(T)$ *without* including additional time-scales. Finding higher-order terms in the *equation* governing the slowly varying amplitude is usually what one desires in physical problems. However, *care must be taken* here, as we will see below. In the example above, we perturb (4.9) as

$$2A_T + A = \varepsilon r_1 + \varepsilon^2 r_2 + \cdots . \tag{4.12}$$

Substituting this into (4.8) shows that

$$y_{1,tt} + y_1 = -i(2A_T + A)e^{it} + \text{c.c.} = \underbrace{-i\varepsilon r_1 e^{it} + \text{c.c.}}_{\varepsilon R_1} + O(\varepsilon^2). \tag{4.13}$$

Notice that there is an $O(\varepsilon)$ residual term on the right-hand side. That term must be added to the right-hand side of the next-order equation. When doing so we can choose $y_1 = 0$ in this case (but see the next example!). Therefore, the $O(\varepsilon^2)$ equation corresponding to (4.7) is

$$\begin{aligned} y_{2,tt} + y_2 &= -y_{0,TT} - y_{0,T} - 2y_{1,tT} - y_{1,t} - R_1, \\ &= -(A_{TT} + A_T)e^{it} - ir_1 e^{it} + \text{c.c.}, \end{aligned}$$

where we have used $y_1 = 0$ and $R_1 = (-ir_1 e^{it} + \text{c.c.})$, the coefficient of the $O(\varepsilon)$ residual term from (4.13). In order to eliminate the secular terms in this equation one requires that

$$r_1 = i(A_{TT} + A_T).$$

Substituting $r_1$ in (4.12) leads to

$$(2A_T + A) = i\varepsilon(A_{TT} + A_T). \tag{4.14}$$

This is a higher-order equation, which at first glance may be a concern. The method for solving this equation, for $\varepsilon \ll 1$, is to solve (4.14) recursively; i.e., replace the $A_{TT}$ term with lower-order derivatives of $A$ by using the equation itself, while maintaining $O(\varepsilon)$ accuracy. Let us see how that works in this example. Equation (4.14) implies that

$$A_T = -\frac{1}{2}A + O(\varepsilon). \tag{4.15}$$

Differentiating with respect to $T$ and using (4.15) to replace $A_T$ gives

$$A_{TT} = -\frac{1}{2}A_T + O(\varepsilon) = -\frac{1}{2}\left[ -\frac{1}{2}A + O(\varepsilon) \right] + O(\varepsilon) = \frac{1}{4}A + O(\varepsilon).$$

We can now substitute $A_{TT}$ using this equation and substitute $A_T$ using (4.15) in (4.14) to find

$$(2A_T + A) = -i\varepsilon\left[\frac{1}{4}A + O(\varepsilon) - \frac{1}{2}A + O(\varepsilon)\right].$$

Simplifying leads to the equation

$$A_T = -\frac{1}{2}\left(1 + \frac{i\varepsilon}{4}\right)A + O(\varepsilon^2),$$

whose solution to $O(\varepsilon)$ is

$$A(T) = A(0)e^{-\left(\frac{1}{2}+\frac{i\varepsilon}{8}\right)T}.$$

The above equation is a higher-order improvement of (4.10). Substituting $A(T)$ and using $T = \varepsilon t$ in (4.11) gives

$$y \sim A(0)e^{(1-\varepsilon^2/8)it}e^{-\varepsilon t/2} + \text{c.c.} \qquad (4.16)$$

This approximate solution is valid for times $t = O(1/\varepsilon^2)$.

We can now compare our result to the exact solution that can be found by substituting $y_s = e^{rt}$ into (4.6). Doing so gives $r^2 + \varepsilon r + 1 = 0$ and therefore

$$y_{\text{exact}}(t) = A(0)e^{\left(-\frac{\varepsilon}{2}+\frac{i}{2}\sqrt{4-\varepsilon^2}\right)t} + \text{c.c.}$$

If we now expand the exponent in the exact solution in powers of $\varepsilon$, we find that it agrees with the exponent in (4.16) to $O(\varepsilon^2)$, which is the order of accuracy we kept in the perturbation analysis:

$$-\frac{\varepsilon}{2} + \frac{i}{2}\sqrt{4-\varepsilon^2} = -\frac{\varepsilon}{2} + i\left(1 - \frac{\varepsilon^2}{8}\right) + O(\varepsilon^3).$$

This agrees with our approximate solution in (4.16).

## 4.4 Method of multiple scales: Nonlinear example

Below we give an example of using the method of multiple scales for the nonlinear ODE

$$\frac{d^2y}{dt^2} + y - \varepsilon y^3 = 0.$$

The first steps are similar to the previous example. We set $T = \varepsilon t$ and obtain the equation

$$y_{tt} + y + \varepsilon(2y_{tT} - y^3) + \varepsilon^2 y_{TT} = 0.$$

The leading-order equation and solution are the same as in the previous example,

$$y_0(t, T) = A(T)e^{it} + \text{c.c.}$$

The next-order equation is

$$O(\varepsilon): \qquad y_{1,tt} + y_1 = -2y_{0,tT} - y_0^3.$$

Substituting the leading-order solution in the right-hand side and expanding the nonlinear term leads to

$$
\begin{aligned}
y_{1,tt} + y_1 &= -(2iA_T e^{it} + \text{c.c.}) + (Ae^{it} + A^*e^{-it})^3 \\
&= -(2iA_T e^{it} + \text{c.c.}) + (A^3 e^{3it} + 3A^2 A^* e^{it} + \text{c.c.}) \\
&= \underbrace{-(2iA_T - 3|A|^2 A)e^{it}}_{\text{secular}} + \underbrace{A^3 e^{3it}}_{\text{non-secular}} + \text{c.c.}
\end{aligned}
$$

It is important to distinguish the terms that multiply $e^{it}$ from those that multiply $e^{3it}$, because the former is a secular term, i.e., it satisfies the homogeneous solution for $y_1$, whereas the second one does not. Therefore, we must remove the secular terms, which multiply $e^{it}$, but keep the non-secular terms, which multiply $e^{3it}$, in the equation for $y_1$ – the non-secular term results in a bounded contribution to $y_1$. To remove secularity, we require that

$$2iA_T = 3|A|^2 A \tag{4.17}$$

and the equation for $y_1$ becomes

$$y_{1,tt} + y_1 = A^3 e^{3it} + \text{c.c.} \tag{4.18}$$

The solution of the (complex-valued) ODE (4.17) can be found by noting that $|A|^2$ is a conserved quantity in time. One way to see this is by multiplying (4.17) by $A^*$, subtracting the complex conjugate, and integrating:

$$2iA_T A^* + 2iA_T^* A = 3|A|^2 AA^* - 3|A|^2 A^* A$$

and therefore $i(|A|^2)_T = 0$. Hence $|A|^2(T) = |A|^2(0) = |A_0|^2$, where $A_0$ is our arbitrary constant. Using this conservation relation we can rewrite (4.17) as

$$2iA_T = 3|A_0|^2 A.$$

This is now a *linear equation* for $A$ and its solution is given by

$$A(T) = A_0 e^{-\frac{3}{2}|A_0|^2 T} = A_0 e^{-\frac{3}{2}|A_0|^2 \varepsilon t}.$$

Note $A(T)$ depends on $T = \varepsilon t$, so we can assume that $A(T)$ is constant when integrating the equation for $y_1$ in (4.18). Assuming that $y_1 = Be^{3it} +$ c.c. gives $B = -A^3/8$, which is bounded, and therefore

$$y_1(t) = -\frac{1}{8}A^3(T)e^{3it} + \text{c.c.} + O(\varepsilon) = -\frac{1}{8}A_0^3 e^{-\frac{9i}{2}|A_0|^2 \varepsilon t}e^{3it} + \text{c.c.} + O(\varepsilon).$$

Finally, the perturbed solution we find is

$$y(t) = A_0 e^{\left(1-\frac{3\varepsilon}{2}|A_0|^2\right)it} + \varepsilon y_1(t) + \text{c.c.},$$

$$= A_0 e^{\left(1-\frac{3\varepsilon}{2}|A_0|^2\right)it} - \frac{\varepsilon}{8}A_0^3 e^{-\frac{9i}{2}|A_0|^2 \varepsilon t}e^{3it} + \text{c.c.}$$

Inspecting our solution we note that the second term (the one multiplied by $\varepsilon$) plays a minor role compared with the first term, since the amplitude of the second term is bounded and so remains $O(\varepsilon)$ small relative to the first term for all values of $t$. We therefore focus our attention on the first term: its effective frequency is given by

$$\Omega = 1 - \frac{3\varepsilon}{2}|A_0|^2. \tag{4.19}$$

Therefore, when $\varepsilon > 0$, the additional frequency contribution decreases with amplitude $|A_0|$ (beyond the linear solution) and the period of oscillations increases (since $T = 2\pi/\Omega$). This is sometimes called a "soft spring", in analogy with a spring whose period is elongated compared with a linear spring. Conversely, when $\varepsilon < 0$ the frequency increases and the period of oscillations decreases. This is called a "hard spring".

Since the equation is conservative and there is only a (nonlinear) frequency shift, this problem can also be done by the frequency-shift method. This is left as an exercise.

## 4.5 Method of multiple scales: Linear and nonlinear pendulum

For our final application of multiple scales to ODEs, we will look at a nonlinear pendulum with a slowly varying length; see Figure 4.1. Newton's second law of motion gives

$$ml\ddot{y} + mg\sin(y) = 0$$

as the equation of motion, where $m$ is the pendulum mass, $g$ the gravitational constant of acceleration, and $l$ the slowly varying length. This implies that

$$\ddot{y} + \rho^2(\varepsilon t)\sin(y) = 0, \tag{4.20}$$

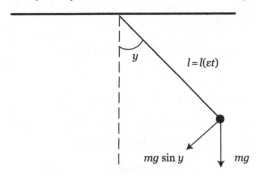

Figure 4.1 Nonlinear pendulum.

where for convenience we denote $\rho^2(\varepsilon t) \equiv g/l(\varepsilon t)$, or more simply $\rho^2(T)$ with $T \equiv \varepsilon t$, which we assume is smooth. If the length were constant, an exact solution would be available in terms of elliptic functions.

Associated with this pendulum is what is usually called an *adiabatic invariant*, i.e., a quantity in the system that is asymptotically invariant when a system parameter is slowly (adiabatically) changed. While analyzing the pendulum problem, we will look for adiabatic invariants. Adiabatic invariants received much attention during the development of the theory of quantum mechanics, cf. Goldstein (1980) and Landau and Lifshitz (1981), and more recently in such areas as plasma physics, particle accelerators, and galactic dynamics, etc.

### 4.5.1 Linear pendulum

To start, we will consider the linear problem

$$\ddot{y} + \rho^2(\varepsilon t)y = 0. \tag{4.21}$$

Suppose we assume a "standard" multi-scale solution in the form $y(t) = y(t, T; \varepsilon)$. Expand $d/dt$ and $y$ as

$$\frac{d}{dt} = \frac{\partial}{\partial t} + \varepsilon \frac{\partial}{\partial T},$$
$$y(t) = y_0(t, T) + \varepsilon y_1(t, T) + \varepsilon^2 y_2(t, T) + \cdots$$

and substitute them into (4.21). Collecting like powers of $\varepsilon$ gives
$O(1)$:

$$Ly_0 \equiv \partial_t^2 y_0 + \rho^2(T)y_0 = 0.$$

$O(\varepsilon)$:

$$Ly_1 = -2\frac{\partial^2 y_0}{\partial t \partial T}.$$

The leading-order solution is

$$y_0 = A(T)\exp(i\rho t) + \text{c.c.}$$

Notice that

$$\partial_T y_0 = (A_T + it\rho_T)\exp(i\rho t) + \text{c.c.}$$

grows as $t \to \infty$; i.e., it is secular! The problem is that we chose the wrong fast time-scale. We modify the fast scale in $y$ as follows:

$$y(t) = y(\Theta(t), T; \varepsilon),$$

where $\Theta_t = \omega(\varepsilon t)$ and $\omega$ will be chosen so that the leading-order solution is not secular. Expand $d/dt$ as

$$\frac{d}{dt} = \Theta_t \frac{\partial}{\partial \Theta} + \varepsilon \frac{\partial}{\partial T} = \omega(T)\frac{\partial}{\partial \Theta} + \varepsilon \frac{\partial}{\partial T}.$$

A little care must be taken in calculating $d^2/dt^2$, since $\omega$ is a function of the slow variable: we get

$$\frac{d^2}{dt^2} = \left(\omega(T)\frac{\partial}{\partial \Theta} + \varepsilon \frac{\partial}{\partial T}\right)\left(\omega(T)\frac{\partial}{\partial \Theta} + \varepsilon \frac{\partial}{\partial T}\right)$$

$$= \omega^2 \frac{\partial^2}{\partial \Theta^2} + \varepsilon\left[2\omega\frac{\partial^2}{\partial \Theta \partial T} + \omega_T \frac{\partial}{\partial \Theta}\right] + \varepsilon^2 \frac{\partial^2}{\partial T^2}. \qquad (4.22)$$

Substituting this into (4.21) and expanding $y = y_0 + \varepsilon y_1 + \cdots$ we get the first two equations

$O(1)$:

$$\omega^2(T)\frac{\partial^2 y_0}{\partial \Theta^2} + \rho^2(T)y_0 = 0.$$

$O(\varepsilon)$:

$$\omega^2(T)\frac{\partial^2 y_1}{\partial \Theta^2} + \rho^2(T)y_1 = -\left(\omega_T \frac{\partial y_0}{\partial \Theta} + 2\omega\frac{\partial^2 y_0}{\partial T \partial \Theta}\right).$$

The leading-order solution is

$$y_0 = A(T)\exp\left(i\frac{\rho}{\omega}\Theta\right) + \text{c.c.}$$

To prevent secularity at this order we require that

$$\frac{\partial y_0}{\partial T} = (A_T + i(\rho/\omega)_T \theta A)e^{i(\rho/\omega)\theta}$$

is bounded in $\theta$. Hence we must take $\omega/\rho$ to be constant in order to remove the secular term. The choice of constant does not affect the final result, so for convenience we take $\omega = \rho$. The order $\varepsilon$ equation then becomes, after substituting in the expression for $y_0$,

$$\rho^2(T)\left(\frac{\partial^2 y_1}{\partial \Theta^2} + y_1\right) = -\left(i\rho_T A e^{i\Theta} + 2iA_T\rho e^{i\Theta} + \text{c.c.}\right).$$

To remove secular terms, we require

$$2\rho A_T + A\rho_T = 0,$$
$$2\rho A_T^* + A^*\rho_T = 0.$$

Multiplying the first equation by $A^*$, the second by $A$, and then adding, we find that

$$\frac{\partial}{\partial T}\left(\rho|A|^2\right) = 0 \quad \Rightarrow \quad \rho(T)|A(T)|^2 = \rho(0)|A(0)|^2 = \frac{E}{\rho},$$

where $E = \rho^2 A^2$ is related to the unperturbed energy $E = \dot{y}^2/2 + \rho^2 y^2/2$, from (4.21); thus $\rho(T)|A(T)|^2 = E/\rho$ is constant in time. This is usually called the adiabatic invariant. Notice that $\rho(T)|A(T)|^2$ is constant on the same time-scale as the length is being varied. Also, using separation of variables on the equation for $A_T$,

$$\frac{A_T}{A} = \frac{-\rho_T}{2\rho} \quad \Rightarrow \quad \log(A\rho^{1/2}) = \text{constant} \quad \Rightarrow \quad A(T) = \frac{C}{\rho^{1/2}(T)}$$

where $C$ is constant. Since $\omega = \rho$,

$$\Theta(t) = \int_0^t \rho(\varepsilon t')\, dt' = \frac{1}{\varepsilon}\int_0^{\varepsilon t} \rho(s)\, ds.$$

The leading-order solution is then

$$y(t) \sim \frac{C}{\sqrt{\rho(\varepsilon t)}} \exp\left\{\frac{i}{\varepsilon}\int_0^{\varepsilon t}\rho(s)\, ds\right\} + \text{c.c.}$$

Alternatively, we can arrive at the same approximate solution using the so-called WKB method,[1] cf. Bender and Orszag (1999). Instead of introducing multiple time-scales, let $T = \varepsilon t$ and simply change variables in (4.21) to get

$$\varepsilon^2 \frac{d^2 y}{dT^2} + \rho^2(T)y(T) = 0.$$

---

[1] The WKB method is named after Wentzel, Kramers and Brillouin who used the method extensively. However, these ideas were used by others including Jeffries and so is sometimes referred to as the WKBJ method.

If $\rho$ were constant, the solution would be $y = e^{i\rho/\varepsilon} + $ c.c. This suggests that we look for a solution in the form $y \sim e^{i\phi(T;\varepsilon)/\varepsilon} + $ c.c. Using this ansatz, we find

$$-\phi_T^2 + i\varepsilon\phi_{TT} + \rho^2 = 0.$$

We now expand $\phi$ as $\phi = \phi_0 + \varepsilon\phi_1 + \varepsilon^2\phi_2 + \cdots$ to get the leading-order equation

$$\frac{\partial\phi_0}{\partial T} = \pm\rho(T) \Rightarrow \phi_0 = \pm\int_0^T \rho(s)\,ds + \mu_0,$$

where $\mu_0$ is the constant of integration. The $O(\varepsilon)$ equation is $-2\phi_{0T}\phi_{1T} + i\phi_{0TT} = 0$, so

$$\frac{\partial\phi_1}{\partial T} = \frac{i}{2}\frac{\phi_{0TT}}{\phi_{0T}} = \frac{i}{2}\frac{\partial}{\partial T}\log|\phi_{0T}|,$$

which gives

$$\phi_1 = \frac{i}{2}\log|\phi_{0T}| = \frac{i}{2}\log(\rho).$$

Thus, in terms of the original variables:

$$y(t) \sim \exp\left[i\left(\phi_0 + \varepsilon\phi_1\right)/\varepsilon\right],$$

$$= \exp\left[\frac{i}{\varepsilon}\left(\int_0^{\varepsilon t} \rho(s)\,ds + \mu_0\right) - \frac{1}{2}\log(\rho(\varepsilon t))\right] + \text{c.c.}$$

Setting $C = \exp(i\mu_0/\varepsilon)$,

$$y(t) \sim \frac{C}{\sqrt{\rho(\varepsilon t)}}\exp\left[\frac{i}{\varepsilon}\int_0^{\varepsilon t} \rho(s)\,ds\right] + \text{c.c.},$$

which is identical to the multiple-scales result.

## 4.5.2 Nonlinear pendulum

Now we will analyze the nonlinear equation, (4.20),

$$\frac{d^2y}{dt^2} + \rho^2(T)\sin(y) = 0.$$

These and more general problems were analyzed by Kuzmak (1959), see also Luke (1966). We will see that this problem is considerably more complicated than the linear problem: multiple scales are required. As in the linear problem, assuming $y = y(\Theta, T; \varepsilon) = y_0 + \varepsilon y_1 + \cdots$, $T = \varepsilon t$ and using (4.22) to expand $d^2/dt^2$, the leading- and first-order equations are, respectively,

$$\omega^2 y_{0\Theta\Theta} + \rho^2 \sin(y_0) = 0,$$
$$L(y_1) = \omega^2 y_{1\Theta\Theta} + \rho^2 \cos(y_0) y_1 = -(\omega_T y_{0\Theta} + 2\omega y_{0\Theta T}) = F_1.$$

The crucial part of the perturbation analysis is to understand the leading-order equation. Multiplying the leading-order equation by $y_{0\Theta}$, we find the "energy integral":

$$\frac{\partial}{\partial \Theta}\left(\frac{\omega^2}{2}y_{0\Theta}^2 - \rho^2 \cos(y_0)\right) = 0 \quad \Rightarrow \qquad (4.23a)$$

$$\frac{\omega^2}{2}y_{0\Theta}^2 - \rho^2 \cos(y_0) = E(T) \quad \Rightarrow \qquad (4.23b)$$

$$\omega^2 y_{0\Theta}^2 = 2(E + \rho^2 \cos(y_0)). \qquad (4.23c)$$

Notice (i) that the coefficient $\rho$ is constant with respect to $\Theta$ and so is $E$; (ii) the left-hand side of the integral equation (4.23b) is not necessarily positive. If we redefine $\tilde{E} = E + \rho^2$ so that $\omega^2 y_{0\Theta}^2/2 + \rho^2(1 - \cos y_0) = \tilde{E}$, then $\tilde{E} \geq 0$ and $\tilde{E}$ is an energy. However, we use $E$ to simplfy our notation. Solving for $dy_0/d\Theta$ and separating variables gives

$$\int \frac{dy_0}{\sqrt{2(E + \rho^2 \cos y_0)}} = \int \frac{d\Theta}{\omega}.$$

We will be concerned with periodic solutions in $\Theta$; see also the phase plane in Figure 4.2. Periodic solutions are obtained for values $|y| \leq y_*$ obtained from $\cos y_* = -E/\rho^2$.

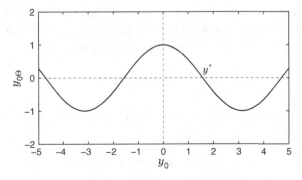

Figure 4.2 Phase plane.

The period of motion, $P$, is

$$P = \omega \oint \frac{dy_0}{\sqrt{2\,(E + \rho^2 \cos y_0)}},$$

where the loop integral is taken over one period of motion. Since $\cos(y_0)$ is even and symmetric, we can also calculate the period from

$$P = 4\omega \int_0^{y_*} \frac{dy_0}{\sqrt{2\,(E + \rho^2 \cos y_0)}}. \tag{4.24}$$

Since the motion is periodic, for any integer $n$

$$y_0(\Theta + nP, T) = y_0(\Theta, T).$$

At this stage $P$ is a function of $T$, thus

$$\frac{\partial y_0(\Theta + nP, T)}{\partial T} = n \frac{\partial P}{\partial T} \frac{\partial y_0(\Theta, T)}{\partial \Theta} + \frac{\partial y_0(\Theta, T)}{\partial T}. \tag{4.25}$$

For a well-ordered perturbation expansion, the first derivative $\partial y_0(\Theta + nP, T)/\partial T$ must be bounded. However, from (4.25), $\partial y_0(\Theta + nP, T)/\partial T \to \infty$ as $n \to \infty$. We can eliminate this divergence if we set $\partial P/\partial T = 0$, i.e., we require the period to be constant. The choice of constant does not affect the final result, so we can take $P = 2\pi$ and then we have from (4.24),

$$\omega = \omega(E) = \frac{\pi}{2 \displaystyle\int_0^{y_*} \frac{dy_0}{\sqrt{2\,(E + \rho^2 \cos y_0)}}}. \tag{4.26}$$

To find out how $E$ varies, we must go to the next-order equation, namely,

$$L y_1 = \left[\omega^2 \partial_\Theta^2 + \rho^2(T) \cos(y_0)\right] y_1 = F_1 = -(\omega_T y_{0\Theta} + 2\omega y_{0\Theta T}). \tag{4.27}$$

We will not actually solve the $O(\varepsilon)$ equation here, but instead we only need to derive a so-called *solvability condition* using a Fredholm alternative. We start by examining the homogeneous equation adjoint to (4.27): $L^*W = 0$, where $L^*$ is the adjoint to the operator $L$, with respect to the inner product

$$\langle Ly, W \rangle = \int_0^{2\pi} (Ly)\, W\, d\Theta = \langle y, L^*W \rangle;$$

integration by parts shows that $L = L^*$, i.e., the operator $L$ is self-adjoint. Now consider

$$L y_1 = \left[\omega^2 \partial_\Theta^2 + \rho^2(T) \cos(y_0)\right] y_1 = F_1,$$
$$L^*W = LW = 0,$$

where $W$ is the periodic solution of the adjoint equation. There is another, non-periodic solution to the adjoint problem (obtained later for completeness); but since we are only interested in periodic solutions, we do not need to consider it in the secularity analysis. Multiply the first equation by $W$, the second by $y_1$, and subtract to find

$$\omega^2 \frac{\partial}{\partial \Theta} (Wy_{1\Theta} - W_\Theta y_1) = WF_1.$$

Integrating this over one period and using the periodicity of the solution $y_1$ gives

$$\int_0^{2\pi} WF_1 \, d\Theta = 0 \qquad (4.28)$$

as a necessary condition for a periodic solution of the $O(\varepsilon)$ equation to exist. This orthogonality condition must hold for the periodic solution, $W$, of the homogeneous problem, which we will now determine. Operate on the leading-order equation with $\partial/\partial \Theta$ to find

$$\omega^2 y_{0\Theta\Theta\Theta} + \rho^2 \cos(y_0) y_{0\Theta} = 0,$$

i.e., $y_{0\Theta}$ is a solution of $LW = 0$; hence we set $W_1 = y_{0\Theta}$. This is the periodic, homogeneous solution. To find the non-periodic homogeneous solution operate on the leading-order equation with $\partial/\partial E$ to find

$$Ly_{0E} = \omega^2 y_{0E\Theta\Theta} + \rho^2 \cos(y_0) y_{0E} = -(\omega^2)_E y_{0\Theta\Theta}.$$

On the other hand, for any constant $\alpha$,

$$L(\alpha\Theta y_{0\Theta}) = \omega^2 \frac{\partial^2}{\partial \Theta^2} (\alpha\Theta y_{0\Theta}) + \rho^2 \cos(y_0)\alpha\Theta y_{0\Theta}$$
$$= \alpha \left[ 2\omega^2 y_{0\Theta\Theta} + \Theta \left( \omega^2 y_{0\Theta\Theta\Theta} + \rho^2 \cos(y_0) y_{0\Theta} \right) \right].$$

The term in parentheses vanishes and we have

$$L(\alpha\Theta y_{0\Theta}) = 2\alpha\omega^2 y_{0\Theta\Theta}.$$

Hence for the combination, $y_{0E} + \alpha\Theta y_{0\Theta}$:

$$L(y_{0E} + \alpha\Theta y_{0\Theta}) = L(y_{0E}) + L(\alpha\Theta y_{0\Theta})$$
$$= -\left( \omega^2 \right)_E y_{0\Theta\Theta} + 2\alpha\omega^2 y_{0\Theta\Theta}.$$

Thus, if we set $\alpha = \omega_E/\omega$,

$$L(y_{0E} + \alpha\Theta y_{0\Theta}) = 0$$

and we have the second solution:

$$W_2 = y_{0E} + \frac{\omega_E}{\omega}\Theta y_{0\Theta},$$

cf. Luke (1966). This solution, however, is not generally periodic since $\omega_E \neq 0$ in the nonlinear problem. We therefore need only one homogeneous solution $W_1$ in the Fredholm alternative (4.28). But the non-periodic solutions are useful if one wishes to find the higher-order solutions. The solution $y_1$ to $Ly_1 = F_1$ can be obtained by using the method of variation of parameters. The solvability condition (4.28) now becomes

$$\int_0^{2\pi} \frac{\partial y_0}{\partial\Theta} F_1 \, d\Theta = 0,$$

$$\int_0^{2\pi} \frac{\partial y_0}{\partial\Theta} (\omega_T y_{0\Theta} + 2\omega y_{0\Theta T}) \, d\Theta = 0,$$

$$\frac{\partial}{\partial T} \int_0^{2\pi} \omega \left(\frac{\partial y_0}{\partial\Theta}\right)^2 d\Theta = 0.$$

We have therefore found an adiabatic invariant for this nonlinear problem:

$$\mathcal{A}(T) \equiv \omega(T) \int_0^{2\pi} \left(\frac{\partial y_0}{\partial\Theta}\right)^2 d\Theta = \mathcal{A}(0).$$

The invariant can also be written as

$$\mathcal{A}(0) = \omega(T) \int_0^{2\pi} \left(\frac{\partial y_0}{\partial\Theta}\right)^2 d\Theta = \omega(T) \oint \left(\frac{\partial y_0}{\partial\Theta}\right)^2 \frac{dy_0}{\frac{dy_0}{d\Theta}},$$

$$\mathcal{A}(0) = 4\omega(T) \int_0^{y_*} \frac{\partial y_0}{\partial\Theta} \, dy_0,$$

$$= 4 \int_0^{y_*} \sqrt{2(E + \rho^2 \cos(y_0))} \, dy_0 \equiv F(E, \rho^2). \qquad (4.29)$$

Thus $E \equiv E(T)$ is a function of $\rho \equiv \rho(T)$ in terms of the const $\mathcal{A}(0)$. Recall that (4.26) gives us $\omega \equiv \omega(E(T))$, in particular this leads to the convenient formula

$$\omega = \pi \left| \frac{\partial}{\partial E} F(E, \rho^2). \right.$$

Hence the solution of the problem is determined in principle; i.e., these relations give $E = E(\rho(T)) = E(T)$ and $\omega = \omega(E(\rho(T))) = \omega(T)$. So, we have determined the leading-order solution: $y \sim y_0(\Theta, E(T), \omega(T))$.

As an aside, the leading-problem can be solved explicitly in terms of ellip-tic functions. For completeness we now outline the method to do this. First, consider

$$\left(\frac{du}{dx}\right)^2 = P(u),$$

where $P$ is a polynomial. If $P$ is of second degree, we can express the solution of the above equation in terms of trigonometric functions. If $P$ is of third or fourth degree, the solution can be expressed in terms of elliptic functions (Byrd and Friedman, 1971).

Now recall the energy integral:

$$\frac{\omega^2}{2} y_{0\Theta}^2 = E(T) + \rho^2 \cos(y_0).$$

To show that $y_0$ can be related to elliptic problems, let $z = \tan(y_0/2)$ so $dz/dy_0 = \frac{1}{2} \sec^2(y_0/2)$. Then

$$\frac{dy_0}{d\Theta} = \frac{dy_0}{dz}\frac{dz}{d\Theta} = 2\cos^2(y_0/2)\frac{dz}{d\Theta}.$$

Using

$$\cos(y_0/2) = \frac{1}{\sqrt{1+z^2}},$$

$$\cos(y_0) = 2\cos(y_0/2)^2 - 1 = \frac{1-z^2}{1+z^2},$$

we substitute these results into the energy integral yielding

$$\frac{4\omega^2}{2(1+z^2)^2}\left(\frac{dz}{d\Theta}\right)^2 = \left(E + \rho^2\frac{1-z^2}{1+z^2}\right),$$

and hence

$$2\omega^2\left(\frac{dz}{d\Theta}\right)^2 = (1+z^2)^2 E + \rho^2(1-z^2)(1+z^2).$$

The right-hand side is a fourth-order polynomial and thus the solutions are elliptic functions. One can also transform the above integrals for $\omega$ and $E$ into standard elliptic integrals, see Exercise 4.6.

## Exercises

4.1  (a) Obtain the nonlinear change in the frequency given by (4.19) by
         applying the frequency-shift method to the equation

$$\frac{d^2y}{dt^2} + y - \varepsilon y^3 = 0, \quad |\varepsilon| \ll 1.$$

    (b) Use the multiple-scales method to find the leading-order approxima-
        tion to the solution of

$$\frac{d^2y}{dt^2} + y - \varepsilon\left(y^3 + \frac{dy}{dt}\right) = 0, \quad 0 < \varepsilon \ll 1.$$

    (c) Find the next-order (first-order) approximation, valid for times $t = o(1/\varepsilon^2)$, to the solution of part (b).

4.2  Apply both the frequency-shift method and the method of multiple scales
     ($|\varepsilon| \ll 1$) to find the solution of:
     (a) $\dfrac{dA}{dt} - iA + \varepsilon A = 0,$
     (b) $\dfrac{dA}{dt} - iA + \varepsilon|A|^2 A = 0.$

4.3  Use the method of multiple scales ($|\varepsilon| \ll 1$) to find an approximation to
     the periodic solutions of:
     (a) $\dfrac{d^4y}{dt^4} - y - \varepsilon y = 0,$
     (b) $\dfrac{d^4y}{dt^4} - y - \varepsilon y^3 = 0.$

4.4  Use the WKB method to find the leading-order approximation to the
     solutions of

$$\frac{d^4y}{dt^4} - \rho^4(\varepsilon t)y = 0, \quad |\varepsilon| \ll 1.$$

4.5  Use the multiple-scale method to find:
     (a) the adiabatic invariant and leading-order solution of

$$\frac{d^2y}{dt^2} - \rho^2(\varepsilon t)(y + y^3) = 0, \quad |\varepsilon| \ll 1;$$

     (b) the leading-order approximation to the solution of

$$\frac{d^2y}{dt^2} - \rho^2(\varepsilon t)(y + y^3) = \varepsilon\frac{dy}{dt}, \quad 0 < \varepsilon \ll 1.$$

4.6  Consider

$$\frac{d^2y}{dt^2} + \omega^2 \sin y = 0,$$

     where $\omega > 0$ is a constant.

(a) Find the energy integral by multiplying the equation with $dy/dt$ and integrating.

(b) Let $y = 2\tan^{-1}(z)$ (inverse tan-function) and find an equation for $z$. Solve the equation in terms of the Jacobi elliptic functions.

(c) Use part (b) to find all bounded solutions $y$.

4.7 Suppose we are given

$$\frac{d^2y}{dt^2} + \rho^2(\varepsilon t)\sin(y) = \varepsilon\frac{dy}{dt}, \quad 0 < \varepsilon \ll 1$$

Find the leading-order solution and contrast it with the asymptotic solution for the pendulum discussed in Section 4.5.

# 5

## Water waves and KdV-type equations

Fluid dynamics is concerned with behavior on scales that are large compared to the distance between the molecules that they consist of. Physical quantities, such as mass, velocity, energy, etc., are usually regarded as being spread continuously throughout the region of consideration; this is often termed the continuum assumption and continuum derivations are based on conservation principles. From these concepts, the equations of fluid dynamics are derived in many books, cf. Batchelor (1967).

A microscopic description of fluids is provided by the Boltzmann equation. When the average distance between collisions of the fluid molecules becomes small relative to the macroscopic dimensions of the fluid, the Boltzmann equation can be simplified to the fluid equations (Chapman and Cowling, 1970).

In this chapter, we focus on the most important results derived from conservation laws and, in particular, how they relate to water waves.

We will use $\rho = \rho(x, t)$ to denote the fluid mass density, $v = v(x, t)$ is the fluid velocity, $P$ is the pressure, $F$ is a given external force, and $v_*$ is the kinematic viscosity that is due to frictional forces. In vector notation, the relevant equations of fluid dynamics we will consider are:

- conservation of mass:

$$\frac{\partial \rho}{\partial t} + \nabla \cdot (\rho v) = 0,$$

- conservation of momentum:

$$\rho \left[ \frac{\partial v}{\partial t} + (v \cdot \nabla) v \right] = F - \nabla P + v_* \Delta v,$$

where $\Delta \equiv \nabla^2$, is the Laplacian.[1] We omit the equation of energy that describes the temperature ($T$) variation of the fluid, so an equation of state, such as

[1] In three-dimensional Cartesian coordinates, the Laplacian is $\Delta \equiv \partial^2/\partial x^2 + \partial^2/\partial y^2 + \partial^2/\partial z^2$.

$P = P(\rho, T)$, is added to close the system of equations. These equations correspond to the first three moments of the Boltzmann equation. When $\rho = \rho_0$ is constant, the first equation then describes an incompressible fluid: $\boldsymbol{\nabla} \cdot \boldsymbol{v} = 0$, also called the divergence equation. The divergence and the momentum equations are often called the incompressible Navier–Stokes equations. The energy equation is not necessary to close this system of equations, which (in three dimensions) are four equations in four unknowns: $\boldsymbol{v} = (u, v, w)$ and $P$.

In this chapter, we will consider the free surface water wave problem, which (interior to the fluid) is the inviscid reduction ($\nu_* = 0$) of the above equations; these equations are called the Euler equations. Supplemented with appropriate boundary conditions, we have the Euler equations with a free surface. We will derive the shallow-water or long wave limit of this system and discuss certain approximate equations: the Korteweg–de Vries (KdV) equation in $1 + 1$ dimensions[2] and the Kadomtsev–Petviashvili (KP) equation in $2 + 1$ dimensions.

## 5.1 Euler and water wave equations

For our discussion of water waves, we will use the above incompressible Navier–Stokes description with constant density $\rho = \rho_0$ and we will assume an ideal fluid: that is, a fluid with zero viscosity ($\nu_* = 0$). Thus, an ideal, incompressible fluid is described by the following Euler equations:

$$\boldsymbol{\nabla} \cdot \boldsymbol{v} = 0,$$

$$\frac{\partial \boldsymbol{v}}{\partial t} + (\boldsymbol{v} \cdot \boldsymbol{\nabla}) \boldsymbol{v} = \frac{1}{\rho_0} (\boldsymbol{F} - \boldsymbol{\nabla}P).$$

Suppose now that the external force is conservative, i.e., we can write $\boldsymbol{F} = -\boldsymbol{\nabla}U$, for some scalar potential $U$. We can then write the momentum equation as

$$\frac{\partial \boldsymbol{v}}{\partial t} + (\boldsymbol{v} \cdot \boldsymbol{\nabla}) \boldsymbol{v} = -\boldsymbol{\nabla} \left( \frac{U + P}{\rho_0} \right).$$

Using the vector identity

$$(\boldsymbol{v} \cdot \boldsymbol{\nabla}) \boldsymbol{v} = \frac{1}{2} \boldsymbol{\nabla} (\boldsymbol{v} \cdot \boldsymbol{v}) - \boldsymbol{v} \times (\boldsymbol{\nabla} \times \boldsymbol{v}),$$

gives

$$\frac{\partial \boldsymbol{v}}{\partial t} - \boldsymbol{v} \times (\boldsymbol{\nabla} \times \boldsymbol{v}) = -\boldsymbol{\nabla} \left( \frac{1}{2} \boldsymbol{v} \cdot \boldsymbol{v} + \frac{U + P}{\rho_0} \right). \tag{5.1}$$

---

[2] The notation "$n + 1$ dimensions" means there are $n$ space dimensions and one time dimension.

Now define the vorticity to be $\omega \equiv \nabla \times v$, which is a local measure of the degree to which the fluid is spinning; more precisely, $\frac{1}{2}\|\nabla \times v\|$ (note that $\|v\|^2 = v \cdot v$) is the angular speed of an infinitesimal fluid element. Taking the curl of the last equation and noting that the curl of a gradient vanishes,

$$\frac{\partial \omega}{\partial t} - \nabla \times (v \times \omega) = 0.$$

Finally, using the vector identity $\nabla \times (F \times G) = (G \cdot \nabla)F - (F \cdot \nabla)G + (\nabla \cdot G)F - (\nabla \cdot F)G$ for vector functions $F$ and $G$ and recalling that the divergence of the curl vanishes, we arrive at the so-called vorticity equation:

$$\frac{\partial \omega}{\partial t} = (\omega \cdot \nabla)v - (v \cdot \nabla)\omega \quad \text{or} \tag{5.2a}$$

$$\frac{D\omega}{Dt} = \omega \cdot \nabla v, \tag{5.2b}$$

where we have used the notation

$$\frac{D}{Dt} = \frac{\partial}{\partial t} + (v \cdot \nabla)$$

to signify the so-called convective or material derivative that moves with the fluid particle ($v = (u, v, w)$). Hence $\omega = 0$ is a solution; moreover, from (5.2b), it can be proven that if the vorticity is initially zero, then (if the solution exists) it is zero for all times. Such a flow is called irrotational. Physically, in an ideal fluid there is no mechanism that will produce "local rotation" if the fluid is initially irrotational. Often it is a good approximation to assume that a fluid is irrotational, with viscosity effects occurring only in thin regions of the fluid flow called boundary layers. In this chapter, we will consider water waves and will assume that the flow is irrotational. In such circumstances, it is convenient to introduce a velocity potential $v = \nabla \phi$. Notice that the vorticity equation (5.2b) is trivially satisfied since

$$\nabla \times (\nabla \phi) = 0.$$

The Euler equations inside the fluid region can now also be simplified:

$$\nabla \cdot v = \nabla \cdot \nabla \phi = \Delta \phi = 0,$$

which is Laplace's equation; it is to be satisfied internal to the fluid, $-h < z < \eta(x, y, t)$, where we denote the height of the fluid free surface above the mean level $z = 0$ to be $\eta(x, y, t)$ and the bottom of the fluid is at $z = -h$. See Figure 5.1 on page 103.

Next we discuss the boundary conditions that lead to complications; i.e., an unknown free surface and nonlinearities. We assume a flat, impenetrable bottom at $z = -h$, so that no fluid can flow through. This results in the condition

$$w = \frac{\partial \phi}{\partial z} = 0, \quad z = -h,$$

where $w$ represents the vertical velocity. On the free surface $z = \eta(x, y, t)$ there are two conditions. The first is obtained from (5.1). Using the fact that $\nabla$ and $\partial/\partial t$ commute,

$$\nabla \left( \frac{\partial \phi}{\partial t} + \frac{1}{2} \|v\|^2 + \frac{U + P}{\rho_0} \right) = 0 \quad \Rightarrow$$

$$\frac{\partial \phi}{\partial t} + \frac{1}{2} \|v\|^2 + \frac{U + P}{\rho_0} = f(t),$$

where we recall $v = (u, v, w)$ and $\|v\|^2 = u^2 + v^2 + w^2 = \phi_x^2 + \phi_y^2 + \phi_z^2$. Since the physical quantity is $v = \nabla \phi$, we can add an arbitrary function of time (independent of space) to $\phi$,

$$\phi \to \phi + \int_0^t f(t')\, dt',$$

to get the so-called Bernoulli, dynamic, or pressure equation,

$$\frac{\partial \phi}{\partial t} + \frac{1}{2} \|v\|^2 + \frac{U + P}{\rho_0} = 0.$$

For now, we will neglect surface tension and assume that the dominant force is the buoyancy force, $F = -\nabla(\rho_0 g z)$, which implies that $U = \rho_0 g z$, where $g$ is the gravitational constant of acceleration. For convenience, we take the pressure to vanish (i.e., $P = 0$) on the free surface, yielding:

$$\frac{\partial \phi}{\partial t} + \frac{1}{2} \|\nabla \phi\|^2 + g\eta = 0, \quad z = \eta(x, y, t)$$

on the free surface.

The second equation governing the free surface is derived from the assumption that if a fluid packet is initially on the free surface, then it will stay there. Mathematically, this implies that if $F = F(x, y, z, t)$, where $(x, y, z)$ is a point on the free surface, then

$$\frac{DF}{Dt} \equiv \frac{\partial F}{\partial t} + v \cdot \nabla F = 0.$$

On the surface, $F = z - \eta(x, y, t) = 0$. Then

$$\frac{Dz}{Dt} = \frac{D\eta}{Dt} \quad \Rightarrow \quad w = \frac{\partial \eta}{\partial t} + v \cdot \nabla \eta,$$

where we have used $v = \left(\dfrac{Dx}{Dt}, \dfrac{Dy}{Dt}, \dfrac{Dz}{Dt}\right) = (u, v, w)$. So, using $w = \partial\phi/\partial z$,

$$w = \frac{\partial\phi}{\partial z} = \frac{\partial\eta}{\partial t} + v \cdot \nabla\eta, \qquad z = \eta(x, y, t), \tag{5.3}$$

on the free surface. Equation (5.3) is often referred to as the kinematic condition. It should be noted that on a single-valued surface $z = \eta(x, y, t)$, equation (5.3) can be written as

$$\frac{\partial\phi}{\partial n} = \frac{\partial\eta}{\partial t},$$

where $\partial\phi/\partial n = (n \cdot \nabla)\phi$ and $n = (-\nabla\eta, 1)$ is the outward normal. Thus, the free surface $z = \eta(x, y, t)$ moves in the direction of the normal velocity.

To summarize, the free-surface water wave equations with a flat bottom are:
- Euler ideal flow

$$\Delta\phi = 0, \qquad -h < z < \eta(x, y, t). \tag{5.4}$$

- No flow through the bottom

$$\frac{\partial\phi}{\partial z} = 0, \qquad z = -h. \tag{5.5}$$

- Bernoulli's or the pressure equation

$$\frac{\partial\phi}{\partial t} + \frac{1}{2}\|\nabla\phi\|^2 + g\eta = 0, \qquad z = \eta(x, y, t). \tag{5.6}$$

- Kinematic condition

$$\frac{\partial\phi}{\partial z} = \frac{\partial\eta}{\partial t} + v \cdot \nabla\eta, \qquad z = \eta(x, y, t). \tag{5.7}$$

These four equations constitute the equations for water waves with the unknowns $\phi(x, y, z, t)$ and $\eta(x, y, t)$. This is a free-boundary problem. In contrast to a Dirichlet or a Neumann boundary value problem where the boundary is fixed and known, in free-boundary problems part of solving the problem is to determine the dynamics of the boundary. This aspect makes the solution to the water wave equations particularly difficult. We also note that if we were given an $\eta$ that satisfies Bernoulli's equation (5.6), the remaining three equations would satisfy a Neumann boundary value problem. The geometry of the problem is shown in Figure 5.1.

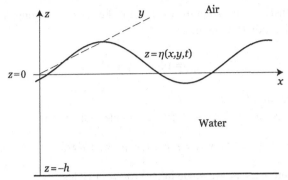

Figure 5.1 Geometry of water waves. The bottom of the water is at a constant level $z = -h$, while the free surface of the water is denoted by $z = \eta(x, y, t)$, where $\eta$ is to be determined. The undisturbed fluid is at a level $z = 0$.

## 5.2 Linear waves

We will first look at the linear case by assuming $|\eta| \ll 1$ and $\|\nabla\phi\| \ll 1$. The first two equations remain unchanged, except that they are to be satisfied on the known surface $z = 0$. The last two equations, using

$$\phi(x, y, \eta, t) = \phi(x, y, 0, t) + \eta \frac{\partial \phi}{\partial z}(x, y, 0, t) + \cdots,$$

$$= \phi_0(x, y, t) + \eta \phi_{0z}(x, y, t) + \cdots,$$

become

$$\frac{\partial \phi_0}{\partial t} = -g\eta, \qquad z = 0,$$

$$\frac{\partial \eta}{\partial t} = \phi_{0z}(x, y, t), \qquad z = 0.$$

We now look for special solutions to the Euler and free-surface equations of the form

$$\phi_s(x, y, z, t) = A(k, l, z, t) \exp(ikx + ily),$$

i.e., we decompose the solution into Fourier modes. Substituting this ansatz into Laplace's equation (5.4),

$$A_{zz} - (k^2 + l^2)A = 0.$$

Calling $\kappa^2 = k^2 + l^2$, the solution is given by

$$A = \tilde{A}(k, l, t) \cosh[\kappa(z + h)] + \tilde{B}(k, l, t) \sinh[\kappa(z + h)];$$

note we translated the solution by $h$ units because satisfying the bottom boundary condition (5.5) requires

$$\frac{\partial A}{\partial z} = 0, \qquad z = -h,$$

which implies that $\tilde{B} = 0$.

If we assume that the free surface has a mode of the form

$$\eta(x, y, t) = \tilde{\eta}(k, l, t) \exp(ikx + ily),$$

then, substituting the above mode into (5.6)–(5.7) yields,

$$\frac{\partial \tilde{A}}{\partial t} \cosh(\kappa h) + g\tilde{\eta} = 0,$$

$$\frac{\partial \tilde{\eta}}{\partial t} - \kappa \sinh(\kappa h)\tilde{A} = 0.$$

Taking the time derivative of the second equation and substituting into the first gives

$$\frac{\partial^2 \tilde{\eta}}{\partial t^2} + g\kappa \tanh(\kappa h)\tilde{\eta} = 0.$$

Assuming $\tilde{\eta}(k, l, t) = \tilde{\eta}(k, l, 0)e^{-i\omega t}$, we find the dispersion relationship:

$$\omega^2 = g\kappa \tanh(\kappa h), \qquad (5.8)$$

which has two branches. "Adding up" all such special solutions implies that the general Fourier solution for a rapidly decaying solution for $\eta$ is

$$\eta = \frac{1}{(2\pi)^2} \iint \{\tilde{\eta}_+ \exp[i(kx + ly - \omega_+ t)]$$

$$+ \tilde{\eta}_- \exp[i(kx + ly + \omega_- t)]\} \, dk \, dl, \quad (5.9)$$

with $\tilde{\eta}_+, \tilde{\eta}_-$ determined by the initial data that is assumed to be real and decaying sufficiently rapidly in space.

The dispersion relation has several interesting limits. The so-called deep-water dispersion relationship, i.e., when $\kappa h \gg 1$, is

$$\omega^2 \simeq g|\kappa|,$$

while for shallow water, i.e., when $\kappa h \ll 1$, it is

$$\omega^2 \simeq g\kappa \left(\kappa h - \frac{(\kappa h)^3}{3} + \cdots\right).$$

The leading term in the shallow-water dispersion relationship is $\omega^2 \simeq gh\kappa^2$. Alternatively, as discussed in earlier chapters, we have that to leading order the amplitude $\eta$ satisfies the wave equation,

$$\eta_{tt} - c_0^2 \eta_{xx} = 0,$$

with a propagation speed $c_0^2 = gh$. We will come back to this point when we discuss the KdV equation later in this chapter.

## 5.3 Non-dimensionalization

To analyze the full equations, we will first non-dimensionalize them. By doing this, it will be easier to compare the "size" of each term. This is convenient since we will then be working with "pure" or dimensionless numbers. For example, a velocity can be large if measured in units of, say, microns per second but in units of light-years per hour it would be small. Indeed small or large in terms of coefficients in an equation are relative concepts and are meaningful only in a comparative sense. By studying non-dimensional equations we can more easily decide which terms are negligible.

For shallow-water waves it is convenient to use the following non-dimensionalization:

$$x = \lambda_x x', \quad y = \lambda_y y', \quad z = hz',$$

$$t = \frac{\lambda_x}{c_0} t', \quad \eta = a\eta', \quad \phi = \frac{\lambda_x ga}{c_0} \phi',$$

where $c_0 = \sqrt{gh}$ is the shallow-water wave speed, $\lambda_x, \lambda_y$ are typical wavelengths of the initial data (in the $x$ or $y$ direction), and $a$ is the maximum or typical amplitude of the initial data. The primed variables are dimensionless. For Laplace's equation:

$$\frac{1}{\lambda_x^2} \phi'_{x'x'} + \frac{1}{\lambda_y^2} \phi'_{y'y'} + \frac{1}{h^2} \phi'_{z'z'} = 0.$$

For the sake of notational simplicity, we will drop the primed notation so that

$$\phi_{zz} + \left(\frac{h}{\lambda_x}\right)^2 \phi_{xx} + \left(\frac{h}{\lambda_y}\right)^2 \phi_{yy} = 0, \qquad -1 < z < \frac{a\eta}{h}.$$

The no-flow condition becomes

$$\frac{\partial \phi}{\partial z} = 0, \qquad z = -1.$$

Bernoulli's equation becomes

$$\phi_t + \frac{a}{2h}\left[\phi_x^2 + \left(\frac{\lambda_x}{\lambda_y}\right)^2 \phi_y^2 + \left(\frac{\lambda_x}{h}\right)^2 \phi_z^2\right] + \eta = 0, \qquad z = \frac{a\eta}{h}.$$

And the kinematic condition becomes

$$\eta_t + \frac{a}{h}\left[\phi_x\eta_x + \left(\frac{\lambda_x}{\lambda_y}\right)^2 \phi_y\eta_y\right] = \left(\frac{\lambda_x}{h}\right)^2 \phi_z, \qquad z = \frac{a\eta}{h}.$$

Now we define the dimensionless parameters $\epsilon \equiv a/h$, $\delta \equiv \lambda_x/\lambda_y$, and $\mu \equiv h/\lambda_x$: $\epsilon$ is a measure of the nonlinearity, or amplitude, of the wave; $\mu$ is a measure of the depth relative to the characteristic wavelength, sometimes called the dispersion parameter; and $\delta$ measures the size of the transverse variations. In summary, the dimensionless equations for water waves propagating over a flat bottom are

- Euler ideal flow, Laplace's equation

$$\phi_{zz} + \mu^2\phi_{xx} + \mu^2\delta^2\phi_{yy} = 0, \quad -1 < z < \epsilon\eta.$$

- No flow through the bottom

$$\frac{\partial\phi}{\partial z} = 0, \quad z = -1.$$

- Bernoulli's or the pressure equation

$$\phi_t + \frac{\epsilon}{2}\left[\phi_x^2 + \delta^2\phi_y^2 + \frac{1}{\mu^2}\phi_z^2\right] + \eta = 0, \quad z = \epsilon\eta.$$

- Kinematic condition:

$$\mu^2\left[\eta_t + \epsilon\left(\phi_x\eta_x + \delta^2\phi_y\eta_y\right)\right] = \phi_z, \quad z = \epsilon\eta.$$

We note that the linear, or small-amplitude, case arises when $\epsilon \ll 1, \mu \sim O(1)$. Then the above equations reduce, by dropping terms of order $\epsilon$, to the linear equations that are equivalent to those we analyzed earlier, i.e.,

$$\phi_t + \eta = 0, \quad z = 0,$$
$$\mu^2\eta_t = \phi_z, \quad z = 0,$$

but now in non-dimensional form.

## 5.4 Shallow-water theory

We will make some simplifying assumptions about the sizes of $\epsilon$, $\mu$, and $\delta$ in order to obtain some interesting limiting equations:

- We will consider shallow water waves (sometimes called long water waves). This regime corresponds to small depth relative to the water wavelength. In our system of parameters, this corresponds to $\mu = h/\lambda_x \ll 1$.
- We will also assume that the wavelength in the transverse direction is much larger than the propagation wavelength, so $\delta = \lambda_x/\lambda_y \ll 1$.
- As before in the linear regime, we assume small-amplitude waves $|\varepsilon| = a/h \ll 1$. Recall that a similar assumption was made the Fermi–Pasta–Ulam (FPU) problem involving coupled nonlinear springs studied in Chapter 1.
- We will make the assumption of "maximal balance" (the small terms, i.e., nonlinearity and dispersion, are of the same order) as was made in the continuum approximation of the FPU problem to the Boussinesq model in order to derive the KdV equation. We will do the same thing now by assuming maximal balance with $\varepsilon = \mu^2$. This reflects a balance of weak nonlinearity and weak dispersion.

### 5.4.1 Neglecting transverse variations

First, we will consider the special case of no transverse waves and later we will incorporate them into our model. Let us rewrite the fluid equations with the simplifying assumptions $\varepsilon = \mu^2$ – this is our "maximal balance" assumption. Let us also consider the one-dimensional case, i.e., we remove the terms involving derivatives with respect to $y$. The four equations then become

$$\varepsilon\phi_{xx} + \phi_{zz} = 0, \qquad\qquad -1 < z < \varepsilon\eta \qquad (5.10a)$$

$$\phi_z = 0, \qquad\qquad z = -1 \qquad (5.10b)$$

$$\phi_t + \frac{\varepsilon}{2}(\phi_x^2 + \frac{1}{\varepsilon}\phi_z^2) + \eta = 0, \qquad\qquad z = \varepsilon\eta \qquad (5.10c)$$

$$\varepsilon(\eta_t + \varepsilon\phi_x\eta_x) = \phi_z, \qquad\qquad z = \varepsilon\eta. \qquad (5.10d)$$

These are coupled, nonlinear partial differential equations in $\phi$ and $\eta$ with a free boundary, and are very difficult to solve exactly. We will use perturbation theory to obtain equations that are more tractable. We will asymptotically expand $\phi$ as

$$\phi = \phi_0 + \varepsilon\phi_1 + \varepsilon^2\phi_2 + \cdots.$$

Substituting this expansion into (5.10a) gives

$$\phi_{0zz} + \varepsilon(\phi_{0xx} + \phi_{1zz}) + \varepsilon^2(\phi_{1xx} + \phi_{2zz}) + \cdots = 0.$$

Equating terms with like powers of $\varepsilon$, we find that $\phi_{0zz} = 0$. This implies that $\phi_0 = A + B(z + 1)$, where $A, B$ are functions of $x, t$. But the boundary condition (5.10b) forces $B = 0$. Hence the leading-order solution for the velocity potential is independent of $z$,

$$\phi_0 = A(x, t).$$

Proceeding to the next order in $\varepsilon$, we see that $\phi_{1zz} = -A_{xx}$. Again using the boundary condition (5.10b),

$$\phi_1 = -A_{xx}(z + 1)^2/2.$$

As before, we absorb any homogeneous solutions that arise in higher-order terms into the leading-order term (that is $\phi_0$). Similarly we find $\phi_2 = A_{xxxx}(z + 1)^4/4!$. Thus, we have the approximation

$$\phi = A - \frac{\varepsilon}{2}A_{xx}(z + 1)^2 + \frac{\varepsilon^2}{4!}A_{xxxx}(z + 1)^4 + \cdots, \tag{5.11}$$

which is valid on the interval $-1 < z < \varepsilon\eta$, in particular, right up to the free boundary $\varepsilon\eta$. This expansion can be carried out to any order in $\varepsilon$, but the first three terms are sufficient for our purpose.

Substituting (5.11) into Bernoulli's equation (5.10c) along the free boundary $z = \varepsilon\eta$ gives

$$A_t - \frac{\varepsilon}{2}A_{xxt}(1 + \varepsilon\eta)^2 + \frac{\varepsilon^2}{4!}A_{xxxxt}(1 + \varepsilon\eta)^4 + \cdots$$
$$+ \frac{\varepsilon}{2}\left(A_x - \frac{\varepsilon}{2}A_{xxx}(1 + \varepsilon\eta)^2 + \cdots\right)^2$$
$$+ \frac{1}{2}(-\varepsilon A_{xx}(1 + \varepsilon\eta) + \cdots)^2 + \eta = 0.$$

Retaining only the first two terms leads to

$$\eta = -A_t + \frac{\varepsilon}{2}\left(A_{xxt} - A_x^2\right) + \cdots. \tag{5.12}$$

Now let us do the same thing for the kinematic equation (5.10d). Substituting in our expansion for $\phi$ and retaining the two lowest-order terms gives

$$\varepsilon\eta_t + \varepsilon^2\eta_x A_x = -\varepsilon A_{xx}(1 + \varepsilon\eta) + \frac{\varepsilon^2}{3!}A_{xxxx} + \cdots. \tag{5.13}$$

We wish to decouple the system of equations involving $A$ and $\eta$. Since we have an expression for $\eta$ in terms of $A$, we substitute (5.12) into (5.13) and retain only the two lowest-order terms:

$$\varepsilon\left(-A_{tt} + \frac{\varepsilon}{2}(A_{xxtt} - 2A_x A_{xt}) + \varepsilon^2(-A_{xt}A_x)\right)$$

$$= -\varepsilon A_{xx}(1 - \varepsilon A_t) + \frac{\varepsilon^2}{3!}A_{xxxx} + \cdots.$$

Dividing through by $\varepsilon$ and keeping order $\epsilon$ terms gives

$$A_{tt} - A_{xx} = \varepsilon\left(\frac{A_{xxtt}}{2} - \frac{A_{xxxx}}{6} - 2A_x A_{xt} - A_{xx}A_t\right). \tag{5.14}$$

Within this approximation, this equation is asymptotically the same as the one Boussinesq derived in 1871, cf. Ablowitz and Segur (1981). Another form, equally valid up to order $\varepsilon$, is derived by noting that on the left-hand side of (5.14) we have $A_{tt} = A_{xx} + O(\varepsilon)$. Then, in particular, we can take two $x$ derivatives to obtain $A_{xxtt} = A_{xxxx} + O(\varepsilon)$. Now we replace the $A_{xxtt}/2$ term in the above Boussinesq model to get

$$A_{tt} - A_{xx} = \varepsilon\left(\frac{A_{xxxx}}{3} - 2A_x A_{xt} - A_{xx}A_t\right), \tag{5.15}$$

which is still valid up to $O(\varepsilon)$.

Now we will make a remark about the linearized model. We have seen that there are various ways to write the equation; for example,

$$A_{tt} - A_{xx} = \varepsilon\left(\frac{1}{2}A_{xxtt} - \frac{1}{6}A_{xxxx}\right) \tag{5.16}$$

$$A_{tt} - A_{xx} = \frac{\varepsilon}{3}A_{xxxx}. \tag{5.17}$$

But only one of these two gives rise to a well-posed problem. We can see this from their dispersion relations: assume a wave solution $A(x, t) = \exp(i(kx - \omega t))$ and substitute it into (5.16) (or recalling the correspondences $k \to -i\partial_x$ and $\omega \to i\partial_t$), then

$$-\omega^2 + k^2 = \varepsilon\left(\frac{\omega^2 k^2}{2} - \frac{k^4}{6}\right) \quad \Longrightarrow \quad \omega^2 = \frac{k^2 + \varepsilon k^4/6}{1 + \varepsilon k^2/2}.$$

Note $\omega^2 > 0$ and for very large $k$, $\omega^2 \sim k^2/3$ . If we substitute this back into our wave solution, then we see that $A(x, t) = \exp(i(kx \pm kt/\sqrt{3}))$ is bounded for all time. But, if we do the same for (5.17), we get

$$-\omega^2 + k^2 = \varepsilon\frac{k^4}{3}.$$

The large $k$ limit of $\omega$ in this case is $\omega_\pm \sim \pm i\sqrt{\varepsilon}k^2/\sqrt{3}$. Substituting the negative root of $\omega$ into the wave solution we end up with

$$A(x, t) \sim \exp(ikx)\exp\left(\sqrt{\tfrac{\varepsilon}{3}}k^2 t\right).$$

This solution blows up arbitrarily fast as $k \to \infty$ so we do not have a convergent Fourier integral solution for "reasonable" (non-analytic) initial data (see also the earlier discussion of well-posedness). If we wish to do numerical calculations then (5.16) is preferred as it is well-posed, as opposed to (5.17). The ill-posedness of (5.17) is due to the long-wave approximation. This is a common difficulty associated with long-wave expansions; see also the discussion in Benjamin et al. (1972).

Nevertheless since there is still a small parameter, $\varepsilon$, in the equation, we can and should do further asymptotics on these equations. This will also remove the ill-posedness and is done in the next section.

First we make another remark about different Boussinesq models. We have derived an approximate equation for the leading term in the expansion of the velocity potential $\phi$ (5.14) or (5.15). Combined with (5.11) it determines the velocity potential, accurate to $O(\epsilon)$. We can also determine an equation for the wave amplitude $\eta$. First we take an $x$ derivative of (5.12) and then use the substitution $u = A_x$ in (5.12)–(5.13) to find

Bernoulli: 
$$\eta_x = -u_t + \frac{\varepsilon}{2}(u_{xxt} - 2uu_x) + \cdots , \qquad (5.18)$$

Kinematic: 
$$\eta_t = -u_x + \varepsilon\left(-\eta_x u - \eta u_x + \frac{u_{xxx}}{6}\right) + \cdots . \qquad (5.19)$$

This is called the coupled Boussinesq model for $\eta$ and $u$.

On the other hand, by differentiating (5.19) with respect to $t$, and by differentiating (5.18) with respect to $x$, we get

$$\eta_{tt} - \eta_{xx} = \varepsilon\left(-\eta_{xt}u - \eta_x u_t - \eta_t u_x - \eta u_{xt}\right.$$
$$\left. + \frac{u_{xxxt}}{6} - \frac{u_{xxxt}}{2} + \frac{(u^2)_{xx}}{2}\right) + O(\varepsilon).$$

Now we use the relations $u_x = -\eta_t + O(\varepsilon)$ and $u_t = -\eta_x + O(\varepsilon)$ from the kinematic and Bernoulli equations, respectively, and make the replacement $u = -\int_{-\infty}^{x} \eta_t \, dx'$. Hence we find

$$\eta_{tt} - \eta_{xx} = \varepsilon\left[\frac{1}{3}\eta_{xxxx} + \eta_{xt}\int_{-\infty}^{x} \eta_t \, dx' + \eta_x^2 + \eta_t^2 \right.$$
$$\left. + \eta\eta_{xx} + \frac{1}{2}\frac{\partial^2}{\partial x^2}\left(\int_{-\infty}^{x} \eta_t \, dx'\right)^2\right] \qquad (5.20)$$

$$= \varepsilon \left[ \frac{\eta_{xxxx}}{3} + \frac{\partial}{\partial x} \left( \eta_t \int_{-\infty}^{x} \eta_t \, dx' + \eta \eta_x \right) \right.$$
$$\left. + \frac{1}{2} \frac{\partial^2}{\partial x^2} \left( \int_{-\infty}^{x} \eta_t \, dx' \right)^2 \right]$$
$$= \varepsilon \left[ \frac{\eta_{xxxx}}{3} + \frac{\partial^2}{\partial x^2} \left( \frac{\eta^2}{2} + \left( \int_{-\infty}^{x} \eta_t \, dx' \right)^2 \right) \right]. \tag{5.21}$$

The last equation (5.21) is the Boussinesq model for the wave amplitude. If we omit the non-local term (which can be done via a suitable transformation) (5.21) is reduced to the Boussinesq model discussed in Chapter 1 [see (1.2), where $\eta = y_x$].

### 5.4.2 Multiple-scale derivation of the KdV equation

Now let us return to the Boussinesq model for the velocity potential equation (5.15) to do further asymptotics

$$A_{tt} - A_{xx} = \varepsilon \left( \frac{A_{xxxx}}{3} - 2A_x A_{xt} - A_{xx} A_t \right). \tag{5.22}$$

Assume an asymptotic expansion for $A$:

$$A = A_0 + \varepsilon A_1 + \cdots ;$$

substituting this expansion into (5.22) gives

$$A_{0tt} + \varepsilon A_{1tt} - A_{0xx} - \varepsilon A_{1xx} + O(\varepsilon^2) = \varepsilon \left( \frac{A_{0xxxx}}{3} - \right.$$
$$\left. - 2A_{0x} A_{0xt} - A_{0xx} A_{0t} + O(\varepsilon) \right).$$

Equating the leading-order terms yields the wave equation: $A_{0tt} - A_{0xx} = 0$. We solved this equation earlier and found $A_0(x, t) = F(x - t) + G(x + t)$ where $F$ and $G$ are determined by the initial conditions. Anticipating secular terms that will need to be removed in the next-order equation, we employ multiple scales; i.e., $A_0 = A_0(\xi, \zeta, T)$:

$$A_0 = F(\xi, T) + G(\zeta, T),$$

where

$$\xi = x - t, \quad \zeta = x + t, \quad T = \varepsilon t.$$

These new variables imply

$$\partial_t = -\partial_\xi + \partial_\zeta + \varepsilon \partial_T \quad \text{and} \quad \partial_x = \partial_\xi + \partial_\zeta.$$

Substituting the expressions for the differential operators into (5.22) leads to

$$((-\partial_\xi + \partial_\zeta + \varepsilon\partial_T)^2 - (\partial_\xi + \partial_\zeta)^2)A =$$
$$\varepsilon\left(\frac{(\partial_\xi + \partial_\zeta)^4}{3}A - 2(\partial_\xi + \partial_\zeta)A(\partial_\xi + \partial_\zeta)(-\partial_\xi + \partial_\zeta + \varepsilon\partial_T)A\right.$$
$$\left. - (\partial_\xi + \partial_\zeta)^2 A(-\partial_\xi + \partial_\zeta + \varepsilon\partial_T)A\right). \qquad (5.23)$$

First we will assume unidirectional waves and only work with the right-moving wave. So $A_0 = F(\xi, T)$ for now. Substituting

$$A = A_0 + \varepsilon A_1 + \cdots$$

into (5.23) and keeping all the $O(\varepsilon)$ terms we get

$$-4A_{1\xi\zeta} = 2F_{\xi T} + \frac{1}{3}F_{\xi\xi\xi\xi} + 3F_{\xi\xi}F_\xi.$$

This equation can be integrated directly to give

$$A_1 \sim -\frac{1}{4}\left(2F_T + \frac{1}{3}F_{\xi\xi\xi} + \frac{3}{2}F_\xi^2\right)\zeta + \cdots,$$

remembering that we absorb homogeneous terms in the $A_0$ solution. In order to remove secular terms, in particular the $\zeta$ term, we require that

$$2F_T + \frac{1}{3}F_{\xi\xi\xi} + \frac{3}{2}F_\xi^2 = 0;$$

or, if we take a derivative with respect to $\xi$ and make the substitution $U = F_\xi$ then we have the Korteweg–de Vries equation

$$2U_T + \frac{1}{3}U_{\xi\xi\xi} + 3UU_\xi = 0.$$

Next we discuss the more general case of waves moving to the left and right; hence $A_0 = F(\xi, T) + G(\zeta, T)$, and we get the following expression for $A_1$

$$-4A_{1\xi\zeta} = 2(F_{\xi T} - G_{\zeta T}) + \frac{1}{3}(F_{\xi\xi\xi\xi} + +G_{\zeta\zeta\zeta\zeta})$$
$$+ 3(F_\xi F_{\xi\xi} + \frac{1}{3}F_{\xi\xi}G_\zeta - \frac{1}{3}G_{\zeta\zeta}F_\xi - G_\zeta G_{\zeta\zeta}).$$

When we integrate this expression, secular terms arise from the pieces that are functions of $\xi$ or $\zeta$ alone, not both. Removal of the secular terms implies the following two equations

$$2F_{\xi T} + \frac{1}{3}F_{\xi\xi\xi\xi} + 3F_{\xi\xi}F_\xi = 0$$
$$-2G_{\zeta T} + \frac{1}{3}G_{\zeta\zeta\zeta\zeta} - 3G_{\zeta\zeta}G_\zeta = 0,$$

which are two uncoupled KdV equations. Simplifying things a bit, we can rewrite the above two equations in terms of $U = F_\xi$ and $V = G_\zeta$:

$$2U_T + \frac{1}{3}U_{\xi\xi\xi} + 3UU_\xi = 0 \tag{5.24}$$

$$2V_T - \frac{1}{3}V_{\zeta\zeta\zeta} + 3VV_\zeta = 0. \tag{5.25}$$

The solution $A_1$ can be obtained by integrating the remaining terms. Since we are only interested in the leading-order asymptotic solution we need not solve for $A_1$.

Hence we have the following conclusion: asymptotic analysis of the fluid equations under shallow-water conditions has given rise to two Korteweg–de Vries (KdV) equations, (5.24) and (5.25), for the right- and left-going waves. Given rapidly decaying initial conditions, we can solve these two PDEs for $U$ and $V$ by the inverse scattering transform (IST). The KdV equation has been studied intensely and its IST solution is described in Chapter 9. Keeping only the leading-order terms, we have an approximate solution for the velocity potential

$$\phi(x, z, t) \sim A_0(x, t) = F(x - t, \varepsilon t) + G(x + t, \varepsilon t),$$

$$= \int_{-\infty}^{x-t} U(\xi', \varepsilon t) \, d\xi' + \int_{-\infty}^{x+t} V(\zeta', \varepsilon t) \, d\zeta',$$

or for the velocity

$$u = \phi_x = F_x + G_x + \cdots = U + V + \cdots .$$

Using the Bernoulli equation (5.12), we have an approximate solution for the wave amplitude of the free boundary

$$\eta(x, t) \sim -A_{0t}(x, t) \sim F_\xi(\xi, T) - G_\zeta(\zeta, T) = U(x - t, \varepsilon t) - V(x + t, \varepsilon t).$$

Thus the wave amplitude $\eta$ has right- and left-going waves that satisfy the KdV equation. Alternatively, we can derive the KdV equation directly by performing an asymptotic expansion in the $\eta$ equation (5.21) though we will not do that here. One can also proceed to obtain higher-order terms and corrections to the KdV equation, though there is much more in the way of details that are required. We will not discuss this here.

### 5.4.3 Dimensional equations

In the previous section, we made our fluid equations non-dimensional for the purpose of determining what terms are "small" in our approximation of shallow-water waves. Now that we have approximate equations (5.24) and (5.25), valid to leading order, we wish to understand the results in dimensional units. We will now transform back to a dimensional equation for (5.24).

Recall that we made two changes of variable. We first changed coordinates $x, t, \eta$ into primed coordinates $x', t', \eta'$ (though we immediately dropped the primes). Then the independent variable substitutions $\xi = x' - t'$ and $T = \varepsilon t'$ were made. First, we remove the $\xi$ and $T$ dependence. Notice that

$$\frac{\partial}{\partial t'} U(\xi, T) = -U_\xi + \varepsilon U_T$$

$$\frac{\partial}{\partial x'} U(\xi, T) = U_\xi.$$

The above two expressions imply that $\partial_T = (\partial_{t'} + \partial_{x'})/\varepsilon$. Making the above substitutions into (5.24), we get

$$2(U_{t'} + U_{x'}) + \varepsilon\left(\frac{1}{3} U_{x'x'x'} + 3UU_{x'}\right) = 0.$$

If we neglect left-traveling waves, $V$, then we have $\eta' \sim U$. Recalling our original non-dimensionalization $x = \lambda_x x'$, $t = \frac{\lambda_x}{c_0} t'$, $\varepsilon = a/h$ and $\eta = a\eta'$, we use $\partial_{x'} = \lambda_x \partial_x$ and $\partial_{t'} = \frac{\lambda_x}{c_0}\partial_t$ to transform the previous expression in $U$ to

$$2\left(\frac{\lambda_x}{c_0}\frac{\eta_t}{a} + \lambda_x\frac{\eta_x}{a}\right) + \varepsilon\left(\frac{\lambda_x^3}{3a}\eta_{xxx} + \frac{3\lambda_x}{a^2}\eta\eta_x\right) = 0 \Rightarrow$$

$$2\left(\frac{\eta_t}{c_0} + \eta_x\right) + \frac{a}{h}\left(\frac{\lambda_x^2}{3}\eta_{xxx} + \frac{3}{a}\eta\eta_x\right) = 0.$$

From maximal balance, we used $\varepsilon = a/h = (h/\lambda_x)^2$ so we can simplify the non-dimensional KdV equation (5.24) to the dimensional form

$$\frac{1}{c_0}\eta_t + \eta_x + \frac{h^2}{6}\eta_{xxx} + \frac{3}{2h}\eta\eta_x = 0, \tag{5.26}$$

where $c_0 = \sqrt{gh}$, $g$ is the gravitational constant of acceleration and $h$ is the water depth. This equation was derived by Korteweg and de Vries (1895).

Let us consider the linear part of (5.26)

$$\frac{1}{c_0}\eta_t + \eta_x + \frac{h^2}{6}\eta_{xxx} = 0.$$

We will interpret the above equation in terms of the dispersion relation for water waves [see (5.8)], noting $\kappa^2 = k^2 + l^2$ with $l = 0$, so $\kappa = k$:

$$\omega^2 = gk\tanh(kh) = gk\left[kh - \frac{1}{3}(kh)^3 + \cdots\right].$$

In shallow-water waves, $kh \ll 1$ so if we retain only the leading-order term in the Taylor series expansion above, we have

$$\omega = \pm k\sqrt{gh} = \pm kc_0.$$

This is the dispersion relation for the linear wave equation and $c_0$ is the wave speed. Retaining the next term in the Taylor series of $\tanh(kh)$ gives

$$\omega \sim \pm k\sqrt{gh(1 - \frac{1}{3}(kh)^2)}$$
$$\sim \pm k\sqrt{gh}(1 - \frac{1}{6}(kh)^2)$$
$$= \pm(kc_0 - \frac{1}{6}c_0 h^2 k^3).$$

Now, we wish to see what linear PDE gives rise to the above dispersion relation. Taking the positive root, replacing $\omega$ with $i\partial_t$ and $k$ with $-i\partial_x$, we find

$$i\partial_t\eta = c_0(-i\partial_x)\eta - \frac{1}{6}c_0 h^2(-i\partial_x)^3\eta \Rightarrow$$
$$\frac{1}{c_0}\eta_t + \eta_x + \frac{h^2}{6}\eta_{xxx} = 0.$$

This is exactly the dimensional linear KdV equation (5.26)! To get the nonlinear term we used the multiple-scales method.

### 5.4.4 Adding surface tension

The model equations (5.4) through (5.7) do not take into account the effects of surface tension. Actually only Bernoulli's equation (5.10c) is affected by surface tension. The modification is due to an additional pressure term from surface tension effects involving the curvature at the surface. We will not go through the derivation here: rather we will just state the result. First, recall Bernoulli's equation used so far

$$\phi_t + \frac{1}{2}|\nabla\phi|^2 + g\eta = 0, \tag{5.27}$$

on $z = \eta$. Adding the surface tension term gives (cf. Lamb 1945 or Ablowitz and Segur 1981)

$$\phi_t + \tfrac{1}{2}|\nabla\phi|^2 + g\eta = \frac{T}{\rho}\nabla \cdot \left(\frac{\nabla\eta}{\sqrt{1+|\nabla\eta|^2}}\right)$$

$$= \frac{T}{\rho}\frac{\left(\eta_{xx}\left(1+\eta_y^2\right)+\eta_{yy}\left(1+\eta_x^2\right)-2\eta_{xy}\eta_x\eta_y\right)}{\left(1+\eta_x^2+\eta_y^2\right)^{\frac{3}{2}}} \qquad (5.28)$$

on $z = \eta$ where $T$ is the surface tension coefficient. Retaining dimensions and keeping linear terms that will affect the previous result to $O(\varepsilon)$, we have:

$$\phi_t + \frac{1}{2}|\nabla\phi|^2 + g\eta - \frac{T}{\rho}(\eta_{xx}+\eta_{yy}) = 0 \qquad z = \eta. \qquad (5.29)$$

Using (5.29) and the same asymptotic procedure as before, the corresponding leading-order asymptotic equation for the free surface is found to be

$$\frac{1}{c_0}\eta_t + \eta_x + \gamma\eta_{xxx} + \frac{3}{2h}\eta\eta_x = 0. \qquad (5.30)$$

The only difference between this equation and (5.26) is the coefficient $\gamma$ of the third-derivative term. This term incorporates the surface tension

$$\gamma = \frac{h^2}{6} - \frac{T}{2\rho g} = \frac{h^2}{6}\left(1 - \frac{3T}{\rho g h^2}\right) = \frac{h^2}{6}(1 - 3\hat{T})$$

$$\hat{T} = \frac{T}{\rho g h^2}.$$

Note (5.30) was also derived by Korteweg and deVries [see Chapter 1, equation (1.4)], and we can see that, depending on the relative sizes of the parameters (in particular $\hat{T} < 1/3$ or $\hat{T} > 1/3$ ), $\gamma$ will be positive or negative. This affects the types of solutions and behavior allowed by the equation and is discussed later. In particular, when we include transverse waves, we will discover the Kadomtsev–Petviashvili (KP) equation.

### 5.4.5 Including transverse waves: The KP equations

Our investigation so far has been for waves in shallow water without transverse modulations. We will now relax our assumptions and include weak transverse variation.

First, let us look at the dispersion relation for water waves. Using the same ideas as previously [see the discussion leading to (5.8)] the reader can verify that the dispersion relation in multidimensions (two space, one time) is given by

$$\omega^2 = \left(g\kappa + \frac{T}{\rho}\kappa^3\right)\tanh(\kappa h), \tag{5.31}$$

where $\kappa^2 = k^2 + l^2$; here $k, l$ are the wavenumbers that correspond to the $x$- and $y$-directions, respectively. Recall that the wavelengths of a wave solution in the form $\alpha e^{i(kx+ly-\omega t)}$ are $\lambda_x = 2\pi/k$ and $\lambda_y = 2\pi/l$. We further assume that (recall our earlier definitions of scales)

$$\delta = \frac{\lambda_x}{\lambda_y} = \frac{l}{k} \ll 1.$$

The above relation says that the wavelength in the $y$-direction is much larger than the wavelength in the $x$-direction. This is what is referred to as weak transverse variation.

We now derive the linear PDE associated with the dispersion relation (5.31) with weak transverse variation. Since we are still in the shallow-water regime, we have $|hk| \propto |h/\lambda_x| \ll 1$ and we assume that $\kappa h = kh\sqrt{1 + l^2/k^2} \ll 1$. Then we expand the hyperbolic tangent term in a Taylor series to find

$$\omega^2 \approx \left(g\kappa + \frac{T}{\rho}\kappa^3\right)\left(\kappa h - \frac{1}{3}(\kappa h)^3\right) + \cdots$$

$$= ghk^2\left(1 + \frac{T}{g\rho}\kappa^2\right)\left(1 - \frac{1}{3}(\kappa h)^2\right) + \cdots$$

$$\omega = \sqrt{gh}\, k\left(1 + \frac{l^2}{k^2}\right)^{\frac{1}{2}}\left(1 + \frac{T}{g\rho}k^2 + O(l^2)\right)^{\frac{1}{2}}$$

$$\times \left(1 - \frac{1}{3}(hk)^2 + O(l^2)\right)^{\frac{1}{2}}.$$

Use of the binomial expansion and assuming $|l/k| \ll 1$ and the maximal balance relation $l^2 \sim O(k^4)$, we end up with ($c_0 = \sqrt{gh}$),

$$\omega k \approx c_0\left(k^2 + k^4\left(\frac{T}{2g\rho} - \tfrac{1}{6}h^2\right) + \tfrac{1}{2}l^2 + O(k^6)\right).$$

Now we can write down the linear PDE associated with this dispersion relation using $\omega \to i\partial_t, k \to -i\partial_x$:

$$\frac{1}{c_0}\eta_{tx} + \eta_{xx} + \tfrac{1}{2}\eta_{yy} + \frac{h^2}{6}(1 - 3\hat{T})\eta_{xxxx} = 0,$$

where we recall $\hat{T} = T/\rho gh^2$. This is the linear Kadomtsev–Petviashvili (KP) equation. To derive the full nonlinear version, we must resort to multiple scales

(Ablowitz and Segur, 1979, 1981). Though we will not do that here, if we take a hint from our work on the KdV equation, we can expect that the nonlinear part of our equation will not change when we add in slow transverse ($y$) dependence. This is true and the full KP equation, first derived in 1970 (Kadomtsev and Petviashvili, 1970), in dimensional form is

$$\partial_x\left(\frac{1}{c_0}\eta_t + \eta_x + \frac{3}{2h}\eta\eta_x + \gamma\eta_{xxx}\right) + \frac{1}{2}\eta_{yy} = 0, \ or$$

$$\frac{1}{c_0}\eta_t + \eta_x + \frac{3}{2h}\eta\eta_x + \gamma\eta_{xxx} = -\frac{1}{2}\int_{-\infty}^{x}\eta_{yy}\ dx', \qquad (5.32)$$

where $\gamma = \dfrac{h^2}{6}(1 - 3\hat{T})$. Usually in water waves surface tension is small and we have $\hat{T} < 1/3$ which gives rise to the so-called KPII equation

$$\frac{1}{c_0}\eta_{xt} + \eta_{xx} + \frac{3}{2h}(\eta\eta_x)_x + \frac{h^2}{6}\eta_{xxxx} = -\frac{1}{2}\eta_{yy}.$$

The equation termed the KPI equation arises when $\hat{T} = T/\rho g h^2 > 1/3$; i.e., when surface tension effects are large. Alternatively, we can rescale the equation into non-dimensional form

$$\partial_x(u_t + 6uu_x + u_{xxx}) + 3\sigma u_{yy} = 0, \qquad (5.33)$$

where $\sigma$ has the following meaning
- $\sigma = +1 \implies$ KPII: typical water waves, small surface tension,
- $\sigma = -1 \implies$ KPI: water waves, large surface tension.

We note that if $\eta_{yy} = 0$ in (5.32) and we rescale, the resulting KP equation can be reduced to the KdV equation in standard form

$$u_t \pm 6uu_x + u_{xxx} = 0. \qquad (5.34)$$

The "+" corresponds to $\hat{T} < 1/3$ whereas the "−" arises when $\hat{T} > 1/3$.

## 5.5 Solitary wave solutions

As discussed in Chapter 1, the KdV and KP equations admit special, exact solutions known as solitary waves. We also mentioned in Chapter 1 that a solitary wave was noted by John Scott Russell in 1834 when he observed a wave detach itself from the front of a boat brought to rest. This wave evolved into a localized rounded hump of water that Russell termed the Great Wave of Translation. He followed this solitary wave on horseback as it moved along the Union Canal between Edinburgh and Glasgow. He noted that it hardly changed its shape or lost speed for over two miles; see Russell (1844); Ablowitz and Segur (1981); Remoissenet (1999) and www.ma.hw.uk/solitons. Today, scientists often use the term soliton instead of solitary wave for localized solutions of many equations despite the fact that the original definition of a soliton reflected the fact that two solitary waves interacted elastically.

### 5.5.1 A soliton in dimensional form for KdV

Recall from above that the KdV equation in standard, non-dimensional form, (5.34), can be written

$$u_t \pm 6uu_x + u_{xxx} = 0, \quad \text{and } \pm \text{ when } \gamma \gtrless 0,$$

where

$$\gamma = \frac{h^2}{6}(1 - 3\hat{T}).$$

A soliton solution admitted by the non-dimensional equation (5.34) is given by

$$u(x, t) = \pm 2\beta^2 \, \text{sech}^2(\beta(x \mp 4\beta^2 t - x_0)). \tag{5.35}$$

Notice that the speed of the wave, $c = 4\beta^2$, is twice the amplitude of the wave. Also, the "+" corresponds to $\gamma > 0$ that is physically a positive elevation wave traveling on the surface like the one first observed by Russell. The "–" case results from $\gamma < 0$ and corresponds to a dip in the surface of the water. In fact, an experiment done only recently with high surface tension produced just such a "depression" wave (Falcon et al., 2002).

The solitary wave or soliton solution equation (5.35) is in non-dimensional form. We can convert this solution of the non-dimensional equation (5.34) directly to a solution of the dimensional equation (5.26). We will only consider here $\gamma > 0$; we leave it as an exercise to find the dimensional soliton when $\gamma < 0$. First consider (5.24) in the form

$$2U'_{T'} + \frac{1}{3}U'_{\xi'\xi'\xi'} + 3U'U'_{\xi'},$$

where we denote all variables with a prime. We will rescale the variables appropriately. Assume the following transformation of coordinates

$$\xi' = l_1\xi, \quad T' = l_2T, \quad U' = l_3U.$$

Then (5.24) becomes

$$\frac{2}{l_2}U_T + \frac{1}{3l_1^3}U_{\xi\xi\xi} + 3\frac{l_3}{l_1}UU_\xi = 0.$$

If we multiply the whole equation by $l_1/l_3$ then we get

$$\frac{2l_1}{l_2 l_3} U_T + \frac{1}{3 l_1^2 l_3} U_{\xi\xi\xi} + 3 U U_\xi = 0.$$

To get (5.24) into standard form, (5.34), we set

$$\frac{2l_1}{l_2 l_3} = \frac{1}{2} \quad \text{and} \quad \frac{1}{3 l_1^2 l_3} = \frac{1}{2}.$$

One solution is $(l_1, l_2, l_3) = (1, 6, 2/3)$. Now our soliton solution takes the form

$$U'(\xi, T) = \frac{4\beta^2}{3} \, \text{sech}^2 \left( \beta(\xi - \tfrac{2}{3}\beta^2 T - x_0) \right).$$

Next use, $\xi' = x' - t'$ and $T' = \varepsilon t'$ to find

$$U'(x', t') = \frac{4\beta^2}{3} \, \text{sech}^2 \left( \beta(x' - (1 + \tfrac{2}{3}\beta^2\varepsilon)t' - x_0) \right).$$

The dimensional solution employs the following change of variables

$$\eta = a U', \quad x = \lambda_x x', \quad t = \frac{\lambda_x}{c_0} t'.$$

Substituting this into our solution, we have

$$\eta(x, t) = \frac{4\beta^2 a}{3} \, \text{sech}^2 \left( \beta \left( \frac{x}{\lambda_x} - \left(1 + \tfrac{2}{3}\beta^2\varepsilon\right) \frac{c_0}{\lambda_x} t - \frac{x_0}{\lambda_x} \right) \right).$$

Now we use the relations $a/h = \varepsilon$, $\mu = h/\lambda_x$, and $\mu^2 = \varepsilon$ to write down the dimensional solution to the KdV equation in the form

$$\frac{\eta}{h} = \tfrac{4}{3}\varepsilon\beta^2 \, \text{sech}^2 \left( \frac{\beta}{h} \sqrt{\varepsilon}(x - (1 + \tfrac{2}{3}\beta^2\varepsilon)c_0 t - x_0) \right).$$

Note that both $\beta$ and $\varepsilon$ are non-dimensional numbers, and $\epsilon$ is related to the size of a typical wave amplitude. Recall that $a = |\eta_{\text{max}}|$ and $c_0^2 = gh$. The excess speed beyond the long wave speed is $\dfrac{2\beta^2}{3} c_0 \varepsilon$. Russell observed this as well! Later, in 1871, Boussinesq found this relationship from the more general point of view of weakly nonlinear waves moving in two directions. Boussinesq also found KdV-type equations (Boussinesq, 1877). Korteweg and de Vries found this result in 1895 concentrating on unidirectional water waves. They also found a class of periodic solutions in terms of elliptic functions and called them "cnoidal" functions (see also the discussion in Chapter 1).

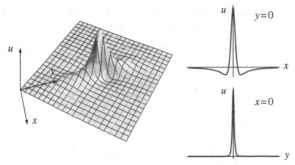

Figure 5.2 Lump solution of KPI, on the left. To better illustrate, we also plot slices along the *x*- and *y*-directions on the right.

### 5.5.2 Solitons in the KP equation

Recall the KPI equation (with "large" surface tension) in non-dimensional form, (5.33), is

$$\partial_x(u_t + 6uu_x + u_{xxx}) - 3u_{yy} = 0.$$

This equation admits the following non-singular traveling "lump" soliton solution (cf. Ablowitz and Clarkson, 1991) (see Figure 5.2)

$$u(x, y, t) = 2\partial_x^2\left[\log\left((\hat{x} - 2k_R\hat{y})^2 + 4k_I^2\hat{y}^2 + \frac{1}{4k_I^2}\right)\right]$$

$$\hat{x} = x - 12\left(k_R^2 + k_I^2\right)t - x_0$$

$$\hat{y} = y - 12k_Rt - y_0.$$

Note that this lump only moves in the positive *x*-direction. Though the one-dimensional solitons in the previous section have been observed, this lump solution has not yet been seen in the laboratory. Presumably, this is because of the difficulties of working with large surface tension fluids. But, as mentioned above, only recently has the one-dimensional KdV "depression" solitary wave been observed for large surface tension (Falcon et al., 2002).

If there exists a solution to (5.33) such that $u, u_x, u_t, u_{xxx} \to 0$ as $x \to -\infty$ then we have

$$u_t + 6uu_x + u_{xxx} = -3\sigma\int_{-\infty}^x u_{yy}\,dx'.$$

In particular, if $u, u_x, u_{xxx} \to 0$ and $u_t \to 0$ as $x \to \infty$, we also have

$$\int_{-\infty}^{\infty} u_{yy}(x, y, t)\,dx = 0.$$

But if we give a general initial condition $u(x, y, t = 0) = u_0(x, y)$, it need not satisfy this constraint at the initial instant. For the linearized version of the KP equation, one can show (Ablowitz and Villarroel, 1991) that if the initial condition satisfies

$$\int_{-\infty}^{\infty} u_0(x, y) \, dx \neq 0,$$

nevertheless the solution obeys

$$\int_{-\infty}^{\infty} u(x, y, t) \, dx = 0, \quad \text{for all } t \neq 0$$

and the function $\partial u / \partial t$ is discontinuous at $t = 0$. Ablowitz and Wang (1997) show why KP models of water waves lead to such constraints asymptotically and how a two-dimensional Boussinesq equation has an "initial value layer" that "smooths" the discontinuity in $\int u_{yy}(x, y, t) \, dx$.

### 5.5.3 KdV and related models

Suppose we consider the KdV equation in the form

$$u_t + u_x + \varepsilon(u_{xxx} + \alpha u u_x) = 0 \tag{5.36}$$

where $\alpha$ is constant. First look at the linear problem and assume the wave solution $u \sim \exp(i(kx - \omega t))$. The dispersion relation is then

$$\omega = k - \varepsilon k^3,$$

which, for $|k| \gg 1$, implies $\omega \sim -\varepsilon k^3$. In particular, note that $\omega \to -\infty$, which can present some numerical difficulties. One way around this issue is to use the relation $u_t = -u_x + O(\varepsilon)$ to alter the equation slightly (Benjamin et al., 1972)

$$u_t + u_x + \varepsilon(-u_{xxt} + \alpha u u_x) = 0. \tag{5.37}$$

Note that (5.36)–(5.37) are asymptotically equivalent to $O(\epsilon)$. But for large $k$, the dispersion relation for the linear problem is now

$$\omega = \frac{k}{1 + \varepsilon k^2} \sim \frac{1}{\varepsilon k}$$

for $|k| \gg 1$. Numerically, (5.37) has some advantages over (5.36). But, we still have a "small" parameter, $\varepsilon$, in both equations so they are not asymptotically reduced. Alternatively, if we use the following transformation in (5.36), as indicated by multiple-scale asymptotics,

$$\xi = x - t, \quad T = \varepsilon t,$$
$$\partial_x = \partial_\xi, \quad \partial_t = -\partial_\xi + \varepsilon \partial_T,$$

then we get the equation

$$u_T + u_{\xi\xi\xi} + \alpha u u_\xi = 0, \tag{5.38}$$

which has no direct dependence on the small parameter $\varepsilon$. Equation (5.38) is also useful for numerical computation since $T = O(1)$ corresponds to $t = O(1/\varepsilon)$.

Finally, recall the Fermi–Pasta–Ulam (FPU) model of nonlinear coupled springs,

$$m\ddot{y}_i = k(y_{i+1} + y_{i-1} - 2y_i) + \alpha((y_{i+1} - y_i)^2 - (y_i - y_{i-1})^2) \tag{5.39}$$

discussed in Chapter 1. In the continuum limit, use $y_{i+1} = y(x) + hy'(x) + h^2 y''(x)/2 + \cdots$, $x = ih$, to find

$$\omega^2 y_{tt} - h^2 y_{xx} = \frac{h^4}{12} y_{xxxx} + 2\alpha h^3 y_x y_{xx} + \cdots,$$

where we have used a dot to denote $\dfrac{d}{dt}$. Making the following substitutions $t = \tau/h\omega$, $\omega^2 = k/m$, $\varepsilon = 2\alpha h$, and $\delta^2 \varepsilon = h^2/12$, where the last two equalities arise from the stipulation of maximal balance, we have

$$y_{\tau\tau} - y_{xx} = \varepsilon(\delta^2 y_{xxxx} + y_x y_{xx}). \tag{5.40}$$

This is a Boussinesq-type equation and is also known to be integrable (Ablowitz and Segur, 1981; Ablowitz and Clarkson, 1991). The term integrable has several interpretations. We will use the notion that if we can exactly linearize (as opposed to solving the equation via a perturbation expansion) the equation then we consider the equation to be integrable (see Chapters 8 and 9). Linearization or direct methods when applied to an integrable equation allow us to find wide classes of solutions.

Performing asymptotic analysis, in particular multiple scales on (5.40) with the expansion

$$y \sim y_0 + \varepsilon y_1 + \cdots$$
$$y_0 = \phi(X; T), \qquad X = x - \tau, \ T = \varepsilon\tau,$$

we find that, after removing secular terms (see also Chapter 1),

$$2\phi_{XT} + \phi_X \phi_{XX} + \delta^2 \phi_{XXXX} = 0.$$

Or, if we make the substitution $\phi_X = u$ then we end up with the integrable KdV equation

$$u_T + \tfrac{1}{2} u u_X + \frac{\delta^2}{2} u_{XXX} = 0.$$

Another lattice equation, called the Toda lattice that is known to be integrable, is given by

$$m\ddot{y}_i = e^{k(y_{i+1}-y_i)} - e^{k(y_i-y_{i-1})}.$$

If we expand the exponential and keep quadratic nonlinear terms, we get the FPU model (5.39), which in turn asymptotically yields the KdV equation. This further shows that the KdV equation (5.34) arises widely in applied mathematics and the FPU problem itself is very "close" to being integrable.

### 5.5.4 Non-local system and the Benney–Luke equations

In a related development, a recent reformulation (see Ablowitz et al., 2006) of the fully nonlinear water wave equations with surface tension leads to two equations for two unknowns: $\eta$ and $q = q(x, y, t) = \phi(x, y, \eta(x, y, t))$. The equations are given by the following "simple looking" system

$$\int_{-\infty}^{\infty} \int_{-\infty}^{\infty} dx\, dy\, e^{i(kx+ly)} \Big( i\eta_t \cosh[\kappa(\eta + h)] $$
$$+ \frac{1}{\kappa}(k, l) \cdot \nabla q \sinh\left[\kappa(\eta + h)\right] \Big) = 0 \qquad \text{(I)}$$

$$q_t + \frac{1}{2}|\nabla q|^2 + g\eta - \frac{(\eta_t + \nabla q \cdot \nabla \eta)^2}{2(1 + |\nabla \eta|^2)} = \sigma \nabla \cdot \left( \frac{\nabla \eta}{\sqrt{1 + |\nabla \eta|^2}} \right) \qquad \text{(II)}$$

where $(k, l) \cdot \nabla q = kq_x + lq_y$, $\kappa^2 = k^2 + l^2$, $\sigma = T/\rho$ where $T$ is the surface tension and $\rho$ is the density; here $\eta$ and derivatives of $q$ are assumed to decay rapidly at infinity. The first of the above two equations is non-local and is written in a "spectral" form.

The non-local equation satisfies the Laplace equation, the kinematic condition and the bottom boundary condition. The second equation is Bernoulli's equation rewritten in terms of the new variable $q$. These variables were introduced by Zakharov (1968) in his study of the Hamiltonian formulation of the water wave problem. Subsequently they were used by Craig and co-workers (Craig and Sulem, 1993; Craig and Groves, 1994) in their discussion of the Dirichlet–Neumann (DN) map methods in water waves. In this regard we note:

$$\eta_t = \phi_z - \nabla\phi \cdot \nabla \eta = \nabla\phi \cdot \vec{n}$$

where $\vec{n} = (-\nabla\eta, 1)$ is the normal on $y - \eta = 0$. Thus finding $q(x, y, t)$, $\eta(x, y, t)$ yields $\eta_t$, which in turn leads to the normal derivative $\partial\phi/\partial n$; hence the DN map is contained in the solution of the non-local system (I–II). Also, solving equations (I)–(II) for $\eta, q$ reduces to solving the remaining water wave equation, Laplace's equation (5.4), which is a linear problem due to the now fixed boundary conditions.

The non-local equation can be derived from Laplace's equation, the kinematic condition and the bottom boundary condition via Green's identity (cf. Haut and Ablowitz, 2009). The essential ideas are outlined below. We also mention that the reader will find valuable discussions of non-local systems with auxiliary parameters in Fokas (2008). Fokas has made numerous important contributions in the study of such non-local nonlinear equations. We define an associated potential $\psi(x, y, z)$ satisfying

$$\Delta\psi(x, y, z) = 0, \quad \text{in } D, \quad \psi_y(x, y, z = -h) = 0. \tag{5.41}$$

Then, from Green's identity,

$$
\begin{aligned}
0 &= \int_{D(\eta)} ((\Delta\psi)\phi - (\Delta\phi)\psi)\, dV \\
&= \int_{\partial D(\eta)} (\phi(\nabla\psi \cdot \hat{n}) - \psi(\nabla\phi \cdot \hat{n}))\, dS
\end{aligned}
$$

where $dV, dS$ are the volume and surface elements respectively, and $\hat{n}$ is the unit normal. It then follows that

$$\int \psi(x, y, \eta)\eta_t\, dxdy = \int q\left(\psi_z(x, \eta) - \nabla_{x,y}\psi(x, y, \eta) \cdot \nabla\eta\right) dx\, dy, \tag{5.42}$$

where we have used $\eta_t = \nabla\phi \cdot \vec{n}$, $\vec{n} = (-\nabla\eta, 1)$ on the free surface as well as the bottom boundary condition and the condition $|\eta|$ decaying at infinity.

We assume a solution of (5.41) can be written in the form

$$\psi(x, y, z) = \int \hat{\xi}(k, l)\hat{\psi}_{k,l}(x, y, z)dkdl, \quad \hat{\psi}_{k,l} = e^{i(kx+ly)}\cosh[\kappa(z + h)].$$

Inserting $\psi_{k,l}(x, y, z)$ into (5.42) yields

$$
\begin{aligned}
\int_{\mathbb{R}^2} & e^{i(kx+ly)}\cosh(\kappa(\eta + h)\eta_t\, dxdy \\
&= \int_{\mathbb{R}^2} q\left(e^{i(kx+ly)}\kappa\sinh(\kappa(\eta + h)) - ie^{i(kx+ly)}\cosh(\kappa(\eta + h)(k \cdot \nabla)\eta\right)dx\, dy.
\end{aligned}
\tag{5.43}
$$

Then, using

$$e^{i(kx+ly)}\kappa \sinh(\kappa(\eta + h)) - ie^{i(kx+ly)} \cosh(\kappa(\eta + h))\,((k, l) \cdot \nabla)\,\eta$$

$$= -i\nabla \cdot \left(e^{ikx}\frac{\sinh(\kappa(\eta + h))}{\kappa}(k, l)\right)$$

in (5.43) and integrating by parts yields the non-local equation (I).

The second equation is essentially a change of variables. Differentiating $q(x, y, t) = \phi(x, y, \eta(x, y, t), t)$ leads to

$$q_x + \phi_x + \phi_z \eta_x, \quad q_y = \phi_y + \phi_z \eta_y.$$

These equations and

$$\eta_t + \nabla\phi \cdot \nabla\eta = \phi_y$$

yield $\phi_x$, $\phi_y$, $\phi_z$ in terms of derivatives of $\eta$, $q$. Then from Bernoulli's equation with surface tension included, (5.28), we obtain equation (II).

If $|\eta|$, $|\nabla q|$ are small then equations (I) and (II) simplify to the linearized water wave equations. Calling the Fourier transform $\hat{\eta} = \int dx\,dy e^{i(kx+ly)}\eta$, and similarly for the derivatives of $q$, we find from equations (I) and (II) respectively

$$i\hat{\eta}_t \cosh \kappa h + \frac{k \cdot \widehat{\nabla q}}{\kappa}\sinh \kappa h = 0$$

$$\hat{q}_t + (g + \sigma\kappa^2)\hat{\eta} = 0.$$

Then from these two equations we get

$$\hat{\eta}_{tt} = -(g\kappa + \sigma\kappa^3)\tanh \kappa h\,\hat{\eta}$$

which is the linearized water wave equation for $\eta$ in Fourier space.

We also remark that one can find integral relations by taking $k, l \to 0$ in equations (I) and (II) above. The first three, corresponding to powers $k^0$, $l^0$, $k, l$, are given by

$$\frac{\partial}{\partial t}\int_{-\infty}^{\infty}\int_{-\infty}^{\infty} dx\,dy\,\eta(x, y, t) = 0 \quad \text{(Mass)}$$

$$\frac{\partial}{\partial t}\int_{-\infty}^{\infty}\int_{-\infty}^{\infty} dx\,dy(x\eta) = \int_{-\infty}^{\infty}\int_{-\infty}^{\infty} dx\,dy\,q_x(\eta + h) \quad \text{(COM}_x)$$

$$\frac{\partial}{\partial t}\int_{-\infty}^{\infty}\int_{-\infty}^{\infty} dx\,dy(y\eta) = \int_{-\infty}^{\infty}\int_{-\infty}^{\infty} dx\,dy\,q_y(\eta + h) \quad \text{(COM}_y).$$

They correspond to conservation of mass and the motion of the center of mass in $x, y$ respectively. The right-hand sides of (COM$_x$, $y$) are related to the $x, y$-momentum respectively. Higher powers of $k, l$ lead to "virial" type identities.

From the above system (I) and (II), one can also derive both deep- and shallow-water approximations to the water wave equations. In deep water, non-linear Schrödinger equation (NLS) systems result (see Ablowitz et al., 2006; Ablowitz and Haut, 2009b). In shallow water the Benney–Luke (BL) equation (Benney and Luke, 1964), properly modified to take into account surface tension, can be obtained.

To do this it is convenient to make all variables non-dimensional:

$$x_1' = \frac{x_1}{\lambda}, \ x_2' = \gamma\frac{x_2}{\lambda}, \ t' = \frac{c_0}{\lambda}t, \ a\eta' = \eta, q' = \frac{a\lambda g}{c_0}q, \ \sigma' = \frac{\sigma}{gh^2}$$

where $c_0 = \sqrt{gh}$ and $\lambda$, $a$ are the characteristic horizontal length and amplitude, and $\gamma$ is a non-dimensional transverse length parameter. The equations are written in terms of non-dimensional variables $\epsilon = a/h \ll 1$, small-amplitude, $\mu = h/\lambda \ll 1$ long waves; $\gamma \ll 1$, slow transverse variation. Dropping $'$ we find

$$\int dx dy e^{i(kx+ly)} \left( i\eta_t \cosh[\tilde{\kappa}\mu(1+\epsilon\eta)] + \sinh[\tilde{\kappa}\mu(1+\epsilon\eta)]\frac{(k,l)\cdot\tilde{\nabla}q}{\tilde{\kappa}\mu} \right) = 0 \quad (5.44)$$

where $(k,l) \cdot \tilde{\nabla} = kq_x + \gamma^2 lq_y$, $\tilde{\kappa}^2 = k^2 + \gamma^2 l^2$.

In (5.44) we use

$$\cosh[\tilde{\kappa}\mu(1+\epsilon\eta)] \sim 1 + \frac{\mu^2}{2}\tilde{\kappa}^2, \quad \sinh[\tilde{\kappa}\mu(1+\epsilon\eta)] \sim \mu\tilde{\kappa} + \frac{\mu^3}{6}\tilde{\kappa}^3 + \epsilon\mu\eta\tilde{\kappa}.$$

Then, after inverse Fourier transforming, and with $(k,l) \rightarrow (i\partial x, i\partial y)$ we find from equations (I) and (II)

$$\left(1 - \frac{\mu^2}{2}\tilde{\Delta}\right)\eta_t + \left(\tilde{\Delta} - \frac{\mu^2}{6}\tilde{\Delta}^2\right)q + \epsilon\left(\tilde{\nabla}\eta \cdot \tilde{\nabla}q\right) + \epsilon\eta\tilde{\Delta}q = 0 \quad \text{(Ia)}$$

$$\eta = -q_t - \frac{\epsilon}{2}|\tilde{\nabla}q|^2 + \tilde{\sigma}\mu^2\tilde{\Delta}\eta \quad \text{(IIa)}$$

where $\tilde{\Delta} = \partial_x^2 + \delta^2\partial_y^2$, $\tilde{\nabla}\eta \cdot \tilde{\nabla}q = \eta_x q_x + \delta^2\eta_y q_y$, $|\tilde{\nabla}q|^2 = \left(q_x^2 + \delta^2 q_y^2\right)$, and $\tilde{\sigma} = \sigma - 1/3$.

Eliminating the variable $\eta$ we find, when $\epsilon, \mu, \delta \ll 1$, the Benney–Luke (BL) equation with surface tension included:

$$q_{tt} - \tilde{\Delta}q + \tilde{\sigma}\mu^2\tilde{\Delta}^2q + \epsilon(\partial_t|\tilde{\nabla}q|^2 + q_t\tilde{\Delta}q) = 0, \quad (5.45)$$

where $\tilde{\sigma} = \sigma - 1/3$. Alternatively, one can derive the BL system from the classical water wave equations via asymptotic expansions as we have done previously for the Boussinesq models. The equations look somewhat different due to different representations of the velocity potential (see also the exercises).

We remark that if we take the maximal balance $\epsilon = \mu^2 = \delta^2$ then using multiple scales the BL equation can be reduced to the unidirectional KP equation, which we have shown can be further reduced to the KdV equation if there is no transverse variation. Namely letting $\xi = x - t$, $T = \epsilon t/2$, $w = q_\xi$ we find the KP equation in the form

$$\partial_\xi(w_T - \tilde{\sigma}w_{\xi\xi\xi} + 3(ww_\xi)) + w_{yy} = 0.$$

Hence the non-local system (I)–(II) contains the well-known integrable water wave reductions. We also note that higher-order asymptotic expansions of solitary waves and lumps can be obtained from equations (I) and (II) using the same non-dimensionalization as we used for the BL equation. These expansions yield solitary waves close to their maximal amplitude (Ablowitz and Haut, 2009a, 2010)

# Exercises

5.1   Derive the KP equation, with surface tension included, from the Benney–Luke (BL) equation (5.45): i.e., letting $w = q_\xi, \varepsilon = \mu^2 = \delta^2$, find

$$w_{T\xi} + \left(1/3 - \hat{T}\right)w_{\xi\xi\xi\xi} + w_{yy} + (3ww_\xi)_\xi = 0 \qquad (KP)$$

   where $T = \epsilon t, \hat{T} = T/\rho g h^2$, which is the well-known Kadomtsev–Petviashvili (KP) equation.

5.2   From the non-dimensional KP equation derived in Exercise 5.1, show that the equation in dimensional form is given by

$$\frac{1}{c_0}\eta_{xt} + \frac{3}{2h}(\eta\eta_x)_x + \gamma\eta_{xxxx} = -\frac{1}{2}\eta_{yy}$$

   where $\gamma = \frac{h^2}{6}(1 - 3\hat{T})$.

5.3   Find the dimensional form of a soliton solution to the KdV equation when $\hat{T} > 1/3$, i.e., $\gamma < 0$.

5.4   Derive the equivalent BL equation for the velocity potential $\phi(x, y, 0, t)$ from the classical water wave equations with surface tension included.

5.5   Rewrite the lump solution given in this section in dimensional form.

5.6   Given the "Boussinesq" model

$$u_{tt} - \Delta u + \Delta^2 u + \|\nabla u\|^2 = 0,$$

   make suitable assumptions about the size of terms, then rescale, and derive a corresponding KP-type equation.

5.7 Given a "modified Boussinesq" model

$$u_{tt} - \Delta u + \Delta^2 u + \|\nabla u\|^{2+n} = 0$$

with $n$ a positive integer, make suitable assumptions about the size of terms, then rescale, and derive a corresponding "modified KP"-type equation.

5.8 Beginning with (5.22) assume right-going waves and obtain the KdV equation with a higher-order correction.

# 6

## Nonlinear Schrödinger models and water waves

A different asymptotic regime will be investigated in this chapter. Since nonlinear, dispersive, energy-preserving systems generically give rise to the nonlinear Schrödinger (NLS) equation, it is important to study this situation. In this chapter first the NLS equation will be derived from a nonlinear Klein–Gordon and the KdV equation. Then a discussion of how the NLS equation arises in the limit of deep water will be undertaken. A number of results associated with NLS-type equations, including Benney–Roskes/Davey–Stewartson systems, will also be obtained and analyzed.

## 6.1 NLS from Klein–Gordon

We start with the so-called "$u$-4" model, which arises in theoretical physics, or the nonlinear Klein–Gordon (KG) equation – with two choices of sign for the cubic nonlinear term:

$$u_{tt} - u_{xx} + u \mp u^3 = 0. \tag{6.1}$$

Note, by multiplying the equation with $u_t$ we obtain the conservation of energy law

$$\partial_t\left(u_t^2 + u_x^2 + u^2 \mp u^4/2\right) - \partial_x(2u_x u_t) = 0$$

and hence the equation has a conserved density $T_1 = u_t^2 + u_x^2 + u^2 \mp u^4/2$. Note the "potential" is proportional to $V(u) = u^2 \mp u^4/2$, hence the term "$u$-4" model is used. We assume a real solution to (6.1) with a "rapid" phase and slowly varying functions of $x, t$:

$$u(x, t) = u(\theta, X, T; \varepsilon); \quad X = \varepsilon x, \ T = \varepsilon t, \ \theta = kx - \omega t, |\varepsilon| \ll 1,$$

where the dispersion relation for the linear problem in (6.1) is $\omega^2 = 1 + k^2$. Based on the form of assumed solution the following operators are introduced in the asymptotic analysis:

$$\partial_t = -\omega\partial_\theta + \varepsilon\partial_T$$
$$\partial_x = k\partial_\theta + \varepsilon\partial_X.$$

The weakly nonlinear asymptotic expansion

$$u = \varepsilon u_0 + \varepsilon^2 u_1 + \varepsilon^3 u_2 + \cdots,$$

and the differential operators are substituted into (6.1) leading to

$$\left[(-\omega\partial_\theta + \varepsilon\partial_T)^2 - (k\partial_\theta + \varepsilon\partial_X)^2\right]u + u \mp u^3 = 0$$

or

$$\left[(\omega^2 - k^2)\partial_\theta^2 - \varepsilon(2\omega\partial_\theta\partial_T + 2k\partial_\theta\partial_X) + \varepsilon^2\left(\partial_T^2 - \partial_X^2\right)\right](\varepsilon u_0 + \varepsilon^2 u_1 + \cdots) +$$
$$+ (\varepsilon u_0 + \varepsilon^2 u_1 + \cdots) \mp (\varepsilon u_0 + \varepsilon^2 u_1 + \cdots)^3 = 0.$$

We will find the equations up to $O(\varepsilon^3)$; the first two are:

$$3 O(\varepsilon): \qquad (\omega^2 - k^2)\partial_\theta^2 u_0 + u_0 = 0$$
$$\Rightarrow u_0 = Ae^{i\theta} + \text{c.c.}$$

$$O(\varepsilon^2): \qquad (\omega^2 - k^2)\partial_\theta^2 u_1 + u_1 = (2\omega i A_T + 2ki A_X)e^{i\theta} + \text{c.c.}$$

where c.c. denotes the complex conjugate. Note: throughout the discussion we will use the dispersion relation $\omega^2 - k^2 = 1$. In order to remove secular terms, we require that the coefficient on the right-hand side of the $O(\varepsilon^2)$ equation be zero. This leads to an equation for $A(X, T)$ that, to go to higher order, we modify by using the asymptotic expansion

$$2i(\omega A_T + k A_X) = \varepsilon g_1 + \varepsilon^2 g_2 + \cdots. \tag{6.2}$$

The quantity $A(X, T)$ is often called the slowly varying envelope of the wave. As usual, we absorb the homogeneous solutions into $u_0$ so $u_1 = 0$. Now let us look at the next-order equation

$$O(\varepsilon^3): \qquad (\omega^2 - k^2)\partial_\theta^2 u_2 + u_2 = -(A_{TT} - A_{XX})e^{i\theta} + \text{c.c.}$$
$$\pm (Ae^{i\theta} + A^*e^{-i\theta})^3 + g_1 e^{i\theta} + \text{c.c.}$$

In order to remove secular terms, we eliminate the coefficient of $e^{i\theta}$. This implies

$$g_1 = (A_{TT} - A_{XX}) \mp 3A^2 A^*.$$

Combining the above relation with the asymptotic expansion in (6.2) we have

$$2i\omega(A_T + \omega'(k)A_X) = \varepsilon(A_{TT} - A_{XX} \mp 3|A|^2 A), \tag{6.3}$$

where we used the dispersion relation to find $\omega'(k) = k/\omega$. After removing the secular terms, we can solve the $O(\varepsilon^3)$ equation for $u_2$ giving

$$u_2 = \mp \frac{1}{8} A^3 e^{3i\theta} + \text{c.c.}$$

Note, in (6.3), we can transform the equation for $A$ and remove the $\varepsilon$ (group velocity) term. To do this, we change the coordinate system so that we are moving with the group velocity $\omega'(k)$ – sometimes this is called the group-velocity frame:

$$\xi = X - \omega'(k)T \qquad\qquad \partial_T = -\omega'(k)\partial_\xi + \varepsilon\partial_\tau$$
$$\tau = \varepsilon T = \varepsilon^2 t \qquad\qquad \partial_X = \partial_\xi.$$

Now, we use (6.3) to find

$$2i\omega(-\omega'A_\xi + \varepsilon A_\tau + \omega'A_\xi) = \varepsilon((\omega')^2 A_{\xi\xi} - A_{\xi\xi} \mp 3|A|^2 A) + O(\varepsilon^2).$$

We can simplify this equation by using the dispersion relation as follows: $\omega' = k/\omega$ and

$$\omega'' = \frac{\omega - \omega'k}{\omega^2} = \frac{1}{\omega} - \frac{k\omega'}{\omega^2} = \frac{1}{\omega}\left[1 - (\omega')^2\right],$$

to arrive at the nonlinear Schrödinger equation in canonical form:

$$iA_\tau + \frac{\omega''}{2} A_{\xi\xi} \pm \frac{3}{2\omega}|A|^2 A = 0. \tag{6.4}$$

It is important to note that, like the derivation of the Korteweg–de Vries equation in the previous chapter, the small parameter, $\varepsilon$, has disappeared from the leading-order NLS equation. The NLS equation is a maximally balanced asymptotic system.

The behavior of (6.4) depends on the plus or minus sign. The NLS equation with a "+", here

$$\omega'' = \frac{\omega^2 - k^2}{\omega^3} = \frac{1}{\omega^3},$$

so $\omega''/\omega > 0$. The NLS with a "+" sign is said to be "focusing" and we will see gives rise to "bright" solitons. Bright solitons have a localized shape and decay at infinity. With a "−", NLS is said to be "defocusing" and we will see admits "dark" soliton solutions. Dark solitons have a constant amplitude at infinity. The terminology "bright" and "dark" solitons is typically used in nonlinear optics. We will discuss these solutions later, and nonlinear optics in a subsequent chapter.

In our derivation of NLS, we transformed our coordinate system to coincide with the group velocity of the wave. An alternative form, useful in nonlinear optics and called the retarded frame, is

$$t' = T - \frac{1}{\omega'(k)}X \qquad\qquad \partial_T = \partial_{t'}$$

$$\chi = \varepsilon X \qquad\qquad \partial_X = -\frac{1}{\omega'}\partial_{t'} + \varepsilon\partial_\chi.$$

Next, we substitute the above relations into (6.3) to find

$$2i\omega\left(A_{t'} + \omega'\left[\left(\frac{-1}{\omega'}\right)A_{t'} + \varepsilon A_\chi\right]\right) = \varepsilon\left(A_{t't'} - \frac{1}{(\omega')^2}A_{t't'} \mp 3|A|^2A\right).$$

Again we simplify this equation into a nonlinear Schrödinger equation

$$iA_\chi + \frac{\omega''}{2(\omega')^3}A_{t't'} \pm \frac{3}{2\omega\omega'}|A|^2A = 0. \qquad (6.5)$$

Note that in the above equation, the "evolution" variable is $\chi$ and the "spatial" variable is $t'$.

## 6.2 NLS from KdV

We can also derive the NLS equation from the KdV equation asymptotically for small amplitudes. To do that, we expand the solution of the KdV equation

$$u_t + 6uu_x + u_{xxx} = 0$$

as

$$u = \varepsilon u_0 + \varepsilon^2 u_1 + \varepsilon^3 u_2 + \cdots,$$

where for convenience the assumption of small amplitude is taken through the expansion of $u$ rather than in the equation itself. Since the nonlinear term, $uu_x$, is quadratic rather than cubic, it will turn out that the reduction to the NLS equation requires more work than in the KG equation. As we will see, in this case one needs to remove secularity at one additional power of $\varepsilon$, which in this case is $O(\varepsilon^3)$. Another difficulty is manifested by the need to consider a *mean term* in the solution. Indeed, the equation to leading order is

$$u_{0,t} + u_{0,xxx} = 0,$$

whose real solution is given by

$$u_0 = A(X, T)e^{i\theta} + \text{c.c.} + M(X, T), \tag{6.6}$$

where again $T = \varepsilon t$ and $X = \varepsilon x$ are the slow variables, $\theta = kx - \omega t$ is the fast variable, with the dispersion relation $\omega = -k^3$, and $M(T, X)$ is a real, slowly varying, mean term, i.e., it corresponds to the coefficient of $e^{in\theta}$ with $n = 0$. The quantity $A(X, T)$ is the slowly varying envelope of the rapidly varying wave contribution. As we will see, in this case the mean term turns out to be $O(\varepsilon)$, however, that is not always the case.[1] Note that the addition of the complex conjugate ("c.c.") in (6.6) corresponds to the first term and not to the mean term, which is real-valued.

Substituting $\partial_t = -\omega\partial_\theta + \varepsilon\partial_T$ and $\partial_x = k\partial_\theta + \varepsilon\partial_X$ in the KdV equation leads to

$$(-\omega\partial_\theta + \varepsilon\partial_T)u + 6u(k\partial_\theta + \varepsilon\partial_X)u + (k\partial_\theta + \varepsilon\partial_X)^3 u = 0.$$

Using the above expansion for the solution, $u = \varepsilon u_0 + \varepsilon^2 u_1 + \varepsilon^3 u_2 + \cdots$ gives

$$(-\omega\partial_\theta + \varepsilon\partial_T)(\varepsilon u_0 + \varepsilon^2 u_1 + \varepsilon^3 u_2 + \cdots)$$
$$+ 6(\varepsilon u_0 + \varepsilon^2 u_1 + \varepsilon^3 u_2 + \cdots)(k\partial_\theta + \varepsilon\partial_X)(\varepsilon u_0 + \varepsilon^2 u_1 + \varepsilon^3 u_2 + \cdots)$$
$$+ (k\partial_\theta + \varepsilon\partial_X)^3(\varepsilon u_0 + \varepsilon^2 u_1 + \varepsilon^3 u_2 + \cdots) = 0.$$

The leading-order solution of $\mathcal{L}u_0 = 0$, $\mathcal{L}$ defined below, is given by (6.6). We recall that the dispersion relation for the exponential term is given by $\omega(k) = -k^3$. The $O(\varepsilon^2)$ equation is

$$\mathcal{L}u_1 = -u_{0,T} - 3k^2 u_{0,\theta\theta X} - 6u_0 k u_{0,\theta}$$
$$= (-A_T e^{i\theta} + \text{c.c.}) - M_T + (3k^2 A_X e^{i\theta} + \text{c.c.})$$
$$- 6(Ae^{i\theta} + \text{c.c.}) + M)(Aike^{i\theta} + \text{c.c.}),$$

where we have defined the linear operator (using $\omega = -k^3$),

$$\mathcal{L}u = k^3(u_\theta + u_{\theta\theta\theta}).$$

Thus, we rewrite the $O(\varepsilon)$ equation as

$$\mathcal{L}u_1 = (-A_T - \omega' A_X - 6iMkA)e^{i\theta} + \text{c.c.} - 6(ikA^2 e^{2i\theta} + \text{c.c.}) - M_T,$$

where we have used the dispersion relation, from which one gets that $\omega'(k) = -3k^2$. In order to remove secular terms we note that such terms arise

---

[1] For example, in finite-depth water waves the mean term is $O(1)$ (Benney and Roskes, 1969; Ablowitz and Segur, 1981).

not only from the coefficients of $e^{i\theta}$, but also from the coefficients of "$e^{i0\theta} = 1$" (which correspond to a mean term). Therefore, according to our methodology, we require that

$$A_T + \omega' A_X + 6iMkA = \varepsilon g_1 + \varepsilon^2 g_2 + \cdots, \tag{6.7}$$

$$M_T = \varepsilon f_1 + \varepsilon^2 f_2 + \cdots \tag{6.8}$$

and solve $u_1$ for what remains, i.e., solve,

$$\mathcal{L}u_1 = -6(ikA^2 e^{2i\theta} + \text{c.c.}).$$

The latter equation can be solved by letting $u = \alpha e^{2i\theta} + \text{c.c.}$, which gives that $\alpha = A^2/k^2$ and (omitting homogeneous terms as was done earlier for perturbations in ODE problems)

$$u_1 = \frac{A^2}{k^2} e^{2i\theta} + \text{c.c.} = \alpha e^{2i\theta} + \text{c.c.},$$

where $\alpha = A^2/k^2$. Inspecting equation (6.8) we see that $M = O(\varepsilon)$. Thus, in retrospect, we could (or should) have introduced the mean term at the $O(\varepsilon)$ solution, i.e., in $u_1$. However, that will not change the result of the analysis, provided we work consistently.

Next we inspect the $O(\varepsilon^3)$ equation:

$$\mathcal{L}u_2 + 6(u_0 k u_{1,\theta} + u_0 u_{0,X} + u_1 k u_{0,\theta}) + 3k^2 u_{1,\theta\theta X} + 3k u_{0,\theta XX} + u_{1T}$$
$$= -f_1 - (g_1 e^{i\theta} + \text{c.c.}),$$

where the $f_1$ and $g_1$ terms arise from equations (6.7) and (6.8). Using (6.6) and the fact that $M = O(\varepsilon)$ we get that to leading order

$$\mathcal{L}u_2 = -6(Ae^{i\theta} + \text{c.c.})(2ik\alpha e^{2i\theta} + \text{c.c.}) - 6(Ae^{i\theta} + \text{c.c.})(A_X e^{i\theta} + \text{c.c.}) -$$
$$- 6(\alpha e^{2i\theta} + \text{c.c.})(ikAe^{i\theta} + \text{c.c.}) - 3k^2[(2i)^2 \alpha_X e^{2i\theta} + \text{c.c.}] -$$
$$- 3k(iA_{XX} e^{i\theta} + \text{c.c.}) - f_1 - (g_1 e^{i\theta} + \text{c.c.}) - (\alpha_T e^{2i\theta} + \text{c.c.}).$$

The reader is cautioned about the "$i$" terms. For example, note that $(iA_{XX}e^{i\theta} + \text{c.c.}) = iA_{XX}e^{i\theta} - iA_{XX}e^{-i\theta}$. Here we need to remove secularity of the $e^{i\theta}$ and mean terms in order to find the $f_1$ and $g_1$ terms. Using $\alpha = A^2/k^2$ this leads to

$$g_1 = -3ikA_{XX} - 6ik\alpha A^* = -3ikA_{XX} - \frac{6i}{k}A^2 A^*,$$

$$f_1 = -6(AA_X^* + \text{c.c.}) = -6(|A|^2)_X,$$

where $A^*$ is the complex conjugate of $A$. Thus, from equations (6.7) and (6.8) using $f_1, g_1$, we obtain to leading order that

$$A_T + \omega' A_X + 6iMkA = \varepsilon\left(-3ikA_{XX} - \frac{6i}{k}A^2A^*\right), \qquad (6.9)$$

$$M_T = -6\varepsilon(|A|^2)_X. \qquad (6.10)$$

As such, we have solved the problem of removing secularity at $O(\varepsilon^3)$. However, in order to obtain the NLS equation one additional step is required. We transform to a moving reference frame, i.e., $A(X, T) = A(\xi, \tau)$, where the new variables are given by

$$\xi = X - \omega'(k)T, \quad \tau = \varepsilon T. \qquad (6.11)$$

Therefore, the derivatives transform according to

$$\partial_X = \partial_\xi, \quad \partial_T = \varepsilon\partial_\tau - \omega'\partial_\xi$$

and equations (6.9) and (6.10) become

$$\varepsilon A_\tau + 6iMkA = \varepsilon\left(-3ikA_{\xi\xi} - \frac{6i}{k}A^2A^*\right), \qquad (6.12)$$

$$\varepsilon M_\tau - \omega' M_\xi = -6\varepsilon(|A|^2)_\xi. \qquad (6.13)$$

The latter equation can be simplified using $M = O(\varepsilon)$, which leads to

$$-\omega' M_\xi = -6\varepsilon(|A|^2)_\xi + O(\varepsilon^2),$$

whose (leading-order) solution is given by (omitting the integration constant)

$$M \sim -\frac{2\varepsilon|A|^2}{k^2},$$

where we have used $\omega' = -3k^2$. Upon substituting the solution for $M$ in (6.12) we have that

$$A_\tau + 3ikA_{\xi\xi} + \frac{6i}{k}(-2|A|^2)A + \frac{6i}{k}|A|^2A = 0.$$

Therefore, we arrive at the following *defocusing* NLS equation

$$iA_\tau - 3kA_{\xi\xi} + \frac{6}{k}|A|^2A = 0.$$

Using $\omega'' = -6k$ leads to the canonical (or generic) form of the NLS equation,

$$iA_\tau + \frac{\omega''(k)}{2}A_{\xi\xi} + \frac{6}{k}|A|^2A = 0.$$

An alternative, more direct, and perhaps "faster", way of obtaining the NLS equation from the KdV equation is described next (Ablowitz et al., 1990, where it was used in the study of the KP equation and the associated boundary conditions). This method consists of making the ansatz (sometimes referred to as the "quasi-monochromatic" assumption. This method can also be used for the Klein–Gordon model in Section 6.1.):

$$u_0 = \varepsilon\left[A(T,X)e^{i\theta} + \text{c.c.} + M(X,T)\right] + \varepsilon^2(\alpha e^{2i\theta} + \text{c.c.}) + \cdots,$$

i.e., we add the $O(\varepsilon^2)$ second-harmonic term to the ansatz that has a fundamental and a mean term. Below we outline the derivation. The KdV terms are then respectively given by

$$u_t = \varepsilon\left[(Aik^3 + \varepsilon A_T)e^{i\theta} + \text{c.c.} + \varepsilon M_T\right] + \varepsilon^2\left[(2ik^3\alpha + \varepsilon\alpha_T)e^{2i\theta} + \text{c.c.}\right] + \cdots,$$

$$6uu_x = 6\varepsilon\left[(Ae^{i\theta} + \text{c.c.} + M) + \varepsilon(\alpha e^{2i\theta} + \text{c.c.})\right]\varepsilon\left[\left((ikA + \varepsilon A_X)e^{i\theta}+\right.\right.$$
$$\left.\left. +\text{c.c.}\right) + \varepsilon M_X + \varepsilon\left((2ik\alpha + \varepsilon\alpha_X)e^{2i\theta} + \text{c.c.}\right)\right] + \cdots,$$

and

$$u_{xxx} = \varepsilon\left\{\left[(ik + \varepsilon\partial_X)^3 A\right]e^{i\theta} + \text{c.c.} + \varepsilon^3 M_{XXX}+\right.$$
$$\left. +\varepsilon^2\left[(2ik + \varepsilon\partial_X)^3 \alpha e^{2i\theta} + \text{c.c.}\right]\right\} + \cdots.$$

We then set the coefficients of the $e^{i\theta}$ and mean terms to zero. The remaining terms are the coefficients of $e^{2i\theta}$, which are also set to zero (similarly if we add terms $e^{in\theta}$ at higher order):

$$\varepsilon^2\left(2ik^3\alpha + 6ikA^2 - 8ik^3\alpha\right) = 0.$$

This equation implies that $\alpha = A^2/k^2$, i.e., the same as in the preceding derivation. Removing the secular mean term gives

$$M_T + 6\varepsilon(|A|^2)_X + 6\varepsilon MM_X = 0.$$

As before, one obtains $M = O(\varepsilon)$, from which it follows that to leading order the equation for $M$ is

$$M_T + 6\varepsilon(|A|^2)_X \sim 0.$$

Then recalling the earlier change of variables, (6.11),

$$\partial_T = \varepsilon\partial_\tau - \omega'\partial_\xi, \partial_X = \partial_\xi$$

we find, after integration (and omitting the integration constant)

$$M \sim -\frac{2\varepsilon|A|^2}{k^2},$$

which agrees with what we found before. Next, removing the $e^{i\theta}$ secular terms leads to

$$\varepsilon^2 \left(A_T - 3k^2 A_X + 6ikMA\right) + \varepsilon^3 \left(3ikA_{XX} + 12ik\alpha A^2 A^* \right.$$
$$\left. - 6ik\alpha A^2 A^* + 6MA_X\right) = 0.$$

Note that the last term in the $O(\varepsilon^3)$ parentheses is negligible, since it is smaller by $M = O(\varepsilon)$ than the other terms. Upon substituting the solution for $M$ one arrives at the equation

$$A_T - 3k^2 A_X - \frac{12i\varepsilon|A|^2}{k}A + \varepsilon\left(3ikA_{XX} + \frac{6i}{k}|A|^2 A\right) = 0.$$

Finally, transforming the coordinates to the moving-frame (6.11) leads to

$$A_\tau + 3ikA_{\xi\xi} - \frac{6i}{k}|A|^2 A = 0,$$

and, using the dispersion relation $\omega'' = -3k$, results in the NLS equation in canonical form

$$iA_\tau + \frac{\omega''(k)}{2}A_{\xi\xi} + \frac{6}{k}|A|^2 A = 0.$$

Note again, given the signs (i.e. the product of nonlinear and dispersive coefficients), the above NLS equation is of defocusing type.

## 6.3 Simplified model for the linear problem and "universality"

It is noteworthy that in the derivations of the NLS equation from both the KG and KdV equations, the linear terms in the NLS equation turn out to be

$$iA_\tau + \frac{\omega''(k)}{2}A_{\xi\xi}.$$

It might seem that the $\omega''(k)/2$ coefficient is a coincidence from these two different derivations. However, as explained below, this is not the case. In fact, this coefficient will always arise in the derivation because it manifests an inherent property of the slowly varying amplitude approximation of a constant coefficient linearized dispersive equation, in the moving-frame system.

To see that concretely, let us recall the linear KdV equation

$$A_t = -A_{xxx},$$

whose dispersion relation is $\omega(k) = -k^3$. We can solve this equation explicitly (e.g., using Fourier transforms), by looking for wave solutions

$$A(x, t) = \tilde{A}(X, T)e^{i(kx - \omega t)} + \text{c.c.}$$

The slowly varying amplitude assumption corresponds to a superposition of waves of the form:

$$\tilde{A}(X, T) = A_0 e^{i(\varepsilon Kx - \varepsilon \Omega t)} = A_0 e^{i(KX - \Omega T)}; \quad A_0 \text{ const.}$$

This gives us

$$A(x, t) = A_0 e^{i(kx + \varepsilon Kx - \omega t - \varepsilon \Omega t)} + \text{c.c.}$$
$$= A_0 e^{i(kx + KX - \omega t - \Omega T)} + \text{c.c.},$$

where $X = \varepsilon x$ and $T = \varepsilon t$. The quantities $K$ and $\Omega$ are sometimes referred to as the sideband wavenumber and frequency, respectively, because they correspond to a small deviation from the central wavenumber $k$ and central frequency $\omega$. It is useful to look at these deviations from the point of view of operators, whereby $\Omega \rightarrow i\partial_T$ and $K \rightarrow -i\partial_X$. Thus,

$$\omega_{\text{tot}} \sim \omega + \varepsilon \Omega \rightarrow \omega + i\varepsilon \partial_T.$$

Similarly,

$$k_{\text{tot}} \sim k + \varepsilon K \rightarrow k - i\varepsilon \partial_X.$$

We can expand $\omega(k)$ in a Taylor series around the central wavenumber as

$$\omega_{\text{tot}}(k + \varepsilon K) \sim \omega(k) + \varepsilon K \omega' + \varepsilon^2 K^2 \frac{\omega''}{2}.$$

Then using the operator $K = -i\partial_X$ we have

$$\omega_{\text{tot}}(k - i\varepsilon \partial_X) \sim \omega(k) - i\varepsilon \omega' \partial_X - \varepsilon^2 \frac{\omega''}{2} \partial_{XX}.$$

Alternatively, if we use the operator $\Omega = i\partial_T$ this yields

$$\omega_{\text{tot}}(k)A = (\omega + \varepsilon \Omega)A \sim [\omega(k) + i\varepsilon \partial_T]A \sim \left(\omega(k) - i\varepsilon \omega' \partial_X - \varepsilon^2 \frac{\omega''}{2} \partial_{XX}^2\right)A.$$

Then to leading order

$$i\varepsilon(A_T + \omega' A_X) + \varepsilon^2 \frac{\omega''}{2} A_{XX} = 0 \tag{6.14}$$

and in the moving frame ("water waves variant"), $\xi = X - \omega'(k)T$, $\tau = \varepsilon T$, this equation transforms to

$$\varepsilon^2\left(iA_\tau + \frac{\omega''}{2}A_{\xi\xi}\right) = 0,$$

which is the linear Schrödinger equation with the canonical $\omega''(k)/2$ coefficient.

If, however, the coordinates transform to the "optics variant" using the "retarded" frame, i.e., $\chi = \varepsilon X$ and $t' = T - X/\omega'$, then equation (6.14) becomes

$$i\varepsilon\left(A_{t'} + \omega'\left(-\frac{1}{\omega'}A_{t'} + \varepsilon A_\chi\right)\right) + \frac{\omega''}{2}\varepsilon^2\frac{1}{(\omega')^2}A_{t't'} = 0,$$

which simplifies to

$$iA_\chi + \frac{\omega''}{2(\omega')^3}A_{t't'} = 0; \tag{6.15}$$

i.e., the "optical" variant of the linear Schrödinger equation has a canonical $\omega''/2(\omega')^3$ coefficient in front of the dispersive term [see (6.5) for a typical NLS equation]. The above interchange of coordinates from $\xi = x - \omega'(k)T$, $T = \varepsilon t$ to $t' = T - x/\omega'$, $\chi = \varepsilon x$ is also reflected in terms of the dispersion relation; namely, instead of considering $\omega = \omega(k)$, let us consider $k = k(\omega)$. Then

$$\frac{d^2k}{d\omega^2} = \frac{d}{d\omega}\frac{dk}{d\omega} = \frac{dk}{d\omega}\frac{d}{dk}\left(\frac{1}{d\omega/dk}\right) = -\frac{1}{\omega'^3}\frac{d^2\omega}{dk^2}$$

so that (6.15) can be written as

$$iA_\chi - \frac{k''}{2}A_{t't'} = 0.$$

On the other hand, suppose we consider rather general conservative nonlinear wave problems with leading quadratic or cubic nonlinearity. We have seen earlier that a multiple-scales analysis, or Stokes–Poincaré frequency-shift analysis (see also Chapter 4), shows, omitting dispersive terms, a wave solution of the form

$$u(x,t) = \varepsilon A(\tau)e^{i(kx-\omega t)} + \text{c.c.},$$

with $\tau = \varepsilon t$ has $A(\tau)$ satisfying

$$i\frac{\partial A}{\partial \tau} + n|A|^2A = 0.$$

The constant coefficient $n$ depends on the particular equation studied. Putting the linear and nonlinear effects together implies that an NLS equation of the form

$$i\frac{\partial A}{\partial \tau} + \frac{\omega''}{2}\frac{\partial^2 A}{\partial \xi^2} + n|A|^2 A = 0$$

is "natural". Indeed, the NLS equation can be viewed as a "universal" equation as it generically governs the slowly varying envelope of a monochromatic wave train (see also Benney and Newell, 1967).

## 6.4 NLS from deep-water waves

In this section we discuss the derivation of the NLS equation from the Euler–Bernoulli equations in the limit of infinitely deep $(1+1)$-dimensional water waves, i.e.,

$$\phi_{xx} + \phi_{zz} = 0, \quad -\infty < z < \varepsilon\eta(x,t) \tag{6.16}$$

$$\phi_z = 0, \quad z \to -\infty \tag{6.17}$$

$$\phi_t + \frac{\varepsilon}{2}\left(\phi_x^2 + \phi_z^2\right) + g\eta = 0, \quad z = \varepsilon\eta \tag{6.18}$$

$$\eta_t + \varepsilon\eta_x\phi_x = \phi_z, \quad z = \varepsilon\eta. \tag{6.19}$$

There are major differences between this model and the shallow-water model discussed in Chapter 5 that require our attention. To begin with, equations (6.16) and (6.17) are defined for $z \to -\infty$, as opposed to $z = -1$. In addition, the parameter $\mu = h/\lambda_x$, which was taken to be very small for shallow-water waves, is not small in this case. In fact, taking $h \to \infty$ in this case would imply $\mu \to \infty$, which is not a suitable limit, so we will not use the parameter $\mu$ (or the previous non-dimensional scaling). We will use (6.16)–(6.19) in dimensional form and begin by only assuming that the nonlinear terms are small.

The idea of the derivation is as follows. We have already seen in the previous section that the linear model always gives rise to the same linear Schrödinger equation. Since water waves have leading quadratic nonlinearity, general considerations mentioned earlier suggest that, for the nonlinear model, we expect to obtain the NLS equation in the form

$$iA_\tau + \frac{\omega''}{2}A_{\xi\xi} + n|A|^2 A = 0,$$

where $n$ is a coefficient (that may depend on $\omega(k)$ and its derivatives) that is yet to be found. Our goal is to obtain the dispersive and nonlinear coefficients. The dispersive term is canonical so we will first find the Stokes frequency shift; i.e., the coefficient $n$ in the above equation.

### 6.4.1 Derivation of the frequency shift

Let us first notice that (6.18) and (6.19) are defined on the free surface, which creates difficulties in the analysis. Therefore, the first step is to approximate the boundary conditions using a Taylor expansion of $z = \varepsilon\eta$ around the stationary limit (i.e., the fixed free surface), which is $z = 0$. Thus, one has that

$$\phi_t(t, x, \varepsilon\eta) = \phi_t(t, x, 0) + \varepsilon\eta\phi_{tz}(t, x, 0) + \frac{1}{2}(\varepsilon\eta)^2\phi_{tzz}(t, x, 0) + \cdots,$$

$$\phi_x(t, x, \varepsilon\eta) = \phi_x(t, x, 0) + \varepsilon\eta\phi_{xz}(t, x, 0) + \frac{1}{2}(\varepsilon\eta)^2\phi_{xzz}(t, x, 0) + \cdots,$$

$$\phi_z(t, x, \varepsilon\eta) = \phi_z(t, x, 0) + \varepsilon\eta\phi_{zz}(t, x, 0) + \frac{1}{2}(\varepsilon\eta)^2\phi_{zzz}(t, x, 0) + \cdots.$$

Using multiple scales in the time variable (only!) and defining $T = \varepsilon t$, we can rewrite (6.18) and (6.19) to $O(\varepsilon^2)$ as

$$\left[\phi_t + \varepsilon\eta\phi_{tz} + \frac{1}{2}(\varepsilon\eta)^2\phi_{tzz} + \varepsilon(\phi_T + \varepsilon\eta\phi_{Tz}) + \cdots\right]$$

$$+ \frac{\varepsilon}{2}\left(\phi_x^2 + \phi_z^2 + 2\varepsilon\eta\phi_x\phi_{xz} + 2\varepsilon\eta\phi_z\phi_{zz} + \cdots\right) + g\eta = 0, \qquad (6.20)$$

and

$$\eta_t + \varepsilon\eta_T + \varepsilon\eta_x(\phi_x + \varepsilon\eta\phi_{xz}) + \cdots = \phi_z + \varepsilon\eta\phi_{zz} + \frac{1}{2}(\varepsilon\eta)^2\phi_{zzz} + \cdots \qquad (6.21)$$

where all the functions in (6.20) and (6.21) are understood to be evaluated at $z = 0$ for all $x$.

Next we expand $\phi$ and $\eta$ as

$$\phi = \phi^{(0)} + \varepsilon\phi^{(1)} + \varepsilon^2\phi^{(2)} + \cdots,$$

$$\eta = \eta^{(0)} + \varepsilon\eta^{(1)} + \varepsilon^2\eta^{(2)} + \cdots,$$

substitute them into the equations, and study the corresponding equations at $O(\varepsilon^0)$, $O(\varepsilon^1)$, and $O(\varepsilon^2)$.

**Leading order, $O(\varepsilon^0)$**

From (6.16) one gets to leading order that

$$\phi_{xx}^{(j)} + \phi_{zz}^{(j)} = 0,$$

i.e., this is Laplace's equation (at every order of $\varepsilon$). Together with the boundary condition

$$\lim_{z \to -\infty} \phi_z^{(j)} = 0$$

(also true at every order of $\varepsilon$). The solution is given by

$$\phi^{(j)} \sim \sum_m A_m^{(j)}(T)e^{im\theta + m|k|z} + \text{c.c.}, \qquad (6.22)$$

where $\theta = kx - \omega t$ and the summation[2] is carried over $m = 0, 1, 2, 3, \ldots$ Note that the choice of $+m|k|z$ in the exponent assures that the solution is decaying as $z$ approaches $-\infty$. In addition, the summation over all possible modes is necessary, since even if the initial conditions excite only a single mode, the nonlinearity will generate the other modes.

To describe the analysis, we will explicitly show (once) the perturbation terms in detail. Keeping up to $O(\varepsilon^2)$ terms in (6.20) and (6.21) on $z = 0$ lead to

$$\left[\left(\phi_t^{(0)} + \varepsilon\phi_t^{(1)} + \varepsilon^2\phi_t^{(2)}\right) + \varepsilon(\eta^{(0)} + \varepsilon\eta^{(1)})\left(\phi_{tz}^{(0)} + \varepsilon\phi_{tz}^{(1)}\right) + \frac{1}{2}\varepsilon^2\eta^{(0)2}\phi_{tzz}^{(0)}\right]$$

$$+ \varepsilon\left(\phi_T^{(0)} + \varepsilon\phi_T^{(1)} + \varepsilon\eta^{(0)}\phi_{Tz}^{(0)}\right) + \frac{1}{2}\varepsilon\left[\left(\phi_x^{(0)} + \varepsilon\phi_x^{(1)}\right)^2 + 2\varepsilon\eta^{(0)}\phi_x^{(0)}\phi_{xz}^{(0)}\right]$$

$$+ \frac{1}{2}\varepsilon\left[\left(\phi_z^{(0)} + \varepsilon\phi_z^{(1)}\right)^2 + 2\varepsilon\eta^{(0)}\phi_z^{(0)}\phi_{zz}^{(0)}\right] + g(\eta^{(0)} + \varepsilon\eta^{(1)} + \varepsilon^2\eta^{(2)}) = 0$$

and

$$\left(\eta_t^{(0)} + \varepsilon\eta_t^{(1)} + \varepsilon^2\eta_t^{(2)}\right) + \varepsilon\left(\eta_T^{(0)} + \varepsilon\eta_T^{(1)}\right)$$

$$\varepsilon\left(\eta_x^{(0)} + \varepsilon\eta_x^{(1)}\right)\left(\phi_x^{(0)} + \varepsilon\phi_x^{(1)}\right) + \varepsilon^2\eta^{(0)}\eta_x^{(0)}\phi_{xz}^{(0)} = \phi_z^{(0)} + \varepsilon\phi_z^{(1)} + \varepsilon^2\phi_z^{(2)}$$

$$+ \varepsilon\left(\eta^{(0)} + \varepsilon\eta^{(1)}\right)\left(\phi_{zz}^{(0)} + \varepsilon\phi_{zz}^{(1)}\right) + \frac{1}{2}\varepsilon^2\eta^{(0)2}\phi_{zzz}^{(0)}.$$

Below we study these equations up to $O(\varepsilon^2)$.

At $O(\varepsilon^0)$ we obtain

$$\phi_t^{(0)} + g\eta^{(0)} = 0, \qquad (6.23)$$

$$\eta_t^{(0)} - \phi_z^{(0)} = 0. \qquad (6.24)$$

---

[2] It will turn out that the mean term $m = 0$ is of low order.

It follows from (6.22), assuming only one harmonic at leading order (recall $\theta = kx - \omega t$),

$$\phi^{(0)} \sim A_1(T)e^{i\theta+|k|z} + \text{c.c.}$$

and that

$$\eta^{(0)} = N_1(T)e^{i\theta} + \text{c.c.}$$

and from (6.23) and (6.24) we get the system

$$-i\omega A_1 + gN_1 = 0, \tag{6.25}$$
$$-i\omega N_1 = |k|A_1.$$

This is a linear system in $A_1$ and $N_1$ that can also be written in matrix form as follows

$$\begin{pmatrix} -i\omega & g \\ -|k| & -i\omega \end{pmatrix} \begin{pmatrix} A_1 \\ N_1 \end{pmatrix} = \begin{pmatrix} 0 \\ 0 \end{pmatrix}.$$

The solution of this linear homogeneous system is unique if and only if its determinant is zero, which leads to the dispersion relation

$$\omega^2(k) = g|k|. \tag{6.26}$$

Note that this dispersion relation can be viewed as the formal limit as $h \to \infty$ of the more general dispersion relation (at any water depth),

$$\omega^2(k) = gk \tanh(kh),$$

since $\tanh(x) \to \text{sgn}(x)$ as $x \to \infty$ and $|k| = k \cdot \text{sgn}(k)$.

### First order, $O(\varepsilon)$

The $O(\varepsilon)$ equation corresponding to the system (6.23), (6.24) reads

$$\phi_t^{(1)} + g\eta^{(1)} = -\left(\eta^{(0)}\phi_{tz}^{(0)} + \phi_T^{(0)}\right) - \frac{1}{2}\left(\phi_x^{(0)2} + \phi_z^{(0)2}\right), \tag{6.27}$$
$$\eta_t^{(1)} - \phi_z^{(1)} = -\eta_x^{(0)}\phi_x^{(0)} - \eta_T^{(0)} + \eta^{(0)}\phi_{zz}^{(0)}. \tag{6.28}$$

It follows from (6.25) that

$$A_1 = -\frac{ig}{\omega}N_1.$$

Using this relation and substituting the solution we found for $\phi^{(0)}$ and $\eta^{(0)}$ into (6.27) and (6.28) leads to[3]

$$\phi_t^{(1)} + g\eta^{(1)} = g|k|N_1^2 e^{2i\theta} - A_{1,T} e^{i\theta} + \text{c.c.,} \qquad (6.29)$$

$$\eta_t^{(1)} - \phi_z^{(1)} = -\frac{2ig}{\omega} k^2 N_1^2 e^{2i\theta} - N_{1,T} e^{i\theta} + \text{c.c.} \qquad (6.30)$$

Removing the secular terms $e^{i\theta}$ requires taking $A_{1,T} = N_{1,T} = 0$ at this order. But, as usual we expand

$$A_{1,T} = \varepsilon f_1 + \varepsilon^2 f_2 + \cdots$$

and

$$N_{1,T} = \varepsilon g_1 + \varepsilon^2 g_2 + \cdots,$$

in which case remaining terms in the equation lead us to a solution of the form

$$\phi^{(1)} = A_2 e^{2i\theta + 2|k|z} + \text{c.c.}$$

and

$$\eta^{(1)} = N_2 e^{2i\theta} + \text{c.c.}$$

Substituting this ansatz into (6.29) and (6.30) yields the system

$$-2i\omega A_2 + gN_2 = g|k|N_1^2,$$

$$-2i\omega N_2 - 2|k|A_2 = -\frac{2ig}{\omega} k^2 N_1^2.$$

This time we arrived at an inhomogeneous linear system for $A_2$ and $N_2$. Note that its solution is unique on account of the fact that its determinant is non-zero: this follows from the dispersion relation (6.26). Using the dispersion relation, the solution is found to be

$$A_2 = 0$$

and

$$N_2 = |k|N_1^2.$$

Therefore,

$$\phi^{(1)} = 0$$

and

$$\eta^{(1)} = |k|N_1^2 e^{2i\theta} + \text{c.c.}$$

---

[3] Note that the mean terms cancel in this case, which is the reason we need not have considered them in the expansion.

**Second order, $O(\varepsilon^2)$**

Now for the $O(\varepsilon^2)$ equation corresponding to (6.20) and (6.21):

$$\phi_t^{(2)} + g\eta^{(2)} = -\eta^{(1)}\phi_{tz}^{(0)} - \eta^{(0)}\phi_{tz}^{(1)} - \frac{1}{2}\eta^{(0)2}\phi_{tzz}^{(0)} - (f_1 e^{i\theta} + \text{c.c.})$$

$$- \left[\phi_x^{(0)}\phi_x^{(1)} + \eta^{(0)}\phi_x^{(0)}\phi_{xz}^{(0)} + \phi_z^{(0)}\phi_z^{(1)} + \eta^{(0)}\phi_z^{(0)}\phi_{zz}^{(0)}\right]$$

$$- \phi_T^{(1)} - \eta^{(0)}\phi_{Tz}^{(0)},$$

$$\eta_t^{(2)} - \phi_z^{(2)} = -\left(\eta_x^{(0)}\phi_x^{(1)} + \eta_x^{(1)}\phi_x^{(0)} + \eta^{(0)}\eta_x^{(0)}\phi_{xz}^{(0)}\right) + \eta^{(0)}\phi_{zz}^{(1)}$$

$$+ \eta^{(1)}\phi_{zz}^{(1)} + \frac{1}{2}\eta^{(0)2}\phi_{zzz}^{(0)} - (g_1 e^{i\theta} + \text{c.c.}) - \eta_T^{(1)}.$$

Substituting $\phi^{(1)} = 0$ and taking into account the residual terms $f_1$ and $g_1$ after the removal of the previous secularities leads to the system

$$\phi_t^{(2)} + g\eta^{(2)} = -\eta^{(1)}\phi_{tz}^{(0)} - \left(f_1 e^{i\theta} + \text{c.c.}\right) - \frac{1}{2}\eta^{(0)2}\phi_{tzz}^{(0)}$$

$$- \left(\eta^{(0)}\phi_x^{(0)}\phi_{xz}^{(0)} + \eta^{(0)}\phi_z^{(0)}\phi_{zz}^{(0)}\right) - \eta^{(0)}\phi_{Tz}^{(0)}$$

$$\eta_t^{(2)} - \phi_z^{(2)} = -\left(\eta_x^{(1)}\phi_x^{(0)} + \eta^{(0)}\eta_x^{(0)}\phi_{xz}^{(0)}\right)$$

$$- \left(g_1 e^{i\theta} + \text{c.c.}\right) + \eta^{(1)}\phi_{zz}^{(0)} + \frac{1}{2}\eta^{(0)2}\phi_{zzz}^{(0)} - \eta_T^{(1)}.$$

As always, we will remove secular terms and solve for the remaining equations. In doing so we will substitute the previous solutions

$$\begin{cases} \phi^{(0)} = A_1 e^{i\theta + |k|z} & + \text{c.c.} \\ \eta^{(0)} = N_1 e^{i\theta} & + \text{c.c.} \\ \eta^{(1)} = |k|^2 N_1^2 e^{2i\theta} & + \text{c.c.} \\ \phi^{(1)} = 0. \end{cases}$$

Thus, by substituting the previous solutions we arrive at

$$\phi_t^{(2)} + g\eta^{(2)} = (C_1 e^{i\theta} + \text{c.c.}) + (C_2 e^{2i\theta} + \text{c.c.} + (C_3 e^{3i\theta} + \text{c.c.}) + C_0,$$

$$\eta_t^{(2)} - \phi_z^{(2)} = (D_1 e^{i\theta} + \text{c.c.}) + (D_2 e^{2i\theta} + \text{c.c.}) + (D_3 e^{3i\theta} + \text{c.c.}),$$

where the coefficients $C_1, C_2, C_3, C_0, D_1, D_2, D_3$ depend on $A_1, N_1, k$ and $\omega$. Removal of secular terms requires that the coefficients of $e^{\pm i\theta}$ be zero. To do this, we will look for a solution of the form

$$\phi^{(2)} = \left(A_1^{(2)} e^{i\theta + |k|z} + A_2^{(2)} e^{2i\theta + |k|z} + A_3 e^{3i\theta + |k|z} + \text{c.c.}\right) + A_0,$$

$$\eta^{(2)} = \left(N_1^{(2)} e^{i\theta} + N_2^{(2)} e^{2i\theta} + N_3 e^{3i\theta} + \text{c.c.}\right) + N_0$$

and use the method of undetermined coefficients. In doing so we call $f_1 = A_{1,\tau}$ and $g_1 = N_{1,\tau}$, where $\tau = \varepsilon T = \varepsilon^2 t$. Therefore, using previous solutions and removing the coefficients of $e^{\pm i\theta}$, we find the system

$$-i\omega A_1^{(2)} + g N_1^{(2)} = -\frac{3}{2} g k^2 |N_1|^2 N_1 - A_{1,\tau}, \qquad (6.31)$$

$$-i\omega N_1^{(2)} - |k| A_1^{(2)} = -\frac{5i}{2} \omega k^2 |N_1|^2 N_1 - N_{1,\tau}. \qquad (6.32)$$

Using (6.31) in (6.32) one arrives at

$$A_1^{(2)} = \frac{g}{i\omega} N_1^{(2)} + \frac{3}{2i\omega} g k^2 |N_1|^2 N_1 + \frac{1}{i\omega} A_{1,\tau}$$

and

$$-i\omega N_1^{(2)} - \frac{g|k|}{i\omega} N_1^{(2)} = \frac{3}{2i\omega} |k| k^2 |N_1|^2 N_1 + \frac{|k|}{i\omega} A_{1,\tau} - \frac{5i}{2} k^2 \omega |N_1|^2 N_1 - N_{1,\tau}.$$

Using $A_1 = -\frac{ig}{\omega} N_1$ and $\omega^2 = g|k|$ leads to

$$-2N_{1,\tau} - 4ik^2 \omega |N_1|^2 N_1 = 0,$$

or

$$N_{1,\tau} = -2ik^2 \omega |N_1|^2 N_1.$$

We have studied this type of equation before in Chapter 5 and have shown that $|N_1|^2(\tau) = |N_1|^2(0)$ and, therefore, that

$$N_1(\tau) = N_1(0) e^{-2ik^2 \omega |N_1(0)|^2 \tau}.$$

Hence the original free-surface solution is given to leading order by

$$\eta = N_1(0) e^{ikx - 2i\omega (1 + 2\varepsilon^2 k^2 |N_1(0)|^2) t} + \text{c.c.},$$

or

$$\eta = a \cos \left[ kx - \omega \left( 1 + \frac{\varepsilon^2 a^2 k^2}{2} \right) t \right],$$

where $a = 2|N_1(0)|$. The total frequency is therefore approximately given by

$$\omega_{\text{new}} = \omega \left( 1 + 2\varepsilon^2 k^2 |N_1(0)|^2 \right)$$

$$= \omega \left( 1 + \frac{\varepsilon^2 a^2 k^2}{2} \right).$$

The $O(\varepsilon^2)$ term, i.e., $2\varepsilon^2 \omega k^2 |N_1(0)|^2$, corresponds to the nonlinear frequency shift.

Note that we can also derive an equation for $A_1$. Indeed, using $N_1 = \dfrac{i\omega}{g}A_1$ gives that

$$A_{1,\tau} = -2ik^2\frac{\omega^3}{g^2}|A_1|^2A_1.$$

Using $\omega^2 = g|k|$ we get that

$$iA_{1,\tau} - \frac{2k^4}{\omega}|A_1|^2A_1 = 0. \tag{6.33}$$

Similar to the derivation of $N_1$, the solution of this equation is given by

$$A_1(\tau) = A_1(0)e^{-2i\frac{k^4}{\omega}|A_1(0)|^2\tau} = A_1(0)e^{-2ik^2\omega|N_1(0)|^2\tau},$$

where in the last equation we have used $A_1 = -\dfrac{ig}{\omega}N_1$. This shows that $A_1$ and $N_1$ have the same nonlinear frequency shift.

It is remarkable that Stokes obtained this nonlinear frequency shift in 1847 (Stokes, 1847)! While his derivation method (a variant of the "Stokes–Poincaré" frequency-shift method we have described) was different from the one we use here in terms of multiple scales, and he used different nomenclature (sines and cosines instead of exponentials), he nevertheless obtained the same result to leading order, i.e.,

$$\omega_{\text{new}} = \omega\left(1 + \frac{\varepsilon^2k^2a^2}{2} + \cdots\right),$$

where $a = 2|N_1(0)|$.

## 6.5 Deep-water theory: NLS equation

In the previous section, we were concerned with deriving the nonlinear term of the NLS equation and thus allowed the slowly varying envelope $A$ to depend only on the slow time $T = \varepsilon t$. Here, however, we outline the calculation when slow temporal and spatial variations are included. Since the water wave equations have an additional depth variable, $z$, we need to take some additional care. Therefore we discuss the calculation in some detail.

We will now use the structure of the water wave equations to suggest an ansatz for our perturbative calculation. The equations we will consider are

$$\phi_{xx} + \phi_{zz} = 0, \qquad -\infty < z < \varepsilon\eta \tag{6.34a}$$

$$\lim_{z \to -\infty} \phi_z = 0 \tag{6.34b}$$

$$\phi_t + \frac{\varepsilon}{2}\left(\phi_x^2 + \phi_z^2\right) + g\eta = 0, \qquad\qquad z = \varepsilon\eta \qquad (6.34c)$$

$$\eta_t + \varepsilon\eta_x\phi_x = \phi_z, \qquad\qquad z = \varepsilon\eta, \qquad (6.34d)$$

i.e., the water wave equations in the deep-water limit. There are three distinct steps in the calculation. First, because of the free boundary, we expand $\phi = \phi(t, x, \varepsilon\eta)$ for $\varepsilon \ll 1$:

$$\phi = \phi(t, x, 0) + \varepsilon\eta\phi_z(t, x, 0) + \frac{(\varepsilon\eta)^2}{2}\phi_{zz}(t, x, 0) + \cdots \qquad (6.35)$$

We similarly expand $\phi_t$, $\phi_x$, and $\phi_z$. Then the free-surface equations (6.34c) and (6.34d) expanded around $z = 0$ take the form:

$$\left[\phi_t + \varepsilon\eta\phi_{tz} + \frac{1}{2}(\varepsilon\eta)^2\phi_{tzz}\right] + \frac{\varepsilon}{2}\left(\phi_x^2 + \phi_z^2 + 2\varepsilon\eta\phi_x\phi_{xz} + 2\varepsilon\eta\phi_z\phi_{zz}\right) + g\eta = 0,$$

and

$$\eta_t + \varepsilon\eta_x(\phi_x + \varepsilon\eta\phi_{xz}) = \phi_z + \varepsilon\eta\phi_{zz} + \frac{1}{2}(\varepsilon\eta)^2\phi_{zzz}.$$

Second, introduce slow temporal and spatial scales:

$$\phi(t, x, z) = \phi(t, x, z, T, X, Z; \varepsilon)$$
$$\eta(t, x) = \phi(t, x, \varepsilon\eta, T, X, \varepsilon),$$

where $X = \varepsilon x$, $Z = \varepsilon z$, and $T = \varepsilon t$. Finally, because of the quadratic nonlinearity, we expect second harmonics and mean terms to be generated. This suggests the ansatz

$$\phi = \left(Ae^{i\theta + |k|z} + \text{c.c.}\right) + \varepsilon\left(A_2e^{2i\theta + 2|k|z} + \text{c.c.} + \overline{\phi}\right) \qquad (6.36a)$$
$$\eta = \left(Be^{i\theta} + \text{c.c.}\right) + \varepsilon\left(B_2e^{2i\theta} + \text{c.c.} + \overline{\eta}\right). \qquad (6.36b)$$

The coefficients $A$, $A_2$ and $\overline{\phi}$ depend on $X$, $Z$, and $T$ while $B$, $B_2$ and $\overline{\eta}$ depend on $X, T$. The rapid phase is given by $\theta = kx - \omega t$, with the dispersion relation $\omega^2 = g|k|$. Substituting the ansatz for $\phi$ into Laplace's equation (6.34a) we find

$$e^{i\theta}\left[2\varepsilon k\left(iA_X + \text{sgn}(k)A_Z\right) + \varepsilon^2\left(A_{XX} + A_{ZZ}\right) + \cdots\right] = 0,$$
$$e^0\left[\overline{\phi}_{XX} + \overline{\phi}_{ZZ}\right] = 0.$$

The first equation implies

$$A_Z = -i\,\text{sgn}(k)A_X - \frac{\varepsilon\,\text{sgn}(k)}{2k}\left(A_{XX} + A_{ZZ}\right) + O(\epsilon^2) \qquad (6.37)$$
$$= -i\,\text{sgn}(k)A_X + O(\varepsilon).$$

Taking the derivative of the above expression with respect to the slow variable $Z$ gives

$$
\begin{aligned}
A_{ZZ} &= -i\,\mathrm{sgn}(k)A_{XZ} + O(\varepsilon) \\
&= -i\,\mathrm{sgn}(k)(-i\,\mathrm{sgn}(k))A_{XX} + O(\varepsilon) \\
&= -A_{XX} + O(\varepsilon),
\end{aligned} \tag{6.38}
$$

where we differentiated (6.37) with respect to $X$ and substituted the resulting expression for $A_{XZ}$ to get the second line. Since the $O(\varepsilon)$ term in (6.37) is proportional to $A_{XX} + A_{ZZ}$, we can use (6.38) to obtain

$$
A_Z = -i\,\mathrm{sgn}(k)A_X + O(\varepsilon^2).
$$

Substituting our ansatz (6.36) into the Bernoulli equation (6.34c) and kinematic equation (6.34d) with (6.35), we find, respectively,

$$
e^{i\theta}\Big\{(-i\omega A + gB) + \varepsilon A_T + \varepsilon^2\Big[-i\omega k^2 A|B|^2 + 4k^2|k||A|^2 B + + 2k^2|k|A^2 B^*
$$
$$
+\frac{i}{2}\omega k^2 B^2 A^* + 4k^2 A_2 A^* - i\omega|k|A\bar\eta + i\omega|k|B_2 A^*
$$
$$
-4i\omega|k|A_2 B^*\Big] + \cdots\Big\} = 0, \tag{6.39}
$$

$$
e^{2i\theta}\Big\{A_2 - \varepsilon\frac{4k^2 A}{2|k|g}\Big(A_T + \frac{\omega}{2k}A_X\Big) + \cdots\Big\} = 0, \tag{6.40}
$$

$$
e^0\Big\{\bar\phi_Z - \bar\eta_T - \frac{2\omega k}{g}\frac{\partial}{\partial X}|A|^2 + \cdots\Big\} = 0, \tag{6.41}
$$

and

$$
e^{i\theta}\Big\{(-i\omega B - |k|A) + \varepsilon[B_T + i\,\mathrm{sgn}(k)A_X] + \varepsilon^2\Big[\frac{k^2|k|}{2}\Big(B^2 A^* - 2|B|^2 A\Big)
$$
$$
+k^2\left(B_2 A^* - 2B^* A_2\right) - k^2\bar\eta A\Big] + \cdots\Big\} = 0, \tag{6.42}
$$

$$
e^{2i\theta}\Big\{B_2 + \frac{k^2 A^2}{g} - \varepsilon\frac{2ik}{g}AA_X + \cdots\Big\} = 0, \tag{6.43}
$$

$$
e^0\{\bar\eta + O(\varepsilon)\} = 0. \tag{6.44}
$$

Note that we used the result $A_Z = -i\,\mathrm{sgn}(k)A_X$ found earlier. Using (6.44), we find from (6.41) that

$$
\bar\phi_Z = \frac{2\omega k}{g}\frac{\partial}{\partial X}|A|^2,
$$

i.e., up to $O(\varepsilon^2)$ the mean velocity potential depends explicitly on $|A|^2$. We also note that if $A$ is independent of $X$ these results agree with those from the previous section. Setting the coefficients of each power of $\varepsilon$ to zero in (6.39) and (6.42) we get to leading order

$$-i\omega A + gB = 0$$
$$-|k|A - i\omega B = 0,$$

which, since the dispersion relationship $\omega^2 = g|k|$ is satisfied, has the non-trivial solution $B = \dfrac{i\omega}{g}A$. From (6.39) – (6.44), we now have

$$B = \frac{i\omega}{g}A - \frac{\varepsilon A_T}{g} + \varepsilon^2 \left[ i\frac{\omega k^2}{g}A|B|^2 - 4\frac{k^2|k|}{g}|A|^2 B \right.$$
$$\left. - \frac{i\omega k^2}{2g}B^2 A^* + \frac{i\omega|k|k^2}{g^2}A^2 A^* \right] + O(\varepsilon^3).$$

Substituting this into (6.42), and with (6.39)–(6.44), yields

$$2i\omega \left( A_T + v_g A_X \right) - \varepsilon \left( A_{TT} + 4k^4|A|^2 A \right) + O(\varepsilon^2) = 0,$$

where we have defined the group velocity as $v_g = \omega'(k) = \omega/2k$. From this and (6.40), we see that $A_2 \sim O(\varepsilon^2)$. If we neglect the $O(\varepsilon^2)$ terms in the above equation and make the change of variables $\tau = \varepsilon T$, $\xi = X - v_g T$, we get the focusing NLS equation

$$iA_\tau + \frac{\omega''}{2}A_{\xi\xi} - \frac{2k^4}{\omega}|A|^2 A = 0. \qquad (6.45a)$$

With $\omega'' = -v_g^2/\omega$, the above equation can be written as

$$iA_\tau - \left( \frac{v_g^2}{2\omega}A_{\xi\xi} + \frac{2k^4}{\omega}|A|^2 A \right) = 0, \qquad (6.45b)$$

which is the typical formulation of the focusing NLS equation found in water wave theory. We also note that in terms of $B$, which is associated with the wave elevation $\eta$, using $A = \dfrac{g}{i\omega}B$, equation (6.45b) becomes

$$iB_\tau + \frac{\omega''}{2}B_{\xi\xi} - 2k^2\omega|B|^2 B = 0.$$

We note the important point that the coefficient of the nonlinear term in (6.45b) agrees with the Stokes frequency shift discussed in the previous section. As discussed earlier an alternative change of coordinates to a retarded time frame is to let $t' = T - X/v_g$ and $\chi = \varepsilon X$. We then get

$$iA_\chi + \frac{\omega''}{2(\omega')^3}A_{t't'} - \frac{2k^4}{\omega\omega'}|A|^2 A = 0. \qquad (6.46)$$

This formulation is more commonly found in the context of nonlinear optics.

This derivation of the NLS equation for deep-water waves was done in 1968 by Zakharov for deep water, including surface tension (Zakharov, 1968) and in the context of finite depth by Benney and Roskes (1969). It took more than a century from Stokes' (Stokes, 1847) initial discovery of the nonlinear frequency shift until these NLS equations were derived.

We remark that this NLS equation is called the "focusing" NLS because the signs of the dispersive and nonlinear terms are the same in (6.45b). To see that in (6.46), we recall that $\omega^2(k) = g|k|$ and therefore, for positive $k$, one gets that $\omega = \sqrt{gk}$, $v_g = \omega' = \sqrt{g/4k}$, and $\omega'' = -\sqrt{g}/4k^{3/2} = -v_g^2/\omega$; we also note $\frac{\omega''}{2(\omega')^2} = -\frac{1}{\omega}$. These results imply that the coefficient of the second derivative term and the nonlinear coefficient have the same sign. As we will see, the focusing NLS equation admits "bright" soliton solutions, i.e., solutions that are traveling localized "humps".

## 6.6 Some properties of the NLS equation

Note that the linear operator in (6.45a) is

$$\widehat{L} = i\partial_\tau + \frac{\omega''}{2}\partial_\xi^2,$$

and is what we expected to find from our earlier considerations (see Section 6.3). Similarly, the nonlinear part:

$$iA_\tau - \frac{2k^4}{\omega}|A|^2 A = 0,$$

is what we expected from the frequency shift analysis in Section 6.4.1; see (6.33). We can rescale (6.45b) by $\xi = \frac{v_g}{\sqrt{2}}x$, $A = k^2 u$, and $\tau = -2\omega^2 t$ to get the focusing NLS equation in standard form:

$$iu_t + u_{xx} + 2|u|^2 u = 0. \qquad (6.47)$$

Remarkably, this equation can be solved exactly using the so-called inverse scattering transform (Zakharov and Shabat, 1972); see also (Ablowitz et al., 2004b). One special solution is a "bright" soliton:

$$u = \eta \operatorname{sech} \left[ \eta \left( x + 2\xi t - x_0 \right) \right] e^{-i\Theta},$$

where $\Theta = \xi x + (\xi^2 - \eta^2)t + \Theta_0$. The parameters $\xi$ and $\eta$ are related to an eigenvalue from the inverse scattering transform analysis via $\lambda = \xi/2 + i\eta/2$ where $\lambda$ is the eigenvalue. If, instead of (6.47), we had the defocusing NLS equation,

$$iu_t + u_{xx} - 2|u|^2 u = 0, \tag{6.48}$$

then we can find "dark" – "black" or more generally "gray" – soliton solutions. Letting $t \to -t/2$, (6.48) goes to

$$iu_t - \frac{1}{2}u_{xx} + |u|^2 u = 0$$

and has a black soliton solution whose amplitude vanishes at the origin

$$u = \eta \tanh (\eta x) \, e^{i\eta^2 t}.$$

Note that $u \to \pm\eta$ as $x \to \pm\infty$. A gray soliton solution is given by

$$u(x, t) = \eta e^{2i\eta^2 t + i\psi_0} \left[ \cos \alpha + i \sin \alpha \, \tanh \left[ \sin \alpha \, \eta(x - 2\eta \cos \alpha \, t - x_0) \right] \right]$$

with $\eta, \alpha, x_0, \psi_0$ arbitrary real parameters. In Figure 6.1 a "bright" and the two dark (black and gray) are depicted.

These solutions satisfy the boundary conditions

$$u(x, t) \to u_{\pm}(t) = \eta e^{2i\eta^2 t + i\psi_0 \pm i\alpha} \qquad \text{as} \quad x \to \pm\infty$$

and appear as localized dips of intensity $\eta^2 \sin^2 \alpha$ on the background field $\eta$. The gray soliton moves with velocity $2\eta \cos \alpha$ and reduces to the dark (black) soliton when $\alpha \to \pi/2$ with $\psi_0 = -\pi/2$ (Hasegawa and Tappert, 1973b; Zakharov and Shabat, 1973); see also Prinari et al. (2006) where the vector IST problem for non-decaying data is discussed in detail.

A property of the NLS equation we will investigate next is its Galilean invariance. That is, if $u_1(x, t)$ is a solution of (6.47) then so is

$$u_2(x, t) = u_1(x - vt, t)e^{i(kx - \omega t)},$$

with $k = v/2$ and $\omega = k^2$. Substituting $u_2$ into (6.47) we find $u_1$ satisfies:

$$iu_{1,t} + \omega u_1 - iv u_{1,x} + (u_{1,xx} + 2ik u_{1,x} - k^2 u_1) + 2|u_1|^2 u_1 = 0.$$

Using the fact that $u_1$ is assumed to be a solution of (6.47) and using the values for $k, \omega$, this implies $u_2$ also satisfies (6.47).

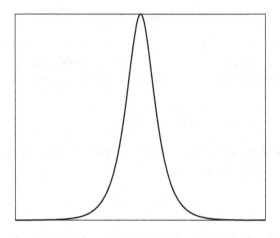

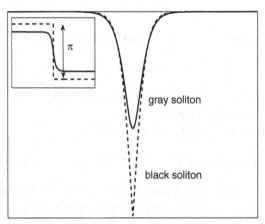

Figure 6.1  Bright (top) and dark (bottom) solitons of the NLS. In the inset we plot the relative phases.

Another important result involves the linear stability of a special periodic solution of (6.47). In (6.45a), i.e., the standard water wave formulation, take $A$ independent of $\xi$ to get

$$iA_\tau = \frac{2k^4}{\omega}|A|^2A.$$

Note that this agrees with the results of the Stoke's frequency shift calculation, (6.33). With the change of variables mentioned above in terms of the standard NLS, (6.47), this means:

$$iu_t = -2|u|^2u,$$

which has the plane wave solution $u = ae^{2ia^2t}$, $a = u(0)$; for convenience take $a$ real. We now perturb this solution: $u = ae^{2ia^2t}(1 + \varepsilon(x,t))$, where $|\varepsilon| \ll 1$. Substituting this into (6.47) and linearizing (i.e., assuming $|\varepsilon| \ll 1$) we find,

$$i\varepsilon_t + \varepsilon_{xx} + 2a^2(\varepsilon + \varepsilon^*) = 0.$$

We will consider the linearized problem on the periodic spatial domain $0 < x < L$. Thus, $\varepsilon(x,t)$ has the Fourier expansion

$$\varepsilon(x,t) = \sum_{-\infty}^{\infty} \widehat{\varepsilon}_n(t)e^{i\mu_n x},$$

where $\mu_n = 2\pi n/L$. Note that $\widehat{\varepsilon}_{-n}(t)$ is not the complex conjugate of $\widehat{\varepsilon}_n(t)$, since $\varepsilon$ is not necessarily real. Since the PDE is linear, it is sufficient to consider $\varepsilon = \widehat{\varepsilon}_n(t)e^{i\mu_n x} + \widehat{\varepsilon}_{-n}(t)e^{-i\mu_n x}$. Thus, with $\varepsilon'_n \equiv \partial \varepsilon_n/\partial t$,

$$i\left(\widehat{\varepsilon}'_n e^{i\mu_n x} + \widehat{\varepsilon}'_{-n} e^{-i\mu_n x}\right) - \mu_n^2\left(\widehat{\varepsilon}_n e^{i\mu_n x} + \widehat{\varepsilon}_{-n} e^{-i\mu_n x}\right)$$
$$+ 2a^2\left(\widehat{\varepsilon}_n e^{i\mu_n x} + \widehat{\varepsilon}_{-n} e^{-i\mu_n x} + \widehat{\varepsilon}^*_n e^{-i\mu_n x} + \widehat{\varepsilon}^*_{-n} e^{i\mu_n x}\right) = 0.$$

Setting to zero the coefficients of $e^{i\mu_n x}$ and $e^{-i\mu_n x}$, we find, respectively,

$$i\widehat{\varepsilon}'_n - \mu_n^2 \widehat{\varepsilon}_n + 2a^2\left(\widehat{\varepsilon}_n + \widehat{\varepsilon}^*_{-n}\right) = 0$$
$$i\widehat{\varepsilon}'_{-n} - \mu_n^2 \widehat{\varepsilon}_{-n} + 2a^2\left(\widehat{\varepsilon}_{-n} + \widehat{\varepsilon}^*_n\right) = 0.$$

Taking the conjugate of the last equation and multiplying it by $-1$, we have the system

$$i\frac{\partial}{\partial t}\begin{pmatrix} \widehat{\varepsilon}_n \\ \widehat{\varepsilon}^*_{-n} \end{pmatrix} + \begin{pmatrix} 2a^2 - \mu_n^2 & 2a^2 \\ -2a^2 & -2a^2 + \mu_n^2 \end{pmatrix}\begin{pmatrix} \widehat{\varepsilon}_n \\ \widehat{\varepsilon}^*_{-n} \end{pmatrix} = 0,$$

to solve. Assuming a solution of the form

$$\begin{pmatrix} \widehat{\varepsilon}_n \\ \widehat{\varepsilon}^*_{-n} \end{pmatrix} = \begin{pmatrix} \alpha \\ \beta \end{pmatrix} e^{i\sigma_n t},$$

we find

$$\det\begin{pmatrix} 2a^2 - \mu_n^2 - \sigma_n & 2a^2 \\ -2a^2 & -2a^2 + \mu_n^2 - \sigma_n \end{pmatrix} = 0$$

must be true, hence with $\mu_n^2 = (2n\pi/L)^2$

$$\sigma_n^2 = \left(\frac{2\pi n}{L}\right)^2\left[\left(\frac{2\pi n}{L}\right)^2 - 4a^2\right].$$

Thus, when

$$\frac{aL}{\pi} > n$$

the system is unstable, since $\sigma_n^2 < 0$ leads to exponential growth. Note that there are only a finite number of unstable modes. In the context of water waves, we have deduced the famous result by Benjamin and Feir (1967) that the Stokes water wave is unstable. Later, Benney and Roskes (1969) (BR) showed that all periodic wave solutions of the slowly varying envelope water wave equations in 2 + 1 dimensions are unstable. The BR equations are also discussed below. We also remark that Zakharov and Rubenchik (1974) showed that solitons are unstable to weak transverse modulations, i.e., one-dimensional soliton solutions of

$$iu_t + u_{xx} + \varepsilon^2 u_{yy} + 2|u|^2 u = 0$$

are unstable. (See Ablowitz and Segur, 1979 for further discussion.)

## 6.7 Higher-order corrections to the NLS equation

As mentioned earlier, the nonlinear Schrödinger equation was first derived in 1968 in deep water with surface tension by Zakharov (1968) and its finite depth analog in 1969 by Benney and Roskes (1969). It took some ten years until in 1979 Dysthe (1979) derived, in the context of deep-water waves, the next-order correction to the NLS equation. We simply quote the $(1 + 1)$-dimensional result:

$$2i\omega\left(A_T + v_g A_X\right) - \varepsilon\left(v_g^2 A_{XX} + 4k^2|A|^2 A\right)$$
$$= \varepsilon^2\left(\frac{i\omega^2}{8k^3}A_{XXX} + 2ik^3 A^2 A_X^* - 12ik^3|A|^2 A_X + 2\omega k\bar\phi_X A\right). \qquad (6.49)$$

The mean term $\bar\phi_X$ is unique to deep-water waves and arises from the quadratic nonlinearity in the water wave equations. The mean field satisfies

$$\bar\phi_{XX} + \bar\phi_{ZZ} = 0, \qquad -\infty < z < 0,$$

with $\bar\phi_Z \to 0$ as $z \to -\infty$ and

$$\bar\phi_Z = \frac{2\omega k}{g}\frac{\partial}{\partial X}|A|^2, \qquad z = 0.$$

Non-dimensionalizing (6.49) by setting $T' = \omega T$, $X' = kX$, $Z' = kZ$, $\eta' = k\eta$, $\bar\phi' = (2k^2/\omega)\bar\phi$, $u = \left(2\sqrt{2}k^2/\omega\right)A$, $\tau = -\varepsilon T'/8$, and $\xi = X' - T'/2$, we get

$$iu_\tau + u_{\xi\xi} + 2|u|^2 u + \varepsilon\left(\frac{i}{2}u_{\xi\xi\xi} - 6i|u|^2 u_\xi + iu^2 u_\xi^* + 2u\mathcal{H}[|u|^2]_\xi\right) = 0, \quad (6.50)$$

where

$$\mathcal{H}[f] = \frac{1}{\pi}\int \frac{f(u)}{u - x}\,du$$

is the so-called Hilbert transform. (The integral is understood in the principle-value sense.) Ablowitz, Hammack, Henderson, and Schober (Ablowitz et al., 2000a, 2001) showed that periodic solutions of (6.50) can actually be chaotic. Experimentally and analytically, a class of periodic solutions is found not to be "repeatable", while soliton solutions are repeatable.

To go from the term including $\bar{\phi}_x$ in (6.49) to the Hilbert transform requires some attention. The velocity potential satisfies Laplace's equation:

$$\bar{\phi}_{xx} + \bar{\phi}_{zz} = 0.$$

In the Fourier domain, with $\widehat{\bar{\phi}} = \int \bar{\phi}e^{-i\xi x}\,dx$, we find

$$\widehat{\bar{\phi}}_{zz} - \xi^2\widehat{\bar{\phi}} = 0.$$

Using the boundary condition $\widehat{\bar{\phi}}_z \to 0$ as $z \to -\infty$, we get

$$\widehat{\bar{\phi}}(\xi, z) = C(\xi)e^{|\xi|z},$$

where $C$ is to be determined. The factor $C$ is fixed by the boundary condition

$$\bar{\phi}_z = \frac{2\omega k}{g}\frac{\partial}{\partial x}|A|^2,$$

on $z = 0$. Hence

$$C(\xi) = \frac{2i\omega k}{g}\frac{\xi}{|\xi|}\mathcal{F}(|A|^2) = \frac{2i\omega k}{g}\mathrm{sgn}(\xi)|\mathcal{F}(|A|^2)$$

where $\mathcal{F}$ represents the Fourier transform. Using a well-known result from Fourier transforms, $\mathcal{F}^{-1}[\mathrm{sgn}\xi\hat{F}(\xi)] = \mathcal{H}[f(x)]$, where $f$ is the inverse Fourier transform of $\hat{F}$ and $\mathcal{H}[\cdot]$ is the Hilbert transform (Ablowitz and Fokas, 2003) and noting that $\mathcal{H}\left[e^{i\xi x}\right] = i\,\mathrm{sgn}(\xi)e^{i\xi x}$, we have

$$\bar{\phi} = \frac{2\omega k}{g}\mathcal{H}\left[|A|^2\right]$$

$$\bar{\phi}_x = \frac{2\omega k}{g}\frac{\partial}{\partial x}\mathcal{H}\left[|A|^2\right].$$

## 6.8 Multidimensional water waves

For multidimensional water waves in finite depth with surface tension, the velocity potential has the form

$$\phi = \varepsilon(\tilde{\phi}(X, Y, T) + \frac{\cosh(k(z + h))}{\cosh(kh)}(\tilde{A}(X, Y, T)e^{i(kx-\omega t)} + \text{c.c.})) + O(\varepsilon^2),$$

where $X = \varepsilon x$, $Y = \varepsilon y$, $T = \varepsilon t$ and $\tilde{\phi}$, $\tilde{A}$ satisfy coupled nonlinear wave equations. Benney and Roskes (1969) derived this system without surface tension. It was subsequently rederived by Davey and Stewartson (1974) who put the system in a simpler form. Later, Djordjevic and Redekopp (1977) included surface tension.

After non-dimensionalization and rescaling, the equations can be put into the following form; we call it a Benney–Roskes (BR) system:

$$iA_t + \sigma_1 A_{xx} + A_{yy} = \sigma_2 |A|^2 A + A\Phi_x$$

$$a\Phi_{xx} + \Phi_{yy} = -b(|A|^2)_x \tag{6.51}$$

$$\sigma_1 = \pm 1, \qquad \sigma_2 = \pm 1.$$

The parameters $\sigma_1$, $\sigma_2$, $a$, and $b$ are dimensionless and depend on the dimensionless fluid depth and surface tension. Here we have presented it in a rescaled, normalized form that helps in analyzing its behavior for different choices of $\sigma_1$ and $\sigma_2$. The quantity $A$ is related to the slowly varying envelope of the first harmonic of the potential velocity field and $\Phi$ is related to the slowly varying mean potential velocity field.

A special solution to (6.51) is the following self-similar solution

$$A = \frac{\Lambda}{t} \exp i\left(\frac{\sigma_1 x^2 + y^2}{4t} + \sigma_2 \frac{\Lambda^2}{t} + B(t) + \phi_0\right)$$

$$\Phi = -B'(t)x + C(t)y + D(t).$$

This is an analog of the similarity solution of the one-dimensional nonlinear Schrödinger (NLS) equation

$$iA_t + A_{xx} + \sigma |A|^2 A = 0$$

$$A = \frac{\Lambda}{t^{1/2}} \exp i\left(\frac{x^2}{4t} + \sigma \Lambda^2 \log(t) + \phi_0\right).$$

Since the above similarity solution of NLS approximates the long-time solution of NLS without solitons in the region $\left|\dfrac{x}{\sqrt{t}}\right| \leq O(1)$, it can be expected that the similarity solution of the BR system is a candidate to approximate long-time "radiative" solutions.

From the water wave equations in the limit $kh \to 0$ (the shallow-water limit) with suitable rescaling, cf. Ablowitz and Segur (1979), the BR system (6.51) reduces to

$$iA_t - \gamma A_{xx} + A_{yy} = A(\gamma|A|^2 + \Phi_x)$$
$$\gamma\Phi_{xx} + \Phi_{yy} = -2(|A|^2)_x, \qquad \gamma = \text{sgn}\left(\tfrac{1}{3} - \hat{T}\right) = \pm 1. \tag{6.52}$$

This is the so-called Davey–Stewartson (DS) equation. As we have mentioned, it describes multidimensional water waves in the slowly varying envelope approximation, with surface tension included, in the shallow-water limit. Here the normalized surface tension is defined as $\hat{T} = \dfrac{T_0}{\rho g h^2}$, $T_0$ being the surface tension coefficient. This system, (6.52), is integrable (Ablowitz and Clarkson, 1991) whereas the multidimensional deep-water limit

$$iA_t + \nabla^2 A + |A|^2 A = 0$$

is apparently not integrable. The concept of integrability is discussed in more detail in Chapters 8 and 9 (cf. also Ablowitz and Clarkson 1991).

The DS equation (6.52) can be generalized by making the following substitutions

$$\phi = \Phi_x, \qquad r = -\sigma q^* = -\sigma A^*, \qquad \sigma = \pm 1$$

and we write

$$iq_t - \gamma q_{xx} + q_{yy} = q(\phi - qr)$$
$$\phi_{xx} + \gamma\phi_{yy} = 2(qr)_{xx}. \tag{6.53}$$

With $q = A$, $\gamma = -1$, and $\sigma = 1$, we get shallow-water waves with "large" surface tension, equation (6.52). It turns out that the generalized DS equation (GDS), (6.53), admits "localized" boundary induced pulse solutions when $\gamma = 1$ (Ablowitz and Clarkson, 1991) and weakly decaying "lump"-type solutions when $\gamma = -1$ (Villarroel and Ablowitz, 2003).

### 6.8.1 Special solutions of the Davey–Stewartson equations

Consider the case where $\gamma = -1$ with $\sigma = 1$. Then the DS system (6.53) becomes

$$2iq_t + q_{xx} + q_{yy} = 2(\phi - qr)q$$
$$\phi_{xx} - \phi_{yy} = 2(qr)_{xx}. \tag{6.54}$$

We call this the DSI system. Note that the DSI system admits an interesting class of solutions, cf. Fokas and Santini (1989, 1990) and Ablowitz et al.

(2001a). In order to write down the pulse solution (also called a "dromion"), we will make a convenient rescaling and change of variable as follows:

$$Q = \phi - qr.$$

Then the second equation in GDS system (6.54) becomes

$$Q_{xx} - Q_{yy} = (qr)_{xx} + (qr)_{yy}. \tag{6.55}$$

Now we make the change of variable

$$\xi = \frac{x+y}{\sqrt{2}}, \quad \eta = \frac{x-y}{\sqrt{2}}$$

$$\partial_x = \frac{1}{\sqrt{2}}(\partial_\xi + \partial_\eta), \quad \partial_y = \frac{1}{\sqrt{2}}(\partial_\xi - \partial_\eta)$$

to transform (6.55) to

$$2Q_{\xi\eta} = (qr)_{\xi\xi} + (qr)_{\eta\eta}. \tag{6.56}$$

Finally, making the substitution

$$Q = -(U_1 + U_2),$$

and integrating with respect to $\xi$ and $\eta$ separates (6.56) into

$$U_1 = u_1(\eta) - \frac{1}{2}\int_{-\infty}^{\xi}(qr)_\eta\, d\xi', \quad U_2 = u_2(\xi) - \frac{1}{2}\int_{-\infty}^{\eta}(qr)_\xi\, d\eta.$$

These two equations along with the transformed second equation in (6.54),

$$2iq_t + q_{\xi\xi} + q_{\eta\eta} + 2(U_1 + U_2)q = 0,$$

give rise to the following "dromion" solution

$$u_1(\eta) = 2\lambda_R^2\ \mathrm{sech}^2(\lambda_R(\hat{\eta} - \eta_0)), \quad \hat{\eta} = \eta - 2\lambda_I t$$

$$u_2(\xi) = 2\mu_R^2\ \mathrm{sech}^2(\mu_R(\hat{\xi} - \xi_0)), \quad \hat{\xi} = \xi - 2\mu_I t$$

$$q = \frac{\rho\,\sqrt{\lambda_R\mu_R}e^{i\theta}}{\cosh(\mu_R(\hat{\xi} - \xi_0))\cosh(\lambda_R(\hat{\eta} - \eta_0)) + (|\rho|/2)^2 e^{(\lambda_R(\hat{\eta} - \eta_0))}e^{\mu_R(\hat{\xi} - \xi_0)}}$$

$$\theta = -(\mu_I\hat{\xi} + \lambda_I\hat{\eta}) + (|\mu|^2 + |\lambda|^2)\frac{t}{2} - \theta_0$$

$$\lambda = \lambda_R + i\lambda_I, \quad \mu = \mu_R + i\mu_I \quad \text{constants}.$$

where $\lambda_R > 0, \mu_R > 0$, $\lambda, \mu, \rho$ are complex constants and $\xi_0$, $\eta_0$ are real constants. See Figure 6.2 where a typical dromion is depicted (with $\lambda_R = 1, \mu_R = 1, \lambda_I = 0, \mu_I = 0, \rho = 1$).

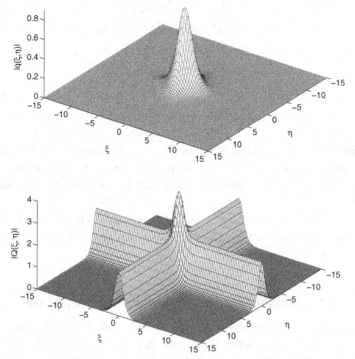

Figure 6.2 The dromion solution for the Benney–Roskes/Davey–Stewartson systems. The top figure represents the field $q$, while the bottom figure represents the mean field $Q = u_1 + u_2$.

### 6.8.2 Lump solution for small surface tension

It turns out that there are lump-type solutions to (6.52) when $\gamma = +1$; this is the so-called DSII system (Villarroel and Ablowitz, 2002). We now consider this case ($\gamma = +1$) corresponding to zero or "small" surface tension $\left(\hat{T} < \dfrac{1}{3}\right)$. Then the GDS system (6.53) is written, after substituting in $r = -\sigma q^*$, $\sigma = \pm 1$,

$$
\begin{aligned}
iq_t - q_{xx} + q_{yy} &= q(\phi + \sigma|q|^2) \\
\phi_{xx} + \phi_{yy} &= -2\sigma(|q|^2)_{xx}.
\end{aligned}
\tag{6.57}
$$

The following is a lump solution

$$
q = 2\rho\sigma \frac{e^{i\theta}}{\hat{x}^2 + \hat{y}^2 + \sigma|\rho|^2}
$$

$$
\phi + \sigma|q|^2 = R_{xx} - R_{yy}
$$

$$R = \log(\hat{x}^2 + \hat{y}^2 + \sigma|\rho|^2) \tag{6.58}$$
$$\theta = a\hat{x} - b\hat{y} + (b^2 - a^2)t - \theta_0$$
$$\hat{x} = x + at - x_0, \quad \hat{y} = y + bt - y_0$$
$$k = \tfrac{1}{2}(a + ib) \quad \text{constant.}$$

This solution is non-singular when $\sigma = 1$, but for $\sigma = -1$, the above solution is singular. Alternatively we have the interesting case of the dark lump-type envelope hole solution (Satsuma and Ablowitz, 1979) when $\sigma = -1$. The case $\sigma = 1$ occurs in water waves; i.e., in this case, (6.57) can be transformed to the reduced water wave equation (6.52). But it has been shown that the lump solution (6.58) is unstable (Pelinovsky and Sulem, 2000).

We also note that Ablowitz et al. (1990) discussed the derivation of the integrable equations (6.54) and (6.57) from the KPI and KPII equations, respectively.

### 6.8.3 Multidimensional problems and wave collapse

As mentioned earlier, in 1969 Benny and Roskes derived a (2+1)-dimensional NLS-type equation for water waves (Benney and Roskes, 1969). In 1977 Djordjevic and Redekopp extended their results by including surface tension effects (Djordjevic and Redekopp, 1977). To get these equations the velocity potential expansion takes the form:

$$\phi = \varepsilon(\tilde{\phi}(X, Y, T) + \frac{\cosh[k(h+z)]}{\cosh kh} \left( \tilde{A}(X, Y, T)e^{i\theta} + \text{c.c.} \right)) + O(\varepsilon^2),$$
$$\theta = kx - \omega t, X = \varepsilon x, Y = \varepsilon y, T = \varepsilon t$$
$$\omega^2 = g\kappa \tanh(\kappa h)(1 + \tilde{T}),$$
$$\kappa^2 = k^2 + l^2,$$
$$\tilde{T} = \frac{k^2 T}{\rho g}.$$

In terms of the redefined functions: $A = \tilde{A}\dfrac{k^2}{\sqrt{gk}}$, $\Phi = \dfrac{k^2}{\sqrt{gk}}\tilde{\phi}$, and variables: $\xi = k(X - \omega'T)$, $\tau = \sqrt{gk}\varepsilon^2 t$, $\eta = \varepsilon kY$, we have that $A$, $\Phi$ satisfy the following coupled system

$$iA_\tau + \lambda A_{\xi\xi} + \mu A_{\eta\eta} = \chi|A|^2 A + \chi_1 A\Phi_\xi \tag{6.59}$$
$$\alpha\Phi_{\xi\xi} + \Phi_{\eta\eta} = -\beta \left( |A|^2 \right)_\xi. \tag{6.60}$$

The parameters $\lambda, \mu, \chi, \chi_1, \alpha,$ and $\beta$ depend on $\omega, k, \widetilde{T},$ and $gh$ (see Ablowitz and Segur, 1979, 1981). As $h \to \infty$, (6.59) reduces to an NLS equation

$$iA_\tau + \lambda_\infty A_{\xi\xi} + \mu_\infty A_{\eta\eta} - \chi_\infty |A|^2 A = 0,$$

where

$$\lambda_\infty = -\frac{\omega_0}{8\omega} \left( \frac{1 - 6\widetilde{T} - 3\widetilde{T}^2}{1 + \widetilde{T}} \right)$$

$$\mu_\infty = \frac{\omega_0}{4\omega} \left( 1 + 3\widetilde{T} \right)$$

$$\chi_\infty = \frac{\omega_0}{4\omega} \left[ \frac{8 + \widetilde{T} + 2\widetilde{T}^2}{\left( 1 - 2\widetilde{T} \right)\left( 1 + \widetilde{T} \right)} \right]$$

$$\omega_0^2 = g\kappa,$$

with $\lambda_\infty > 0$, $\mu_\infty > 0$ and $-\xi_\infty > 0$ for $\widetilde{T}$ large enough. This equation has solutions that blow-up in finite time (Ablowitz and Segur, 1979). When $\widetilde{T} = 1/2$, the expansion breaks down due to second harmonic resonance: $\omega^2(2k) = [2\omega(k)]^2$. However, when $\widetilde{T} = 0$, i.e., when there is no surface tension, some of the coefficients in the above equation change sign and we have $\left( \frac{\omega_0}{\omega} > 0 \right)$: $\lambda_\infty < 0$, $\mu_\infty > 0$ and $\chi_\infty > 0$. To date no blow-up has been found for the above NLS equation with the latter choices of signs.

## Blow-up

One can verify that

$$iA_t + \Delta A + |A|^2 A = 0, \qquad x \in \mathbb{R}^2$$

has the conserved quantities

$$P(t) = \int |A|^2 \, dx,$$

$$\underline{M}(t) = \int A \nabla A \, dx$$

$$H(t) = \int |\nabla A|^2 - \frac{1}{2}|A|^4 \, dx,$$

i.e., mass (power), momentum, and energy (Hamiltonian) are conserved. Define

$$V(t) = \int |\mathbf{r}|^2 |A|^2 \, dx,$$

where $|\mathbf{r}|^2 = x^2 + y^2$. Then one can show that in two dimensions (Vlasov et al., 1970),

$$\frac{d^2V}{dt^2} = 8H.$$

This is often called the virial theorem. Integrating this equation, we get

$$V(t) = 4Ht^2 + c_1 t + c_0.$$

If the initial conditions are such that $H < 0$, then there exists a time $t^*$ when

$$\lim_{t \to t^*} V(t) = 0$$

and for $t > t^*$, $V(t) < 0$. However, $V$ is a positive quantity. Thus a singularity in the solution has occurred in finite time $t = t^*$. Combining this result with the conservation of mass and a bit more analysis, one can show that in fact

$$\int |\nabla A|^2 \, dx$$

becomes infinite as $t \to t^*$, which in turn implies that $A$ also becomes infinite as $t \to t^*$. For an extensive discussion of the NLS equation in one and multidimensions, see Sulem and Sulem (1999).

The more general equation

$$i\psi_t + \Delta_d \psi + |\psi|^{2\sigma} \psi = 0, \qquad x \in \mathbb{R}^d, \tag{6.61}$$

where $\Delta_d$ is the $d$-dimensional Laplacian, has also been studied. There are three cases:

- "supercritical": $\sigma d > 2$ blow-up occurs;
- "critical": $\sigma d = 2$ blow-up can occur; collapse can be arrested with small perturbations;
- "subcritical": $\sigma d < 2$ global solutions exist.

A special solution to (6.61) can be found by assuming a solution of the form $\psi = f(r)e^{i\lambda t}$. In two dimensions, we find the nonlinear eigenvalue problem

$$\frac{1}{r}\frac{\partial}{\partial r}\left(r\frac{\partial f}{\partial r}\right) + f^3 - \lambda f = 0,$$

certain solutions to which are the so-called Townes modes. Asymptotically, $f \sim e^{-r}/\sqrt{r}$, $1 \ll r$ (Fibich and Papanicolaou, 1999). Weinstein (1983) showed that there is a critical energy (found from the Townes mode with the smallest energy),

$$E_c = \int |u|^2 dx dy = 2\pi \int_0^\infty r f^2(r) \, dr \simeq 2\pi(1.86) \simeq 11.68.$$

Pulses with energy below $E_c$ exist globally (Weinstein, 1983). In many cases when $E > E_c$, even by a small amount, blow-up occurs. In the supercritical case, the blow-up solution has a similarity form

$$\psi \sim \frac{1}{t'^\alpha} f\left(\frac{r}{t'^\beta}\right) e^{i\lambda \ln t'},$$

with suitable constants $\alpha$ and $\beta$. But for the critical case the structure is

$$\psi \sim \frac{\sqrt{L(t')}}{\sqrt{t'}} f\left(\frac{r\sqrt{L(t')}}{\sqrt{t'}}\right) e^{i\phi(t')},$$

where $t' = t_c - t$, $L(t) \sim \ln(\ln(1/t'))$ as $t \to t_c$ is the blow-up or collapse time.

The stationary states, collapse and properties of the BR equations (6.59) (sometimes referred to as Davey–Stewartson-type equations) as well as similar ones that arise in nonlinear optics were studied in detail by Papanicolaou et al. (1994) and Ablowitz et al. (2005). More specifically, the equations studied were

$$iU_z + \frac{1}{2}\Delta U + |U|^2 U - \rho U V_x = 0, \quad \text{and}$$

$$V_{xx} + \nu V_{yy} = (|U|^2)_x.$$

The case $\rho < 0$ corresponds to water waves, cf. Ablowitz and Segur (1979, 1981), and the case $\rho > 0$ corresponds to $\chi^{(2)}$ nonlinear optics (Ablowitz et al., 1997, 2001a; Crasovan et al., 2003).

When $\nu > 0$ collapse is possible. That there is a singularity in finite time can be shown by the virial theorem. In analogy with the Townes mode for NLS, there are stationary states (ground states) that satisfy $U = F(x, y)e^{i\mu z}$, $V = G(x, y)$

$$-\mu F + \frac{1}{2}\Delta F + |F|^2 F - \rho F G_x = 0$$

$$G_{xx} + \nu G_{yy} = (|F|^2)_x.$$

The stationary states $F, G$ can be obtained numerically (Ablowitz et al., 2005). In Figure 6.3 slices of the modes along the $y = 0$, $x = 0$ axes, respectively, are given for different values of $\rho$. In Figure 6.3(c) and (d) contour plots of some typical modes $F$ are shown. One can see that the modes are elliptical in nature when $\rho \neq 0$.

### Quasi-self-similar collapse

Papanicolaou et al. (1994) showed that as collapse occurs, i.e., as $z \to z_c$, then $U, V$ have a quasi-self-similar structure:

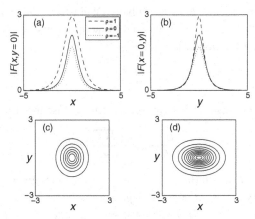

Figure 6.3 $F(x, y)$; $v = 0.5$; top (a), (b): "slices"; bottom: contour plots; (c) $\rho = -1$, (d)$\rho = 1$.

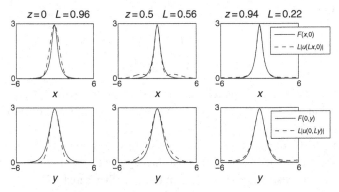

Figure 6.4 Wave collapse to the stationary state as $z \to z_c$.

$$U \sim \frac{1}{L(z)} F\left(\frac{x}{L(z)}, \frac{y}{L(z)}\right), \quad V \sim \frac{1}{L(z)} G\left(\frac{x}{L(z)}, \frac{y}{L(z)}\right)$$

where $L \to 0$. Further, one finds from direct simulation (Ablowitz et al., 2005) that the stationary modes are good approximations of collapse profiles. By taking typical initial conditions (of Gaussian type) that collapse, we find that the pulse structure is well approximated by the stationary modes in the neighborhood of the collapse point, where $L(z) = F(0, 0)/U(0, 0, z) \to 0$. In Figure 6.4 we compare $L(z)|U(Lx, Ly)|$ with $F(x, y)$ where $L(z) = F(0, 0)/U(0, 0, z)$. As collapse occurs, $L(z) \to 0$ for typical values $(v, \rho) = (0.5, 1)$. In Figure 6.4 we provide snapshots of the solution compared to the stationary state $F$ as the wave collapse begins to occur. Along the $y = 0$ axis (top) and the $x = 0$ axis (bottom) at the edges of the solution, radiation is seen. But the behavior at the center of the wave indicates that collapse is well approximated by the stationary state.

Finally, we mention that a similar, but more complicated blow-up or wave collapse phenomenon occurs for the generalized KdV equation

$$u_t + u^p u_x + u_{xxx} = 0$$

when $p \geq 4$ (Ablowitz and Segur, 1981; Merle, 2001; Angulo et al., 2002).

## Exercises

6.1   Derive the NLS equation from the following nonlinear equations.
   (a) $u_{tt} - u_{xx} + \sin u = 0$ (sine–Gordon).
   (b) $u_{tt} - u_{xx} + u_{xxxx} + (u u_x)_x = 0$ (Boussinesq-type).
   (c) $u_t + u^2 u_x + u_{xxx} = 0$ (mKdV) .

6.2   Given the equation

$$i u_t - u_{xxxx} + |u|^{2(n+1)} u = 0$$

   with $n \geq 1$ an integer, substitute the mutli-scale and quasi-monochromatic assumption:

$$u \sim \mu e^{(ikx - \omega t)} A(\epsilon_1 x, \epsilon_2 t),$$

   where $\mu$, $\epsilon_1$, and $\epsilon_2$ are all asymptotically small parameters, into the equation. Choose a maximal balance between the small parameters to find a nonlinear Schrödinger-type wave equation for the slowly varying envelope $A$.

6.3   From the water wave equations with surface tension included, derive the dispersion relation

$$\omega^2 = g\kappa \tanh(\kappa h)(1 + \widetilde{T}), \quad \kappa^2 = k^2 + l^2, \quad \widetilde{T} = \frac{k^2 T_0}{\rho g}$$

   in finite depth water waves with surface tension, where $T_0$ is the surface tension coefficient.

6.4   Beginning with the KP equation

$$\partial_x(u_t + 6 u u_x + u_{xxx}) + 3\sigma u_{yy} = 0, \sigma = \pm 1$$

   derive the integrable Davey–Stewartson equation by employing the slowly varying quasi-monochromatic wave expansion. Hint: see Ablowitz et al. (1990).

6.5     Suppose we are given a "modified KP" equation

$$\partial_x \left( u_t + \frac{3}{n+1}(u^{2(n+1)})_x + u_{xxx} \right) + 3\sigma u_{yy} = 0, \sigma = \pm 1$$

where $n \geq 1$, an integer. Following a similar analysis as in the previous exercise, derive a "modified" Davey–Stewartson equation by employing the slowly varying quasi-monochromatic wave expansion.

6.6     Suppose we are given the "damped NLS" equation

$$iu_t + u_{xx} + 2|u|^2 u = -i\gamma u$$

where $0 < \gamma \ll 1$. Use the integral relation involving mass (power) $\int |u|^2 dx$ to derive an approximate evolution equation for the slowly varying soliton amplitude $\eta$ where the unperturbed soliton is given by

$$u = \eta \text{sech}\left[\eta(x - x_0)\right] e^{i\eta^2 t + i\theta_0}.$$

Hint: see Ablowitz and Segur (1981). Various perturbation problems involving related NLS-type equations are also discussed in Chapter 10.

6.7     Discuss the stability associated with the special solution

$$u = ae^{i(kx - (k^2 - 2\sigma|a|^2)t)}$$

where $k, \sigma$ are constant, associated with the NLS equation

$$iu_t + u_{xx} + 2\sigma|u|^2 u = 0$$

on the infinite interval.

6.8     Given the two-dimensional NLS equation

$$iA_t + \Delta A + |A|^2 A = 0, \qquad x \in \mathbb{R}^2$$

and associated "Townes" mode $A = f(r)e^{i\lambda t}$, where $r^2 = x^2 + y^2$, show that the "virial" equation reduces to

$$\frac{d^2 V}{dt^2} = 0$$

where $V(t) = \int (x^2 + y^2)|A|^2\, dx$. Hint: see the appendix in Ablowitz et al. (2005).

# 7

## Nonlinear Schrödinger models in nonlinear optics

An important and rich area of application of nonlinear wave propagation is the field of nonlinear optics. Asymptotic methods play an important role in this field. For example, since the scales are so disparate in Maxwell's equations, long-distance transmission in fiber optic communications depends critically on asymptotic models. Hence the nonlinear Schrödinger (NLS) equation is central for understanding phenomena and detailed descriptions of the dynamics. In this chapter we will outline the derivation of the NLS equation for electromagnetic wave propagation in bulk optical media. We also briefly discuss how the NLS equation arises as a model of spin waves in magnetic media.

## 7.1 Maxwell equations

We begin by considering Maxwell's equations for electromagnetic waves with no source charges or currents, cf. Landau et al. (1984) and Jackson (1998)

$$\nabla \times H = \frac{\partial D}{\partial t} \tag{7.1a}$$

$$\nabla \times E = -\frac{\partial B}{\partial t} \tag{7.1b}$$

$$\nabla \cdot D = 0 \tag{7.1c}$$

$$\nabla \cdot B = 0 \tag{7.1d}$$

where $H$ is the magnetic field, $E$ is the electromagnetic field, $D$ is the electromagnetic displacement and $B$ is the magnetic induction.

We first consider non-magnetic media so there is no magnetization term in $B$. The magnetic induction $B$ and magnetic field $H$ are then related by

$$B = \mu_0 H. \tag{7.2}$$

169

The constant $\mu_0$ is the magnetic permeability of free space. We allow for induced polarization of the media giving rise to the following relation

$$D = \epsilon_0(E + P), \tag{7.3}$$

where $\epsilon_0$ is the electric permittivity of free space, a constant. When we apply an electric field $E$ to an ideal dielectric material a response of the material can occur and the material is said to become polarized. Typically in these materials the electrons are tightly bound to the nucleus and a displacement of these electrons occurs. The macroscopic effect (summing over all displacements) yields the induced polarization $P$.

Since we are working with non-magnetic media, we can reduce Maxwell's equations to those involving only the electromagnetic field $E$ and the polarization $P$. To that end, we take the curl of (7.1b) and use (7.2) and (7.1a) to find

$$\nabla \times (\nabla \times E) = -\frac{\partial}{\partial t}(\nabla \times B)$$

$$= -\mu_0 \frac{\partial}{\partial t}(\nabla \times H)$$

$$= -\mu_0 \frac{\partial^2}{\partial t^2}D.$$

The vector identity

$$\nabla \times (\nabla \times E) = \nabla(\nabla \cdot E) - \nabla^2 E, \tag{7.4}$$

is useful here. With this, along with relation (7.3), we find the following equations for the electromagnetic field

$$\nabla^2 E - \nabla(\nabla \cdot E) = \frac{1}{c^2}\frac{\partial^2}{\partial t^2}(E + P(E))$$

$$\nabla \cdot (E + P(E)) = 0, \tag{7.5}$$

where the constant $c^2 = \dfrac{1}{\mu_0 \epsilon_0}$ is the square of the speed of light in a vacuum. Notice that we have made explicit the connection between the polarization $P$ and the electromagnetic field $E$.

Before we investigate polarizable, non-magnetic media in detail, let us also write down the dual equation for magnetic media without polarization ($P = 0$) satisfying

$$B = \mu_0(H + M(H)) \tag{7.6}$$

where $M$ is called the magnetization. We see that the magnetization vector $M$ plays a similar role to that of the polarization $P$, i.e., we are assuming that $M = M(H)$. Namely, we assume that the magnetization is related to the magnetic field. The magnetization can be permanent for ferromagnetic materials (permanent magnets such as iron). Taking the curl of (7.1a) then for non-polarized media using $D = \epsilon_0 E$ gives

$$\nabla \times (\nabla \times H) = \frac{\partial}{\partial t}(\nabla \times D)$$

$$= \epsilon_0 \frac{\partial}{\partial t}(\nabla \times E)$$

$$= -\epsilon_0 \frac{\partial^2}{\partial t^2} B.$$

Using (7.4) and (7.6) gives

$$\nabla^2 H - \nabla(\nabla \cdot H) = \frac{1}{c^2} \frac{\partial^2}{\partial t^2}(H + M(H))$$

$$\nabla \cdot (H + M(H)) = 0.$$

Later, in Section 7.4 we will discuss one way $M$ can be coupled to $H$.

## 7.2 Polarization

In homogeneous, non-magnetic media, matter responds to intense electromagnetic fields in a nonlinear manner. In order to model this, we use a well-known relationship between the polarization vector $P$ and the electromagnetic field $E$ that is a good approximation to a wide class of physically relevant media:

$$P(E) = \int \chi^{(1)} * E + \int \chi^{(2)} * EE + \int \chi^{(3)} * EEE. \qquad (7.7)$$

The above equation involves tensors, so the operation $*$ is a special type of convolution that will be defined below.

For notational purposes, we will write the polarization vector as follows:

$$P = \begin{pmatrix} P_x \\ P_y \\ P_z \end{pmatrix} \equiv \begin{pmatrix} P_1 \\ P_2 \\ P_3 \end{pmatrix}.$$

Notice that we can break up (7.7) into a linear and nonlinear part

$$P = P_L + P_{NL}.$$

In glass, and hence fiber optics, the quadratic term $\chi^{(2)}$ is zero. This is cubically nonlinear and is a so-called "centro-symmetric material", cf. Agrawal (2001), Boyd (2003) and also Ablowitz et al. (1997, 2001a) and included references for more information. We write the linear and nonlinear components of the polarization vector, and explicitly define the convolution mentioned above, as follows:

$$P_{L,i} = (\chi^{(1)} * E)_i = \sum_{j=1}^{3} \int_{-\infty}^{\infty} \chi_{ij}^{(1)}(t - \tau) E_j(\tau) \, d\tau \qquad (7.8)$$

$$P_{NL,i} = \sum_{j,k,l} \int_{-\infty}^{\infty} \chi_{ijkl}^{(3)}(t - \tau_1, t - \tau_2, t - \tau_3) E_j(\tau_1) E_k(\tau_2) E_l(\tau_3) \, d\tau_1 d\tau_2 d\tau_3. \quad (7.9)$$

In (7.9), the sums are over the indices $\{1, 2, 3\}$ and the integration is over all of $\mathbb{R}^3$. The matrix $\chi^{(1)}$ is called the linear susceptibility and the tensor $\chi^{(3)}$ is the third-order susceptibility. If the material is "isotropic", $\chi_{ij}^{(1)} = 0$, $i \neq j$ (exhibiting properties with the same values when measured along axes in all directions), then the matrix $\chi^{(1)}$ is diagonal. We will also usually identify subscripts $i = 1, 2, 3$ as $i = x, y, z$. In cubically nonlinear or "Kerr" materials (such as glass), it turns out that the only important terms in the $\chi^{(3)}$ tensor correspond to $\chi_{xxxx}^{(3)}$, which is equal to $\chi_{yyyy}^{(3)}$ and $\chi_{zzzz}^{(3)}$.

From now on, we will suppress the superscript in $\chi$ when context makes the choice clear. For example $\chi_{xx}^{(1)} = \chi_{xx}$ and $\chi_{xxxx}^{(3)} = \chi_{xxxx}$ follow due to the number of entries in the subscript.

We can now pose the problem of determining the electromagnetic field $E$ in Kerr media as solving (7.5) subject to (7.8) and (7.9). Consider the asymptotic expansion

$$E = \varepsilon E^{(1)} + \varepsilon^2 E^{(2)} + \varepsilon^3 E^{(3)} + \cdots, \quad |\varepsilon| \ll 1. \qquad (7.10)$$

Then, to leading order, we assume that the electromagnetic field is initially polarized along the $x$-axis; it propagates along the $z$-axis, and for simplicity we assume no transverse $y$ variations. So

$$E^{(1)} = \begin{pmatrix} E_x^{(1)} \\ 0 \\ 0 \end{pmatrix}, \qquad E_x^{(1)} = A(X, Z, T) e^{i\theta} + \text{c.c.}, \qquad (7.11)$$

where $\theta = kz - \omega t$ and the amplitude $A$ is assumed to be slowly varying in the $x, z$, and $t$ directions:

$$X = \varepsilon x, \quad Z = \varepsilon z, \quad T = \varepsilon t.$$

In order to simplify the calculations, we assume $E$ and hence $A$ are independent of $y$ (i.e., $Y = \varepsilon y$). Derivatives are replaced as follows

$$\partial_x = \varepsilon \partial_X$$
$$\partial_t = -\omega \partial_\theta + \varepsilon \partial_T \qquad (7.12)$$
$$\partial_z = k\partial_\theta + \varepsilon \partial_Z.$$

Substituting in the asymptotic expansion for $E$ we note the important simplification that the assumption of slow variation has on the polarization $P$:

$$P_{L,x} = \sum_{j=1}^{3} \int_{-\infty}^{\infty} \chi_{xj}^{(1)}(t-\tau)E_j(\tau)\, d\tau$$

$$= \varepsilon \int_{-\infty}^{\infty} \chi_{xx}(t-\tau)\left(A(X,Z,T)e^{i(kz-\omega\tau)} + \text{c.c.}\right)d\tau + O(\varepsilon^2)$$

$$= \varepsilon \int_{-\infty}^{\infty} \chi_{xx}(t-\tau)e^{i\omega(t-\tau)}\left(A(X,Z,T)e^{i(kz-\omega t)} + \text{c.c.}\right)d\tau + O(\varepsilon^2).$$

Now make the substitution $t - \tau = u$ to get

$$P_{L,x} = \varepsilon \int_{-\infty}^{\infty} \chi_{xx}(u)e^{i\omega u}\left(A(X,Z,\varepsilon t - \varepsilon u)e^{i(kz-\omega t)} + \text{c.c.}\right)du + O(\varepsilon^2).$$

We expand the slowly varying amplitude $A$ around the point $\varepsilon t$

$$P_{L,x} = \varepsilon \int_{-\infty}^{\infty} du\, \chi_{xx}(u)e^{i\omega u} \times$$

$$\left[\left(1 - \varepsilon u\frac{\partial}{\partial T} + \frac{(\varepsilon u)^2}{2}\frac{\partial^2}{\partial T^2} + \cdots\right)A(X,Z,T)e^{i(kz-\omega t)} + \text{c.c.}\right] + O(\varepsilon^2).$$

Recall that the Fourier transform of $\chi_{xx}$, written as $\hat{\chi}_{xx}$, and its derivatives are

$$\hat{\chi}_{xx}(\omega) = \int_{-\infty}^{\infty} \chi_{xx}(u)e^{i\omega u}\, du$$

$$\hat{\chi}'_{xx}(\omega) = \int_{-\infty}^{\infty} iu\chi_{xx}(u)e^{i\omega u}\, du$$

$$\hat{\chi}''_{xx}(\omega) = \int_{-\infty}^{\infty} -u^2\chi_{xx}(u)e^{i\omega u}\, du.$$

Then we can write the linear part of the polarization as

$$P_{L,x} = \varepsilon\left(\hat{\chi}_{xx}(\omega) + \hat{\chi}'_{xx}(\omega)i\varepsilon\partial_T - \hat{\chi}''_{xx}(\omega)\frac{(\varepsilon\partial_T)^2}{2} + \cdots\right)(Ae^{i\theta} + \text{c.c.}) + O(\varepsilon^2)$$

$$= \varepsilon\hat{\chi}_{xx}(\omega + i\varepsilon\partial_T)(Ae^{i\theta} + \text{c.c.}) + O(\varepsilon^2).$$

The last line is in a convenient notation where the term $\hat{\chi}_{xx}(\omega + i\varepsilon\partial_T)$ is an operator that we can expand around $\omega$ to get the previous line.

In the nonlinear polarization equation (7.9), there will be interactions due to the leading-order mode $Ae^{i\theta}$, for example, the nonlinear term includes $E_x^3$, which leads to terms such as $A^3 e^{3i\theta}$. Then we denote the corresponding linear polarization term (containing interaction terms) as

$$P_{L,x}^{\text{interactions}} = \hat{\chi}_{xx}(\omega_m + i\varepsilon\partial_T)(B_m e^{im\theta} + \text{c.c.}),$$

where $\omega_m = m\omega$ and $B_m$ contains the nonlinear terms generated by the interaction (e.g. $B_3 = A^3$).

Similarly there are nonlinear polarization terms. For example, a typical term in the nonlinear polarization is

$$
\begin{aligned}
P_{\text{NL},x} = \varepsilon^3 \hat{\chi}_{xxxx}(\omega_m + i\varepsilon\partial_{T_1}, \omega_n + i\varepsilon\partial_{T_2}, \omega_l + i\varepsilon\partial_{T_3}) \\
\times B_m(T_1)B_n(T_2)B_l(T_3)\big|_{T_1=T_2=T_3=T}\, e^{i(m+n+l)\theta} + \text{c.c.}
\end{aligned}
\tag{7.13}
$$

All other terms are of smaller order $O(\varepsilon^4)$.

## 7.3 Derivation of the NLS equation

So far, we have derived expressions for the polarization with a cubic nonlinearity (third-order susceptibility). Now we will use (7.5) to derive the NLS equation. This will give the leading-order equation for the slowly varying amplitude of the $x$-component of the electromagnetic field $E_x$.

We begin the derivation of the NLS equation by showing that $E_z \ll E_x$. Recall (7.1c) with (7.3),

$$\nabla \cdot (E + P) = 0.$$

Since there is no $y$ dependence, we get

$$\partial_x(E_x + P_x) + \partial_z(E_z + P_z) = 0. \tag{7.14}$$

We assume

$$
\begin{aligned}
E_x &= \varepsilon A(X, Z, T)e^{i\theta} + \text{c.c.} + O(\varepsilon^2) \\
E_z &= \varepsilon A_z(X, Z, T)e^{i\theta} + \text{c.c.} + O(\varepsilon^2),
\end{aligned}
\tag{7.15}
$$

where as before c.c. denotes the complex conjugate of the preceding term. A further note on notation. Because the $x$-component of the electromagnetic field $E_x$ appears so often, we label its slowly varying amplitude with $A$, whereas for the $z$-component of the $E$-field, $E_z$, we label the slowly varying amplitude with $A_z$ (note: $A_z$ does not mean derivative with respect to $z$).

Substituting the derivatives (7.12) into (7.14) we find

$$(k\partial_\theta + \varepsilon\partial_Z)(E_z + P_z) = -\varepsilon\partial_X(E_x + P_x),$$

which, when we use the ansatz (7.15), yields

$$\varepsilon(k\partial_\theta + \varepsilon\partial_Z)\left(A_z e^{i\theta} + \text{c.c.} + \hat{\chi}_{zz}(\omega + i\varepsilon\partial_T)(A_z e^{i\theta} + \text{c.c.})\right)$$

$$= -\varepsilon\partial_X\left(\varepsilon A e^{i\theta} + \text{c.c.} + \hat{\chi}_{xx}(\omega + i\varepsilon\partial_T)(\varepsilon A e^{i\theta} + \text{c.c.})\right) + \cdots.$$

The leading-order equation is

$$\varepsilon ik A_z(1 + \hat{\chi}_{zz}(\omega)) = -\varepsilon^2\frac{\partial A}{\partial X}(1 + \hat{\chi}_{xx}(\omega)) \qquad \text{implying}$$

$$A_z = -\varepsilon\frac{1 + \hat{\chi}_{xx}(\omega)}{ik(1 + \hat{\chi}_{zz}(\omega))}\frac{\partial A}{\partial X}$$

$$= \frac{-\varepsilon}{ik}\frac{\partial A}{\partial X} = A_X, \quad \text{if} \quad \hat{\chi}_{xx} = \hat{\chi}_{zz}.$$

Assuming $\partial A/\partial X$ is $O(1)$, this implies $A_z = O(\varepsilon)$, which implies $E_z = O(\varepsilon^2)$. While we used the divergence equation in (7.5) with (7.3) to calculate this relationship, we could have proved the same result with the $z$-component of the dynamic equation in (7.5). We also claim that from the dynamic equation of motion in (7.5), if we had studied the $y$-component, then it would have followed that $E_y = O(\varepsilon^3)$. Also, $P_{\text{NL},z} = O(\varepsilon^4)$, which is obtained from the symmetry of the third-order susceptibility tensor $\chi^{(3)}$ since $\chi_{zxxx} = 0$, etc. (e.g., for glass).

Now we will use the dynamic equation in (7.5) to investigate the behavior of the $x$-component of the electromagnetic field via multiple scales. The dynamic equation (7.5) becomes

$$\nabla^2 E_x - \frac{\partial}{\partial x}\left(\frac{\partial E_x}{\partial x} + \frac{\partial E_z}{\partial z}\right) = \frac{1}{c^2}\frac{\partial^2}{\partial t^2}(E_x + P_x). \qquad (7.16)$$

The asymptotic expansions for $E_x$ and $E_z$ take the form

$$E_x = \varepsilon(A e^{i\theta} + \text{c.c.}) + \varepsilon^2 E_x^{(2)} + \varepsilon^3 E_x^{(3)} + \cdots$$
$$E_z = \varepsilon^2(A_z e^{i\theta} + \text{c.c.}) + \varepsilon^3 E_z^{(2)} + \cdots. \qquad (7.17)$$

The polarization in the $x$-direction depends on the linear and nonlinear polarization terms, $P_{L,x}$ and $P_{NL,x}$, but the nonlinear term comes in at $O(\varepsilon^3)$ due to the cubic nonlinearity. Explicitly, we have

$$P_x = \varepsilon(\hat{\chi}_{xx}(\omega) + \hat{\chi}'_{xx}(\omega)i\varepsilon\partial_T - \frac{\hat{\chi}''_{xx}(\omega)}{2}\varepsilon^2\partial_T^2 + \cdots)(Ae^{i\theta} + \text{c.c.}) + P_{NL,x}. \quad (7.18)$$

Expanding the derivatives and using (7.12) in (7.16) gives

$$\left((k\partial_\theta + \varepsilon\partial_Z)^2 + \varepsilon^2\partial_X^2\right)E_x - \varepsilon\partial_X\left(\varepsilon\partial_X E_x + (k\partial_\theta + \varepsilon\partial_Z)E_z\right)$$

$$= \frac{1}{c^2}(-\omega\partial_\theta + \varepsilon\partial_T)^2(E_x + P_{L,x} + P_{NL,x}). \quad (7.19)$$

The leading-order equation is

$$O(\varepsilon): \qquad \left(-k^2 + \left(\frac{\omega}{c}\right)^2\right)A = -\left(\frac{\omega}{c}\right)^2\hat{\chi}_{xx}(\omega)A.$$

Since we assume $A \neq 0$ the above equation determines the dispersion relation; namely solving for $k(\omega)$ we find the dispersion relation

$$k^2 = \left(\frac{\omega}{c}\right)^2(1 + \hat{\chi}_{xx}(\omega))$$

$$k(\omega) = \frac{\omega}{c}(1 + \hat{\chi}_{xx}(\omega))^{1/2} \qquad (7.20)$$

$$\equiv \frac{\omega}{c}n_0(\omega).$$

The term $n_0(\omega)$ is called the linear index of refraction; if it were constant then $c/n_0$ would be the "effective speed of light". The slowly varying amplitude $A(X, Z, T)$ is still free and will be determined by going to higher order and removing secular terms.

From (7.19) and the asymptotic expansions for $E_x$ and $P_{L,x}$ in (7.17) and (7.18) (the nonlinear polarization term, $P_{NL,x}$, does not come in until the next order), we find

$$O(\varepsilon^2): \qquad \left(k^2 - \left(\frac{\omega}{c}\right)^2\right)\partial_\theta^2 E_x^{(2)} = \left[-2ik\partial_Z A - \frac{2i\omega}{c^2}(1 + \hat{\chi}_{xx}(\omega))\partial_T A\right.$$

$$\left. -\left(\frac{\omega}{c}\right)^2 i\hat{\chi}'_{xx}(\omega)\partial_T A\right]e^{i\theta}.$$

To remove secularities, we must equate the terms in brackets to zero. This can be written as

$$2k\frac{\partial A}{\partial Z} + \left(\frac{2\omega}{c^2}(1 + \hat{\chi}_{xx}(\omega)) + \left(\frac{\omega}{c}\right)^2\hat{\chi}'_{xx}(\omega)\right)\frac{\partial A}{\partial T} = 0. \qquad (7.21)$$

Notice that if we differentiate the dispersion relation (7.20) with respect to $\omega$ we find

$$2kk' = \frac{2\omega}{c^2}(1 + \hat{\chi}_{xx}(\omega)) + \left(\frac{\omega}{c}\right)^2 \hat{\chi}'_{xx}(\omega).$$

This means that we can rewrite (7.21) in terms of the group velocity $v_g = 1/k'(\omega)$:

$$\frac{\partial A}{\partial Z} + \frac{1}{v_g}\frac{\partial A}{\partial T} = 0. \tag{7.22}$$

This is a first-order equation that can be solved by the method of characteristics giving, to this order, $A = A(T - z/v_g)$. To obtain a more accurate equation for the slowly varying amplitude $A$, we will now remove secular terms at the next order and obtain the nonlinear Schrödinger (NLS) equation.

As before, we perturb (7.22)

$$2ik\left(\frac{\partial A}{\partial Z} + \frac{1}{v_g}\frac{\partial A}{\partial T}\right) = \varepsilon f_1 + \varepsilon^2 f_2 + \cdots.$$

Using this and (7.19) to $O(\varepsilon^3)$ and then, as we have discussed in earlier chapters, choosing $f_1$ to remove secular terms, we find after some calculation

$$O(\varepsilon^3): \quad 2ik\left(\frac{\partial A}{\partial Z} + k'(\omega)\frac{\partial A}{\partial T}\right)$$

$$+\varepsilon\left[\partial_X^2 + \partial_Z^2 - \frac{1}{c^2}(1 + \hat{\chi}_{xx}(\omega) + 2\omega\hat{\chi}'_{xx}(\omega) + \frac{1}{2}\omega^2\hat{\chi}''_{xx}(\omega))\partial_T^2\right]A$$

$$+3\varepsilon\left(\frac{\omega}{c}\right)^2 \hat{\chi}_{xxxx}(\omega,\omega,-\omega)|A|^2 A = 0. \tag{7.23}$$

We now see in (7.23) a term due to the nonlinear polarization $P_{NL,x}$ i.e., the term with $\hat{\chi}_{xxxx}(\omega,\omega,-\omega)$. In fact, there are three choices for $m,n,p$ in the leading-order term for the nonlinear polarization [recall (7.13)] that give rise to the exponential $e^{i\theta}$ (the coefficient of $e^{i\theta}$ is the only term that induces secularity). Namely, two of these integers are $+1$ and the other one is $-1$. In the bulk media we are working in, the third-order susceptibility tensor is the same in all cases. This means

$$\hat{\chi}_{xxxx}(\omega,\omega,-\omega) + \hat{\chi}_{xxxx}(\omega,-\omega,\omega) + \hat{\chi}_{xxxx}(-\omega,\omega,\omega)$$

$$= 3\hat{\chi}_{xxxx}(\omega,\omega,-\omega)$$

$$\equiv 3\hat{\chi}_{xxxx}(\omega).$$

Let us investigate the coefficient of $A_{TT}$ due to the linear polarization. If we differentiate the dispersion relation (7.20) two times with respect to $\omega$ we get the following result

$$k'^2 + kk'' = \frac{1}{c^2}\left(1 + \hat{\chi}_{xx}(\omega) + 2\omega\hat{\chi}'_{xx}(\omega) + \tfrac{1}{2}\omega^2\hat{\chi}''_{xx}(\omega)\right).$$

Then we can conveniently rewrite the linear polarization term as

$$\varepsilon\left[\partial_X^2 + \partial_Z^2 - (k'^2 + kk'')\partial_T^2\right]A.$$

Now, make the following change of variable

$$\xi = T - k'(\omega)Z, \quad Z' = \varepsilon Z$$
$$\partial_Z = -k'\partial_\xi + \varepsilon\partial_{Z'}, \quad \partial_T = \partial_\xi.$$

Substituting this into the third-order equation (7.23), we find

$$2ik\frac{\partial A}{\partial Z'} + \frac{\partial^2 A}{\partial X^2} - kk''\frac{\partial^2 A}{\partial \xi^2} + 2kv|A|^2 A = 0. \tag{7.24}$$

Here we took

$$v = \frac{3(\omega/c)^2\hat{\chi}_{xxxx}(\omega)}{2k}. \tag{7.25}$$

It is useful to note that had we included variation in the $y$-direction, then the term $\partial^2 A/\partial Y^2$ would be added to (7.24) and we would have found

$$2ik\frac{\partial A}{\partial Z'} + \nabla^2 A - kk''\frac{\partial^2 A}{\partial \xi^2} + 2kv|A|^2 A = 0, \tag{7.26}$$

where

$$\nabla^2 A = \frac{\partial^2 A}{\partial X^2} + \frac{\partial^2 A}{\partial Y^2}.$$

If there is no variation in the $X$-direction then we get the $(1 + 1)$-dimensional (one space dimension and one time dimension) nonlinear Schrödinger (NLS) equation describing the slowly varying wave amplitude $A$ of an electromagnetic wave in cubically nonlinear bulk media

$$i\frac{\partial A}{\partial Z} + \left(\frac{-k''(\omega)}{2}\right)\frac{\partial^2 A}{\partial \xi^2} + v|A|^2 A = 0, \tag{7.27}$$

where we have removed the prime from $Z'$ for convenience. If $k'' < 0$ then we speak of anomalous dispersion. The NLS equation is then called "focusing" and gives rise to "bright" soliton solutions as discussed in Chapter 6 (note: $v > 0$ and $k > 0, \hat{\chi}_{xxxx} > 0$). If $k'' > 0$ then the system is said to have normal dispersion and the NLS equation is called "defocusing". This equation then admits "dark" soliton solutions (see also Chapter 6).

Next we remark about the coefficient of nonlinearity $v$ defined in (7.25). The previous derivation of the NLS equation was for electromagnetic waves in bulk optical media. In the field of optical communications, one is interested in waves propagating down narrow fibers. Perhaps surprisingly, the NLS arises again in exactly the same form as (7.27) with only a small change in the value of $v$:

$$v_{\text{eff}} = \frac{v}{A_{\text{eff}}} = \frac{3(\omega/c)^2 \hat{\chi}_{xxxx}(\omega)}{2k A_{\text{eff}}} \;;$$

thus $v$ only changes from its vacuum value by an additional factor in the denominator. Here, $A_{\text{eff}}$ corresponds to the effective cross-sectional area of the fiber; see Hasegawa and Kodama (1995) and Agrawal (2001) for more information.

We often write $v$ in a slightly different way,

$$v = \frac{\omega}{c} \tilde{n}_2(\omega),$$

so that the nonlinear index of refraction $\tilde{n}_2$ is

$$\tilde{n}_2(\omega) = \frac{3\omega}{2kc} \hat{\chi}_{xxxx}(\omega).$$

We can relate the linear and quadratic indices of refraction ($n_0(\omega)$ and $\tilde{n}_2(\omega)$) to the Stokes frequency shift as follows. We assume the wave amplitude has the following ansatz

$$A(Z, \xi) = \tilde{A}(Z) e^{-ikZ},$$

called a CW or continuous wave, i.e., this is the electromagnetic analog of a Stokes water wave. Substituting this into (7.27) gives

$$\begin{aligned} -i\frac{\partial \tilde{A}}{\partial Z} &= k\tilde{A} + v|\tilde{A}|^2 \tilde{A} \\ &= \frac{\omega}{c}(n_0 + \tilde{n}_2|\tilde{A}|^2)\tilde{A}. \end{aligned} \tag{7.28}$$

If we multiply (7.28) by $\tilde{A}^*$ then add that to the conjugate of (7.28) multiplied by $\tilde{A}$, we see that

$$\frac{\partial}{\partial Z}|\tilde{A}|^2 = 0 \quad \Rightarrow \quad |\tilde{A}(Z)|^2 = |\tilde{A}(0)|^2.$$

Now we can solve (7.28) directly to find

$$\tilde{A}(Z) = \tilde{A}(0) \exp\left(i\frac{\omega}{c}(n_0 + \tilde{n}_2|\tilde{A}(0)|^2)Z\right).$$

The additional "nonlinear" frequency shift is $\tilde{n}_2|\tilde{A}(0)|^2$. This term is often referred to as self-phase-modulation. Hence the wavenumber, $k$, is now a function of the initial amplitude and the frequency

$$k(\omega, |\tilde{A}(0)|) = \frac{\omega}{c}(n_0 + \tilde{n}_2|\tilde{A}(0)|^2).$$

In the literature, the index of refraction combining both the linear and quadratically nonlinear indices is often written in terms of the initial electric field $E$

$$n(\omega, E(0)) = n_0(\omega) + n_2(\omega)|E_x(0)|^2.$$

Since we assumed the form

$$E_x(0) = \varepsilon(A(0)e^{i\theta} + \text{c.c.}) = \varepsilon(\tilde{A}(0)e^{-i\omega t} + \text{c.c.}) = 2\varepsilon|\tilde{A}(0)|\cos(\omega t + \phi_0),$$

for the electric field, we see that

$$|\tilde{A}(0)| = \frac{|E_x(0)|}{2} \quad \Rightarrow \quad n_2 = \frac{\tilde{n}_2}{4}.$$

### Including quadratic nonlinear media; NLS with mean terms

We will briefly discuss the effects of including a non-zero quadratic nonlinear polarization. This implies that the second-order susceptibility tensor $\chi^{(2)}$ in the nonlinear polarization is non-zero.

Recall the definition of the nonlinear polarization for quadratic and cubic media is written schematically as,

$$P_{NL}(E) = \int \chi^{(2)} : EE + \int \chi^{(3)} \vdots EEE + \dots$$

where the symbols ":" and "$\vdots$" represent tensor notation. The $i$th component of the quadratic term in the nonlinear polarization is written

$$P_{NL,i}^{(2)} = \sum_{j,k} \int_{-\infty}^{\infty} \chi_{ijk}(t - \tau_1, t - \tau_2) E_j(\tau_1) E_k(\tau_2) \, d\tau_1 d\tau_2.$$

In the derivation of the nonlinear Schrödinger equation with cubic nonlinearity, the nonlinear terms entered the calculation at third order. The multi-scale procedure for $\chi^{(2)} \neq 0$ follows the same lines as before:

$$E = \varepsilon E^{(1)} + \varepsilon^2 E^{(2)} + \varepsilon^3 E^{(3)} + \cdots$$

$$E^{(1)} = \begin{pmatrix} E_x^{(1)} \\ 0 \\ 0 \end{pmatrix}$$

$$E_x^{(1)} = A(X, Y, Z, T)e^{i\theta} + \text{c.c.},$$

see (7.10) and (7.11); to be general, we allow the amplitude $A$ to depend on $X = \varepsilon x, Y = \varepsilon y, Z = \varepsilon z$ and $T = \varepsilon t$. The new terms of interest due to the quadratic nonlinearity are (Ablowitz et al., 1997, 2001a):

$$E_x^{(2)} = 4\left(\frac{\omega}{c}\right)^2 \frac{\hat{\chi}_{xxx}}{\Delta(\omega)}\left(A^2 e^{2i\theta} + \text{c.c.}\right) + \phi_x$$

$$E_y^{(2)} = \phi_y$$

$$E_z^{(2)} = i\frac{n_x^2}{kn_z^2}\frac{\partial A}{\partial X}e^{i\theta} + \text{c.c.} + \phi_z$$

$$n_x^2 = 1 + \hat{\chi}_{xx}(\omega), \quad n_z^2 = 1 + \hat{\chi}_{zz}(\omega), \quad k = \frac{\omega}{c}n_x(\omega)$$

$$\Delta(\omega) = (2k(\omega))^2 - (k(2\omega))^2 \neq 0.$$

We assume that $\chi_{yy} = \chi_{xx}$. Notice the inclusion of "mean" terms $\phi_x, \phi_y, \phi_z$. If $|\Delta(\omega)| \ll 1$ then second harmonic resonance generation occurs (Agrawal, 2002). We will assume $\Delta(\omega) \neq 0$ and $\Delta(\omega)$ is not small.

In our derivation of NLS for cubic media, the only term at third order that gave rise to secularity was $e^{i\theta}$. Now, at third order, we have two sources for secularity, $e^{i\theta}$ and the mean term $e^{i0}$ (see also the derivation of NLS from KdV in Chapter 6). Removing these secular terms in the usual way as before leads to the following equations:

$$O(\varepsilon^3): \quad e^{i\theta} \rightarrow 2ik\partial_Z A + \left(\partial_X^2 + \partial_Y^2 - kk''\partial_\xi^2\right)A$$

$$+ (M_1|A|^2 + M_0\phi_x)A = 0 \tag{7.29}$$

$$e^{i0} \rightarrow \left(\alpha_x\partial_X^2 + \partial_Y^2 + s_x\partial_\xi^2\right)\phi_x - \left(N_1\partial_\xi^2 - N_2\partial_X^2\right)|A|^2 = 0. \tag{7.30}$$

In this, the change of variable, $\xi = T - k'(\omega)Z$, was made and the following definitions used:

$$M_0 = 2\left(\frac{\omega}{c}\right)^2 \hat{\chi}_{xxx}(\omega, 0), \qquad M_1 = 3\left(\frac{\omega}{c}\right)^2 \hat{\chi}_{xxxx}(\omega, \omega, -\omega)$$

$$+ \frac{8(\omega/c)^4\hat{\chi}_{xxx}(2\omega, -\omega)\hat{\chi}_{xxx}(\omega, \omega)}{\Delta(\omega)},$$

$$N_1 = \frac{2}{c^2}\hat{\chi}_{xxx}(\omega, -\omega), \qquad N_2 = \frac{c^2 N_1}{1 + \hat{\chi}_{xx}(\omega)},$$

$$\alpha_x = \frac{1 + \hat{\chi}_{xx}(\omega)}{1 + \hat{\chi}_{zz}(\omega)}, \qquad s_x = k'(\omega)^2 - \frac{1 + \hat{\chi}_{xx}(\omega)}{c^2}.$$

There are several things to notice. First, if $\hat{\chi}_{xxx} = 0$ then $N_1 = N_2 = M_0 = 0$, $M_1 = 2(\omega/c)^2\hat{\chi}_{xxx}(\omega, -\omega)$. Thus (7.29) reduces to the nonlinear Schrödinger equation obtained earlier for cubic media. Since (7.29) of NLS-type but includes a mean term $\phi_x$ we call it NLS with mean (NLSM). We must solve these coupled equations (7.29) and (7.30) for $\phi_x$ and $A$. It turns out, when $\chi_{xx} = \chi_{yy}$, that the mean terms $\phi_y$, $\phi_z$ decouple from the $\phi_x$ equation and they can be solved in terms of $\phi_x$ and $A$ (Ablowitz et al., 1997, 2001a). These equations, (7.29) and (7.30), are a three-dimensional generalization of the Benney–Roskes (BR) type equations which we discussed in our study of multi-dimensional water waves in Chapter 6. Moreover, when $A$ is independent of time, $t$, i.e., independent of $\xi$, then (7.29) and (7.30) are $(2 + 1)$-dimensional and reduce to the BR form discussed in Section 6.8 and wave collapse is possible (see also Ablowitz et al., 2005).

## 7.4 Magnetic spin waves

In this section, we will briefly discuss a problem involving magnetic materials. As discussed in Section 7.1, in this case, we neglect the polarization, $\boldsymbol{P} = 0$, but we do include magnetization $\boldsymbol{M}(\boldsymbol{H})$.

The typical situation is schematically depicted in Figure 7.1 Kalinikos et al., 1997; Chen et al., 1994; Tsankov et al., 1994; Patton et al., 1999; Kalinikos et al., 2000; Wu et al., 2004, and references therein). Consider a slab of magnetic material with magnetization $\boldsymbol{M}_s$ of thickness $d$. Suppose an external magnetic field $\boldsymbol{G}_s$ is applied from above and below the slab. We wish to determine an equation for the magnetic field $\boldsymbol{H}$ inside the slab. The imposed magnetic field $\boldsymbol{G}_s$ induces a magnetization $\boldsymbol{m}$ inside the material. The total magnetization inside the slab is taken to be

$$\boldsymbol{M} = \boldsymbol{M}_s + \boldsymbol{m}.$$

We write the magnetic fields inside and outside the slab as

$$\boldsymbol{H} = \boldsymbol{H}_s + \boldsymbol{h}, \qquad \boldsymbol{G} = \boldsymbol{G}_s + \boldsymbol{g}.$$

A relation between the steady state values of the internal and external fields as well as the magnetization, $\boldsymbol{H}_s, \boldsymbol{G}_s$, and $\boldsymbol{M}_s$, will be determined by the boundary

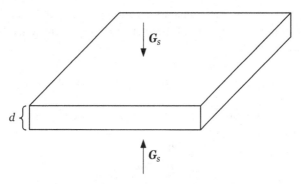

Figure 7.1  Magnetic slab.

conditions. Once a perturbation is introduced into the system, $h$, $g$, $m$ are non-trivial. The goal is to find the equations governing these quantities.

Recall from Section 7.1 the dynamical equations for the magnetic field $H$ and the magnetization $M$ with no source currents or charge:

$$\boldsymbol{\nabla} \times (\boldsymbol{\nabla} \times \boldsymbol{H}) = \nabla^2 \boldsymbol{H} - \boldsymbol{\nabla}(\boldsymbol{\nabla} \cdot \boldsymbol{H}) = \frac{1}{c^2}\frac{\partial^2}{\partial t^2}(\boldsymbol{H} + \boldsymbol{M}(\boldsymbol{H})) \qquad (7.31)$$

$$\boldsymbol{\nabla} \cdot (\boldsymbol{H} + \boldsymbol{M}(\boldsymbol{H})) = 0. \qquad (7.32)$$

We also need a relation between the magnetic field $H$ and the magnetization $M$. The approximation we will work with, assuming no damping, is the so-called "torque" equation

$$\frac{\partial \boldsymbol{M}}{\partial t} = -\gamma \boldsymbol{M} \times \boldsymbol{H}.$$

Since $1/c^2$ is very small given the scales we are interested in, we neglect the time derivative term in (7.31) and get the following quasistatic approximation for the magnetic field inside the slab:

$$\boldsymbol{\nabla} \times \boldsymbol{H} = 0$$
$$\boldsymbol{\nabla} \cdot (\boldsymbol{H} + \boldsymbol{M}) = 0 \qquad (7.33)$$
$$\frac{\partial \boldsymbol{M}}{\partial t} = -\gamma \boldsymbol{M} \times \boldsymbol{H}.$$

Outside the slab we have

$$\boldsymbol{\nabla} \times \boldsymbol{G} = 0$$
$$\boldsymbol{\nabla} \cdot \boldsymbol{G} = 0. \qquad (7.34)$$

The coordinate origin is taken to be in the center of the slab and the $z$-direction is vertical thus $\boldsymbol{G}_s = G_s \hat{\boldsymbol{z}}$. We also take $\boldsymbol{H}_s = H_s \hat{\boldsymbol{z}}$, $\boldsymbol{M}_s = M_s \hat{\boldsymbol{z}}$, with

$G_s$, $H_s$, $M_s$ all constant. The boundary conditions (since we have no sources) are derived from the continuity of the induction and magnetic fields that can be written as

$$H_x = G_x, \quad H_y = G_y, \quad H_s + M_s = G_s \quad \text{all at } z = \pm\frac{d}{2}.$$

It is convenient to introduce the scalar potentials $\psi$ and $\phi$ as

$$h = \nabla\psi,$$
$$g = \nabla\phi.$$

Thus (7.33) and (7.34) can be reduced to

$$\frac{\partial m}{\partial t} = -\gamma(M_s \times h + m \times H_s + m \times h)$$
$$\nabla^2\psi + \nabla \cdot m = 0 \quad \text{(inside slab)} \tag{7.35}$$
$$\nabla^2\phi = 0 \quad \text{(outside slab)}.$$

Suppose we assume a quasi-monochromatic wave expansion for the potentials:

$$\phi = \varepsilon(\phi^{(1)}e^{i\theta} + \text{c.c.}) + \varepsilon^2(\phi^{(0)} + \phi^{(2)}e^{2i\theta} + \text{c.c.}) + \cdots$$

$$\psi = \varepsilon(\psi^{(1)}e^{i\theta} + \text{c.c.}) + \varepsilon^2(\psi^{(0)} + \psi^{(2)}e^{2i\theta} + \text{c.c.}) + \cdots$$

$$m = \varepsilon \begin{pmatrix} m_1^{(1)} \\ m_2^{(1)} \\ 0 \end{pmatrix}(e^{i\theta} + \text{c.c.}) + \cdots.$$

where $\theta = kx - \omega t$ and $\phi^{(j)}$, $\psi^{(j)}$, $m^{(j)}$ are slowly varying functions of $x$, $t$ and are independent of $y$; i.e., $\phi^{(j)} = \phi^{(j)}(X, T)$, $X = \varepsilon x$, $T = \varepsilon t$, etc.

The first of (7.35) show that $m_1^{(1)}$ and $m_2^{(1)}$ satisfy

$$-i\omega m_1^{(1)} = -\omega_H m^{(2)} - i\omega m_1^{(2)} = -ik\omega_M\psi^{(1)} + \omega_H m_1^{(1)}$$

where $\omega_M = \gamma M_s$, $\omega_H = \gamma H_s$. Hence

$$m_1^{(1)} = \chi_1 ik\psi^{(1)}, \qquad m_2^{(1)} = -\chi_2 k\psi^{(1)},$$

where

$$\chi_1 = \frac{\omega_H\omega_M}{\omega_H^2 - \omega^2}, \qquad \chi_2 = \frac{\omega_M\omega}{\omega_H^2 - \omega^2}.$$

To satisfy the latter two equations in (7.35) we find

$$\phi^{(1)} = A(X, T)e^{-k\left(|z|-\frac{d}{2}\right)} \cos(k_1 d/2)$$
$$\psi^{(1)} = A(X, T)\cos(k_1 z),$$

with

$$k_1^2 = k^2(1 + \chi_1).$$

Using continuity of the vertical fields, $h_z + m_z = g_z$ or $\partial_z \psi = \partial_z \phi$ at $z = \pm d/2$, we find the dispersion relation

$$(1 + \chi_1)^{1/2} = \cot(k_1 d/2),$$

which implicitly determines $\omega = \omega(k)$. Finally we remark that at higher order in the perturbation expansion, from a detailed calculation, an NLS equation for $A = A(X, T)$ is determined. For thin films ($d \ll 1$), the equation is found to be (see Zvezdin and Popkov, 1983)

$$i(A_T + v_g A_X) + \varepsilon\left(\frac{\omega''(k)}{2}A_{XX} + n|A|^2 A\right) = 0,$$

where $n$ is a suitable function of $k$, $\gamma$, $H_s$, and $M_s$.

## Exercises

7.1 (a) Derive the NLS equation (7.23) filling in the steps in the derivation.

  (b) Obtain an NLS equation in electromagnetic "bulk" media when the slowly varying amplitude $A$ in ( 7.15) is independent of $T$; here the longitudinal direction plays the role of the the temporal variable.

  (c) Obtain the NLS equation in these (bulk) media when the slowly varying amplitude $A$ in (7.15) is a function of $X$, $Y$, $Z$, $T$. What happens when $A$ is independent of $T$?

7.2 (a) Derive the "NLSM system (7.29)–(7.30) by filling in the steps in the derivation.

  (b) Derive time-independent systems similar to those discussed by Ablowitz et al. (2005) (see also Chapter 6) by assuming the amplitude is steady; i.e., has no time dependence.

7.3 Assume quasi-monchromatic waves in both transverse dimensions and obtain a vector NLSM system in quadratic ($\chi^{(2)} \neq 0$) nonlinear optical media. Hint: see Ablowitz et al. (1997, 2001a).

7.4   Use (7.35), and the assumed form of solution in that subsection, to obtain the linear dispersion relation of a wave $e^{i(kx-\omega t)}$ propagating in a ferromagnetic strip of width $d$ (like that depicted in Figure 7.1) but now with the external magnetic field aligned along the:

(a)  $x$-axis;

(b)  $y$-axis.

# PART II

---

## INTEGRABILITY AND SOLITONS

# 8

## Solitons and integrable equations

In this chapter we begin by finding traveling wave solutions. We then discuss the so-called Miura transformation that is used to find an infinite number of conservation laws and the associated linear scattering problem, the time-independent Schrödinger scattering problem, for the Korteweg–de Vries (KdV) equation. Lax pairs and an expansion method that shows how compatible linear problems are related to other nonlinear evolution equations are described. An evolution operator, sometimes referred to as a recursion operator, is also included. This operator can be used to find general classes of integrable nonlinear evolution equations. We point out that there are a number of texts that discuss these matters (Ablowitz and Segur, 1981; Novikov et al., 1984; Calogero and Degasperis, 1982; Dodd et al., 1984; Faddeev and Takhtajan, 1987); we often follow closely the analytical development Ablowitz and Clarkson (1991).

## 8.1 Traveling wave solutions of the KdV equation

We begin with the Korteweg–de Vries equation in non-dimensional form

$$u_t + 6uu_x + u_{xxx} = 0. \tag{8.1}$$

For the dimensional form of the equation we refer the reader to Chapter 1. An important property of the KdV equation is the existence of traveling wave solutions, including solitary wave solutions. The solutions are written in terms of elementary or elliptic functions. Since a discussion of this topic appears in Chapter 1 we will sometimes call on those results here.

A traveling wave solution of a PDE in one space, one time dimension, such as the KdV equation, where $t \in \mathbb{R}$, $x \in \mathbb{R}$ are temporal and spatial variables and $u \in \mathbb{R}$ the dependent variable, has the form

$$u(x, t) = w(x - Ct) = w(z). \tag{8.2}$$

A solitary wave is a special traveling wave solution, which is bounded and has constant asymptotic states as $z \to \mp\infty$.

To obtain traveling wave solutions of the KdV equation (8.1), we seek a solution in the form (8.2) that yields a third-order ordinary differential equation for $w$

$$\frac{d^3 w}{dz^3} + 6w\frac{dw}{dz} - C\frac{dw}{dz} = 0.$$

Integrating this once gives

$$\frac{d^2 w}{dz^2} + 3w^2 - Cw = E_1, \tag{8.3}$$

with $E_1$ an arbitrary constant. Multiplying (8.3) by $dw/dz$ and integrating again yields

$$\frac{1}{2}\left(\frac{dw}{dz}\right)^2 = f(w) = -(w^3 - Cw^2/2 - E_1 w - E_2), \tag{8.4}$$

with $E_2$ another arbitrary constant. We are interested in obtaining real, bounded solutions for the KdV equation (8.1), so we require that $f(w) \geq 0$. This leads us to study the zeros of $f(w)$. There are two cases to consider: when $f(w)$ has only one real zero or when $f(w)$ has three real zeros.

If $f(w)$ has only one real zero, say $w = \alpha$, so that $f(w) = -(w - \alpha)^3$, we can integrate directly to find that

$$u(x, t) = w(x - Ct) = \alpha - \frac{2}{(x - 6\alpha t - x_0)^2}.$$

So there are no bounded solutions in this case.

On the other hand, if $f(w)$ has three real zeros, which we take without loss of generality to satisfy $\alpha \leq \beta \leq \gamma$, then

$$f(w) = -(w - \alpha)(w - \beta)(w - \gamma), \tag{8.5}$$

where $\gamma = 2(\alpha + \beta + \gamma)$, $E_1 = -2(\alpha\beta + \beta\gamma + \gamma\alpha)$, $E_2 = 2\alpha\beta\gamma$.

Suppose $\alpha$, $\beta$, $\gamma$ are distinct; then the solution of the KdV equation may be expressed in terms of the Jacobian elliptic function $\mathrm{cn}(z; m)$ (see also Chapter 1)

$$u(x, t) = w(x - Ct) = \beta + (\gamma - \beta)\,\mathrm{cn}^2\left\{\left[\tfrac{1}{2}(\gamma - \alpha)\right]^{1/2}[x - Ct - x_0]\,; m\right\}, \tag{8.6}$$

with

$$C = 2(\alpha + \beta + \gamma), \quad m = \frac{\gamma - \beta}{\gamma - \alpha}, \quad \text{where} \quad 0 < m < 1, \qquad (8.7)$$

$x_0$ is a constant and $m$ denotes the modulus of the Jacobian elliptic function. These solutions of the KdV equation were originally found by Korteweg and de Vries; in fact they called these solutions cnoidal waves. Since $cn(z; m)$ has period $[z] = 4K(m)$ where $K$ is the complete elliptic integral of the first kind, we have that the period of $u(x, t)$ is $[x] = L = 2K(m)/\left(\frac{1}{2}(\gamma - \alpha)\right)$.

Noting that $cn(z; m) = \cos(z) + O(m)$ for $m \ll 1$ $(\beta \to \gamma)$, it follows that

$$u(x, t) = w(x - Ct) \sim \tfrac{1}{2}(\beta + \gamma) + \frac{\gamma - \beta}{2} \cos\{2\kappa[x - Ct - x_0]\},$$

where $\kappa^2 = \frac{1}{2}(\gamma - \alpha)$, $C \approx 2(\alpha - 2\beta)$. Thus when $\gamma \to \beta$ we have the limiting case of a sinusoidal wave.

Next suppose in (8.6), (8.7) we take the limit $\beta \to \alpha$ (i.e., $m \to 1$), with $\beta \neq \gamma$; then since $cn(z; 1) = \operatorname{sech}(z)$ it reduces to the solitary wave solution

$$u(x, t) = w(x - Ct) = \beta + 2\kappa^2 \operatorname{sech}^2\left\{\kappa[x - (4\kappa^2 + 6\beta u_0)t + \delta_0]\right\} \qquad (8.8)$$

where $\kappa^2 = \frac{1}{2}(\gamma - \beta)$.

It then follows that the speed of the wave relative to the uniform state $u_0$ is related to the amplitude: bigger solitary waves move faster than smaller ones. Note, if $\beta = 0$ $(\alpha = 0)$, then the wave speed is twice the amplitude. It is also clear that the amplitude $2\kappa^2$ is independent of the uniform background $\beta$. Finally the width of the solitary wave is inversely proportional to $\kappa$; hence narrower solitary waves are taller and faster. We note that these solitary waves are consistent with the observations of J. Scott Russell mentioned in Chapter 1.

If $\alpha < \beta = \gamma$, then the solution of (8.6) is given by

$$u(x, t) = w(x - Ct),$$
$$= \beta - (\beta - \alpha)\sec^2\left\{\left[\tfrac{1}{2}(\beta - \alpha)\right]^{1/2}[x - 2(\alpha + 2\beta)t - x_0]\right\}$$

[this can be obtained by using (8.5) and directly integrating], where $x_0$ is a constant. This solution is unbounded unless $\alpha = \beta$ in which case $u = u_0 = \beta$.

## 8.2 Solitons and the KdV equation

As discussed in Chapter 1, the physical model that motivated the recent discoveries associated with the KdV equation (1.5) was a problem of a one-dimensional anharmonic lattice of equal masses coupled by nonlinear strings that remarkably was studied numerically by Fermi, Pasta and Ulam (FPU).

The FPU model consisted of identical masses, $m$, connected by nonlinear springs with the force law $F(\Delta) = -k(\Delta + \alpha\Delta^p)$, with $k > 0$, $\alpha$ being spring constants:

$$
\begin{aligned}
m\frac{d^2y_n}{dt^2} &= k[y_{n+1} - y_n + \alpha(y_{n+1} - y_n)^p] - k[y_n - y_{n-1} + \alpha(y_n - y_{n-1})^p] \\
&= k(y_{n+1} - 2y_n + y_{n-1}) + k\alpha[(y_{n+1} - y_n)^p - (y_n - y_{n-1})^p]
\end{aligned}
\tag{8.9}
$$

for $n = 1, 2, \ldots, N - 1$, with $y_0 = 0 = y_N$ and initial condition $y_n(0) = \sin(n\pi/N)$, $y_{n,t}(0) = 0$.

In 1965, Zabusky and Kruskal transformed the differential-difference equation (8.9) to a continuum model and studied the case $p = 2$. Calling the length between springs $h$, $t = \omega t$ (with $\omega = \sqrt{K/m}$), $x = x/h$ with $x = nh$ and expanding $y_{n\pm1}$ in a Taylor series, then (dropping the tildes) (8.9) reduces to

$$
y_{tt} = y_{xx} + \varepsilon y_x y_{xx} + \tfrac{1}{12}h^2 y_{xxxx} + O(\varepsilon h^2, h^4),
\tag{8.10}
$$

where $\varepsilon = 2\alpha h$. They found (see Chapter 1) an asymptotic reduction by looking for a solution of the form (unidirectional waves)

$$
y(x, t) \sim \phi(\xi, \tau), \qquad \xi = x - t, \qquad \tau = \tfrac{1}{2}\varepsilon t,
$$

whereupon (8.10), with $u = \phi_\xi$, and ignoring small terms, where $\delta^2 = h^2/12\varepsilon$, reduces to the KdV equation [with different coefficients from (8.1)]

$$
u_\tau + u u_\xi + \delta^2 u_{\xi\xi\xi} = 0.
\tag{8.11}
$$

It is also noted that when the force law is taken as $F(\Delta) = -k(\Delta + \alpha\Delta^p)$ with $p = 3$ then, after rescaling, one obtains the modified KdV (mKdV) equation (see also the next section)

$$
u_\tau + u^2 u_\xi + \delta^2 u_{\xi\xi\xi} = 0.
$$

We saw earlier that the KdV equation possesses a solitary wave solution; see (8.8). Although solitary wave solutions of the KdV equation had been well known for a long time, it was not until Zabusky and Kruskal, in 1965, did extensive numerical studies for the KdV equation (8.11) that the remarkable properties of the solitary waves associated with the KdV equation were discovered. Zabusky and Kruskal considered the initial value problem for the KdV equation (8.11) with ($\delta^2 \ll 1$) the initial condition

$$u(\xi, 0) = \cos(\pi\xi), \qquad 0 \le \xi \le 2,$$

where $u$, $u_\xi$, $u_{\xi\xi}$ are periodic on $[0, 2]$ for all $t$. As also described in Chapter 1, they found that after a short time the wave steepens and almost produces a shock, but the dispersive term $\delta^2 u_{xxx}$ then becomes significant and a balance between the nonlinear and dispersion terms ensues. Later the solution develops a train of (eight) well-defined waves, each like a solitary wave (i.e., sech$^2$ functions), with the faster (taller) waves catching up and overtaking the slower (smaller) waves.

From this they observed that when two solitary waves with different wave speeds are initially well separated, with the larger one to the left (the waves are traveling from left to right), then the faster, taller wave catches up and subsequently overlaps the slower, smaller one; the waves interact nonlinearly. After the interaction, they found that the waves separate, with the larger one in front and slower one behind, but both having regained the same amplitudes they had before the interactions, the only effect of the interaction being a phase shift; that is, the centers of the waves are at different positions than where they would have been had there been no interaction. Thus they deduced that these nonlinear solitary waves essentially interacted elastically. Making the analogy with particles, Zabusky and Kruskal called these special waves *solitons*.

Mathematically speaking, a *soliton* is a solitary wave that asymptotically preserves its shape and velocity upon nonlinear interaction with other solitary waves, or more generally, with another (arbitrary) localized disturbance.

In the physics literature the term soliton is often meant to be a solitary wave without the elastic property. When dealing with physical problems in order to be consistent with what is currently accepted, we frequently refer to solitons in the latter sense (i.e., solitary waves without the elastic property).

Kruskal and Zabusky's remarkable numerical discovery demanded an analytical explanation and a detailed mathematical study of the KdV equation. However the KdV equation is nonlinear and at that time no general method of solution for nonlinear equations existed.

## 8.3 The Miura transformation and conservation laws for the KdV equation

We have seen that the KdV equation arises in the description of physically interesting phenomena; however interest in this nonlinear evolution equation is also due to the fact that analytical developments have led researchers to consider the KdV equation to be an *integrable* or sometimes referred to as an *exactly solvable* equation. This terminology is a consequence of the fact that the initial value problem corresponding to appropriate data (e.g., rapidly decaying initial data) for the KdV equation can be "solved exactly", or more precisely, linearized by a method that employs *inverse scattering*. The inverse scattering method was discovered by Gardner, Greene, Kruskal and Miura (1967) as a means for solving the initial value problem for the KdV equation on the infinite line, for initial values that decay sufficiently rapidly at infinity. Subsequently this method has been significantly enhanced and extended by many researchers and is usually termed the inverse scattering transform (IST) (cf. Ablowitz and Segur, 1981; Ablowitz and Clarkson, 1991; Ablowitz et al., 2004b). We also note that Lax and Levermore (1983a,b,c) and Venakides (1985) have reinvestigated (8.11) with $\delta$ small; that leads to an understanding of the weak limit of KdV.

The KdV equation has several other remarkable properties beyond the existence of solitons (discussed earlier), including the possession of an infinite number of polynomial conservation laws. Discovering this was important in the development of the general method of solution for the KdV equation.

Associated with a partial differential equation denoted by

$$F[x, t; u(x, t)] = 0, \tag{8.12}$$

where $t \in \mathbb{R}$, $x \in \mathbb{R}$ are temporal and spatial variables and $u(x, t) \in \mathbb{R}$ the dependent variable (in the argument of $F$ additional partial derivatives of $u$ are understood) may be a *conservation law*; that is, an equation of the form

$$\partial_t T_i + \partial_x X_i = 0, \tag{8.13}$$

that is satisfied for all solutions of (8.12). Here $T_i(x, t; u)$, called the *conserved density*, and $X_i(x, t; u)$, the *associated flux*, are, in general, functions of $x$, $t$, $u$ and the partial derivatives of $u$ – see also Chapter 2.

If additionally, $u \to 0$ as $|x| \to \infty$ sufficiently rapidly, then by integrating (8.13) we get

$$\frac{d}{dt} \int_{-\infty}^{\infty} T_i(x, t; u) \, dx = 0$$

in which case

$$\int_{-\infty}^{\infty} T_i(x, t; u) \, dx = c_i,$$

where $c_i$, constant, is called the *conserved quantity*.

For the KdV equation (1.5), the first three conservation laws are

$$(u)_t + (3u^2 + u_{xx})_x = 0,$$

$$(u^2)_t + \left(4u^3 + 2uu_{xx} - u_x^2\right)_x = 0,$$

$$\left(u^3 - \tfrac{1}{2}u_x^2\right)_t + \left(\tfrac{9}{2}u^4 + 3u^2 u_{xx} - 6uu_x^2 - u_x u_{xxx} + \tfrac{1}{2}u_{xx}^2\right)_x = 0.$$

The first two of these correspond to conservation of mass (or sometimes momentum) and energy, respectively. The third is related to the Hamiltonian (Whitham, 1965). The fourth and fifth conservation laws for the KdV equation were found by Kruskal and Zabusky. Subsequently some additional ones were discovered by Miura and it was conjectured that there was an infinite number.

After studying these conservation laws of the KdV equation and those associated with another equation, the so-called modified KdV (mKdV) equation that we write as

$$m_t - 6m^2 m_x + m_{xxx} = 0, \tag{8.14}$$

Miura (1968) discovered that with a solution $m$ of (8.14), then $u$ given by

$$u = -m^2 - m_x, \tag{8.15}$$

is a solution of the KdV equation (8.1). Equation (8.15) is known as the *Miura transformation*. Direct substitution of (8.15) into (1.5) yields

$$
\begin{aligned}
u_t + 6uu_x + u_{xxx} &= -\left(m^2 + m_x\right)_t + 6\left(m^2 + m_x\right)\left(m^2 + m_x\right)_x - \left(m^2 + m_x\right)_{xxx} \\
&= -(2mm_t + m_{xt}) + 6\left(m^2 + m_x\right)(2mm_x + m_{xx}) \\
&\quad - (2mm_{xxx} + 6m_x m_{xx} + m_{xxxx}) \\
&= -\left(m_{xt} - 6m^2 m_{xx} - 12mm_x^2 + v_{xxxx}\right) \\
&\quad - (2mm_t - 12m^3 m_x + 2mm_{xxx}) \\
&= -(2m + \partial_x)(m_t - 6m^2 m_x + m_{xxx}) \\
&= -(2m + \partial_x)\left(m_t - 6m^2 m_x + m_{xxx}\right).
\end{aligned}
$$

Note that every solution of the mKdV equation (8.14) leads, via Miura's transformation (8.15), to a solution of the KdV equation, but the converse is not true; i.e., not every solution of the KdV equation can be obtained from a solution of the mKdV equation – see, for example, Ablowitz, Kruskal and Segur (1979). Miura's transformation leads to many other important results related to the KdV equation. Initially it formed the basis of a proof that the KdV and mKdV equations have an infinite number of conserved densities (Miura et al., 1968), which is outlined below. Importantly, Miura's transformation (8.15) was also critical in the development of the *inverse scattering method* or *inverse scattering transform* (IST) for solving the initial value problem for the KdV equation (8.1) – details are in the next chapter.

To show that the KdV equation admits an infinite number of conservation laws, define a variable $w$ by the relation

$$u = w - \varepsilon w_x - \varepsilon^2 w^2, \tag{8.16}$$

which is a generalization of Miura's transformation (8.15).[1] Then the equivalent relation is

$$u_t + 6uu_x + u_{xxx} = \left(1 - \varepsilon\partial_x - 2\varepsilon^2 w\right)\left(w_t + 6(w - \varepsilon^2 w^2)w_x + w_{xxx}\right).$$

Therefore $u$, as defined by (8.16), is a solution of the KdV equation provided that $w$ is a solution of

$$w_t + 6(w - \varepsilon^2 w^2)w_x + w_{xxx} = w_t + \partial_x(3w^2 - 2w^3 + w_{xx}) = 0. \tag{8.17}$$

Since the KdV equation does not contain $\varepsilon$, then its solution $u$ depends only upon $x$ and $t$. However $w$, a solution of (8.17), depends on $x$, $t$ and $\varepsilon$. Since $\varepsilon$ is arbitrary we can seek a formal power series solution of (8.16), in the form

$$w(x, t; \varepsilon) = \sum_{n=0}^{\infty} w_n(x, t)\varepsilon^n. \tag{8.18}$$

The equation (8.17) is in conservation form, thus

$$\int_{-\infty}^{\infty} w(x, t; \varepsilon)\, dx = \text{constant},$$

and so

$$\int_{-\infty}^{\infty} w_n(x, t)\, dx = \text{constant},$$

for each $n = 0, 1, 2, \dots$ . Substituting (8.18) into (8.16), equating coefficients of powers of $\varepsilon$ and solving recursively gives

$$
\begin{aligned}
w_0 &= u, \\
w_1 &= w_{0,x} = u_x, \\
w_2 &= w_{1,x} + w_0^2 = u_{xx} + u^2, \\
w_3 &= w_{2,x} + 2w_0 w_1 = u_{xxx} + 4uu_x = (u_x x + 2u^2)_x, \\
w_4 &= w_{3,x} + 2w_0 w_2 + w_1^2 = u_{xxxx} + 6uu_{xx} + 5u_x^2 + 2u^3,
\end{aligned}
\tag{8.19}
$$

[1] Taking $m = \varepsilon w - 1/\varepsilon$ and performing a Galilean transformation.

etc. Continuing to all powers of $\varepsilon$ gives an infinite number of conserved densities. The corresponding conservation laws may be found by substituting (8.18)–(8.19) into (8.17) and equating coefficients of powers of $\varepsilon$. We further note that odd powers of $\varepsilon$ are trivial since they are exact derivatives. So the even powers give the non-trivial conservation laws for the KdV equation.

## 8.4 Time-independent Schrödinger equation and a compatible linear system

The Miura transformation (8.15) may be viewed as a Riccati equation for $m$ in terms of $u$. It is well known that it may be linearized by the transformation $m = v_x/v$, which yields

$$u + m^2 + m_x = u + \frac{v_x^2}{v^2} + \left( \frac{v_{xx}}{v} - \frac{v_x^2}{v^2} \right) = 0,$$

and so $u = -v_{xx}/v$ or

$$v_{xx} + uv = 0. \tag{8.20}$$

This is reminiscent of the Cole–Hopf transformation (see Section 3.6). However, (8.20) does not yield an explicit linearization for the KdV equation. Much more must be done. Since the KdV equation is Galilean invariant – that is, it is invariant under the transformation

$$(x, t, u(x, t)) \rightarrow (x - 6\lambda t, t, u(x, t) + \lambda),$$

where $\lambda$ is some constant – then it is natural to consider

$$\mathcal{L}v := v_{xx} + u(x, t)v = \lambda v. \tag{8.21}$$

This equation is the time-independent Schrödinger equation that has been extensively studied by mathematicians and physicists; we note that here $t$ plays the role of a parameter and $u(x, t)$ the potential.

The associated time dependence of the eigenfunctions of (8.21) is given by

$$v_t = (\gamma + u_x)v - (4\lambda + 2u)v_x, \tag{8.22}$$

where $\gamma$ is an arbitrary constant. This form is different from the original one found by Gardner, Greene, Kruskal and Miura (1967); see also Gardner et al. (1974): it is simpler and leads to the same results. Then if $\lambda = \lambda(t)$ from (8.21) and (8.22) we obtain

$$v_{txx} = [(\gamma + u_x)(\lambda - u) + u_{xxx} + 6uu_x]v - (4\lambda + 2u)(\lambda - u)v_x,$$
$$v_{xxt} = [(\lambda - u)(\gamma + u_x) - u_t + \lambda_t]v - (\lambda - u)(4\lambda + 2u)v_x.$$

Therefore (8.21) and (8.22) are compatible, i.e., $v_{xxt} = v_{txx}$, if $\lambda_t = 0$ and $u$ satisfies the KdV equation (1.5). Similarly, if (1.5) is satisfied, then necessarily the eigenvalues must be time independent (i.e., $\lambda_t = 0$).

It turns out that the eigenvalues and the behavior of the eigenfunctions of the scattering problem (8.21) as $|x| \to \infty$, determine the *scattering data*, $S(\lambda, t)$, which in turn depends upon the potential $u(x, t)$. The *direct scattering problem* is to map the potential into the scattering data. The time evolution equation (8.22) takes initial scattering data $S(\lambda, t = 0)$ to data at any time $t$, i.e., to $S(\lambda, t)$. The *inverse scattering problem* is to reconstruct the potential from the scattering data. The details of the solution method of the KdV equation (8.1) are described in the next chapter.

## 8.5 Lax pairs

Recall that the operators associated with the KdV equation are

$$\mathcal{L}v = v_{xx} + u(x, t)v = \lambda v,$$
$$v_t = \mathcal{M}v = (u_x + \gamma)v - (2u + 4\lambda)v_x,$$

where $\gamma$ is an arbitrary constant parameter and $\lambda$ is a spectral parameter that is constant ($\lambda_t = 0$). We found that these equations are compatible (i.e., $v_{xxt} = v_{txx}$) provided that $u$ satisfies the KdV equation (8.1).

Lax (1968) put the inverse scattering method for solving the KdV equation into a more general framework that subsequently paved the way to generalizations of the technique as a method for solving other partial differential equations. Consider two operators $\mathcal{L}$ and $\mathcal{M}$, where $\mathcal{L}$ is the operator of the spectral problem and $\mathcal{M}$ is the operator governing the associated time evolution of the eigenfunctions

$$\mathcal{L}v = \lambda v, \tag{8.23}$$
$$v_t = \mathcal{M}v. \tag{8.24}$$

Now take $\partial/\partial t$ of (8.23), giving

$$\mathcal{L}_t v + \mathcal{L}v_t = \lambda_t v + \lambda v_t;$$

hence using (8.24)

$$\mathcal{L}_t v + \mathcal{L}\mathcal{M}v = \lambda_t v + \lambda \mathcal{M}v,$$
$$= \lambda_t v + \mathcal{M}\lambda v,$$
$$= \lambda_t v + \mathcal{M}\mathcal{L}v.$$

Therefore we obtain

$$[\mathcal{L}_t + (\mathcal{L}\mathcal{M} - \mathcal{M}\mathcal{L})]\, v = \lambda_t v,$$

and hence in order to solve for non-trivial eigenfunctions $v(x, t)$

$$\mathcal{L}_t + [\mathcal{L}, \mathcal{M}] = 0, \tag{8.25}$$

where

$$[\mathcal{L}, \mathcal{M}] := \mathcal{L}\mathcal{M} - \mathcal{M}\mathcal{L},$$

if and only if $\lambda_t = 0$. Equation (8.25) is sometime called *Lax's equation* and $\mathcal{L}$, $\mathcal{M}$ are called the Lax pair. Furthermore, (8.25) contains a nonlinear evolution equation for suitably chosen $\mathcal{L}$ and $\mathcal{M}$. For example, if we take [suitably replacing $\lambda$ by $\partial_x^2 + u$ in (8.24)]

$$\mathcal{L} = \frac{\partial^2}{\partial x^2} + u, \tag{8.26}$$

$$\mathcal{M} = \gamma - 3u_x - 6u\frac{\partial}{\partial x} - 4\frac{\partial^3}{\partial x^3}, \tag{8.27}$$

then $\mathcal{L}$ and $\mathcal{M}$ satisfy (8.25) and $u$ satisfies the KdV equation (8.1). Therefore, the KdV equation may be thought of as the compatibility condition of the two linear operators given by (8.26), (8.27). As we will see below, there is a general class of equations that are associated with the Schrödinger operator (8.26).

## 8.6 Linear scattering problems and associated nonlinear evolution equations

Following the development of the method of inverse scattering for the KdV equation by Gardner, Greene, Kruskal and Miura (1967), it was then of considerable interest to determine whether the method would be applicable to other physically important nonlinear evolution equations. The method was thought to possibly only apply to one physically important equation; perhaps it was analogous to the Cole–Hopf transformation (Cole, 1951; Hopf, 1950) that linearizes Burgers' equation (see Chapter 3)

$$u_t + 2uu_x - u_{xx} = 0.$$

Namely, if we make the transformation $u = -\phi_x/\phi$, then $\phi(x, t)$ satisfies the linear heat equation $\phi_t - \phi_{xx} = 0$. [We remark that Forsyth (1906) first pointed out the relationship between Burgers' equation and the linear heat equation]. But Burgers' equation is the only known physically significant equation linearizable by the above transformation.

However, Zakharov and Shabat (1972) proved that the method was indeed more general. They extended Lax's method and related the nonlinear Schrödinger equation

$$iu_t + u_{xx} + \kappa u^2 u^* = 0, \qquad (8.28)$$

where $*$ denotes the complex conjugate and $\kappa$ is a constant, to a certain linear scattering problem or Lax pairs where now $\mathcal{L}$ and $\mathcal{M}$ are $2\times2$ matrix operators. Using these operators, Zakharov and Shabat were able to solve (8.28), given initial data $u(x, 0) = f(x)$, assuming that $f(x)$ decays sufficiently rapidly as $|x| \to \infty$. Shortly thereafter, Wadati (1974) found the method of solution for the modified KdV equation

$$u_t - 6u^2 u_x + u_{xxx} = 0, \qquad (8.29)$$

and Ablowitz, Kaup, Newell and Segur (1973a), motivated by some important observations by Kruskal, solved the sine–Gordon equation:

$$u_{xt} = \sin u. \qquad (8.30)$$

Ablowitz, Kaup, Newell and Segur (1973b, 1974) then developed a general procedure, which showed that the initial value problem for a remarkably large class of physically interesting nonlinear evolution equations could be solved by this method. There is also an analogy between the Fourier transform method for solving the initial value problem for linear evolution equations and the inverse scattering method. This analogy motivated the term: *inverse scattering transform* (IST) (Ablowitz et al., 1974) for the new method.

### 8.6.1 Compatible equations

Next we outline a convenient method for finding "integrable" equations. Consider two linear equations

$$\mathbf{v}_x = \mathbf{X}\mathbf{v}, \qquad \mathbf{v}_t = \mathbf{T}\mathbf{v}, \qquad (8.31)$$

where $\mathbf{v}$ is an $n$-dimensional vector and $\mathbf{X}$ and $\mathbf{T}$ are $n\times n$ matrices. If we require that (8.31) are compatible – that is, requiring that $\mathbf{v}_{xt} = \mathbf{v}_{tx}$ – then $\mathbf{X}$ and $\mathbf{T}$ must satisfy

$$\mathbf{X}_t - \mathbf{T}_x + [\mathbf{X}, \mathbf{T}] = 0. \qquad (8.32)$$

Equation (8.32) and Lax's equation (8.26), (8.27) are similar; equation (8.31) is somewhat more general as it allows more general eigenvalue dependence other than $\mathcal{L}v = \lambda v$.

The method is sometimes referred to as the AKNS method (Ablowitz et al., 1974). As an example, consider the $2 \times 2$ scattering problem [this is a generalization of the Lax pair studied by Zakharov and Shabat (1972)] given by (with the eigenvalue denoted as $k$)

$$
\begin{aligned}
v_{1,x} &= -ikv_1 + q(x,t)v_2, \\
v_{2,x} &= ikv_2 + r(x,t)v_1,
\end{aligned} \tag{8.33}
$$

with the linear time dependence given by

$$
\begin{aligned}
v_{1,t} &= Av_1 + Bv_2, \\
v_{2,t} &= Cv_1 + Dv_2,
\end{aligned} \tag{8.34}
$$

where $A$, $B$, $C$ and $D$ are scalar functions of $q(x,t)$, $r(x,t)$, and their derivatives, and $k$. In terms of $\mathbf{X}$ and $\mathbf{T}$, we specify that

$$
\mathbf{X} = \begin{pmatrix} -ik & q \\ r & ik \end{pmatrix}, \qquad \mathbf{T} = \begin{pmatrix} A & B \\ C & D \end{pmatrix}.
$$

Note that if there were any $x$-derivatives on the right-hand side of (8.34) then they could be eliminated by use of (8.33). Furthermore, when $r = -1$, then (8.33) reduces to the Schrödinger scattering problem

$$
v_{2,xx} + (k^2 + q)v_2 = 0. \tag{8.35}
$$

It is interesting to note that physically interesting nonlinear evolution equations arise from this procedure when either $r = -1$ or $r = q^*$ (or $r = q$ if $q$ is real).

This method provides an elementary technique that allows us to find nonlinear evolution equations. The compatibility of (8.33)–(8.34) – that is, requiring that $v_{j,xt} = v_{j,tx}$, for $j = 1, 2$ – and the assumption that the eigenvalue $k$ is time-independent – that is, $dk/dt = 0$ – imposes a set of conditions that $A$, $B$, $C$ and $D$ must satisfy. Sometimes the nonlinear evolution equations obtained this way are referred to as *isospectral* flows. Namely,

$$
\begin{aligned}
v_{1,xt} &= -ikv_{1,t} + q_t v_2 + q v_{2,t}, \\
&= -ik(Av_1 + Bv_2) + q_t v_2 + q(Cv_1 + Dv_2), \\
v_{1,tx} &= A_x v_1 + A v_{1,x} + B_x v_2 + B v_{2,x}, \\
&= A_x v_1 + A(-ikv_1 + qv_2) + B_x v_2 + B(ikv_2 + rv_1).
\end{aligned}
$$

Hence by equating the coefficients of $v_1$ and $v_2$, we obtain

$$
\begin{aligned}
A_x &= qC - rB, \\
B_x + 2ikB &= q_t - (A - D)q,
\end{aligned} \tag{8.36}
$$

respectively. Similarly

$$v_{2,xt} = ikv_{2,t} + r_t v_1 + r v_{1,t},$$
$$= ik(Cv_1 + Dv_2) + r_t v_1 + a(A_1 + Bv_2),$$
$$v_{2,tx} = C_x v_1 + C v_{1,x} + D_x v_2 + D v_{2,x},$$
$$= C_x v_1 + C(-ikv_1 + qv_2) + D_x v_2 + D(ikv_2 + rv_1),$$

and equating the coefficients of $v_1$ and $v_2$ we obtain

$$C_x - 2ikC = r_t + (A - D)r,$$
$$(-D)_x = qC - rB. \tag{8.37}$$

Therefore, from (8.36), without loss of generality, we may assume $D = -A$, and hence it is seen that $A$, $B$ and $C$ necessarily satisfy the compatibility conditions

$$A_x = qC - rB,$$
$$B_x + 2ikB = q_t - 2Aq, \tag{8.38}$$
$$C_x - 2ikC = r_t + 2Ar.$$

We next solve (8.38) for $A$, $B$ and $C$, thus ensuring that (8.33) and (8.34) are compatible. In general, this can only be done if another condition (on $r$ and $q$) is satisfied, this condition being the evolution equation. Since $k$, the eigenvalue, is a free parameter, we may find solvable evolution equations by seeking finite power series expansions for $A$, $B$ and $C$:

$$A = \sum_{j=0}^{n} A_j k^j, \qquad B = \sum_{j=0}^{n} B_j k^j, \qquad C = \sum_{j=0}^{n} C_j k^j. \tag{8.39}$$

Substituting (8.39) into (8.38) and equating coefficients of powers of $k$, we obtain $3n + 5$ equations. There are $3n + 3$ unknowns, $A_j$, $B_j$, $C_j$, $j = 0, 1, \ldots, n$, and so we also obtain two nonlinear evolution equations for $r$ and $q$. Now let us consider some examples.

**Example 8.1** $n = 2$. Suppose that $A$, $B$ and $C$ are quadratic polynomials, that is

$$A = A_2 k^2 + A_1 k + A_0,$$
$$B = B_2 k^2 + B_1 k + B_0, \tag{8.40}$$
$$C = C_2 k^2 + C_1 k + C_0.$$

Substitute (8.40) into (8.38) and equate powers of $k$. The coefficients of $k^3$ immediately give $B_2 = C_2 = 0$. At order $k^2$, we obtain $A_2 = a$, a constant (we could have allowed $a$ to be a function of time, but for simplicity we do not do so), $B_1 = iaq$, $C_1 = iar$. At order $k^1$, we obtain $A_1 = b$, constant, and for simplicity we set $b = 0$ (if $b \neq 0$ then a more general evolution equation is obtained); then $B_0 = -\frac{1}{2}aq_x$ and $C_0 = \frac{1}{2}ar_x$. Finally, at order $k^0$, we obtain $A_0 = \frac{1}{2}arq + c$, with $c$ a constant (again for simplicity we set $c = 0$). Therefore we obtain the following evolution equations

$$-\frac{1}{2}aq_{xx} = q_t - aq^2r,$$
$$\frac{1}{2}ar_{xx} = r_t + aqr^2. \tag{8.41}$$

If in (8.41) we set $r = \mp q^*$ and $a = 2i$, then we obtain the nonlinear Schrödinger equation

$$iq_t = q_{xx} \pm 2q^2q^*. \tag{8.42}$$

for both focusing (+) and defocusing (−) cases. In summary, setting $r = \mp q^*$, we find that

$$A = 2ik^2 \mp iqq^*,$$
$$B = -2qk - iq_x, \tag{8.43}$$
$$C = \pm 2q^*k \mp iq_x^*$$

satisfy (8.38) provided that $q(x,t)$ satisfies the nonlinear Schrödinger equation (8.42).

**Example 8.2** $n = 3$. If we substitute the following third-order polynomials in $k$

$$A = a_3k^3 + a_2k^2 + \frac{1}{2}(a_3qr + a_1)k + \frac{1}{2}a_2qr - \frac{1}{4}ia_3(qr_x - rq_x) + a_0,$$
$$B = ia_3qk^2 + \left(ia_2q - \frac{1}{2}a_3q_x\right)k + \left[ia_1q - \frac{1}{2}a_2q_x + \frac{1}{4}ia_3(2q^2r - q_{xx})\right],$$
$$C = ia_3rk^2 + \left(ia_2r + \frac{1}{2}a_3r_x\right)k + \left[ia_1r + \frac{1}{2}a_2r_x + \frac{1}{4}ia_3(2r^2q - r_{xx})\right],$$

into (8.38), with $a_3$, $a_2$, $a_1$ and $a_0$ constants, then we find that $q(x,t)$ and $r(x,t)$ satisfy the evolution equations

$$q_t + \frac{1}{4}ia_3(q_{xxx} - 6qrq_x) + \frac{1}{2}a_2(q_{xx} - 2q^2r) - ia_1q_x - 2a_0q = 0,$$
$$r_t + \frac{1}{4}ia_3(r_{xxx} - 6qrr_x) - \frac{1}{2}a_2(r_{xx} - 2qr^2) - ia_1r_x + 2a_0r = 0. \tag{8.44}$$

For special choices of the constants $a_3$, $a_2$, $a_1$ and $a_0$ in (8.44) we find physically interesting evolution equations. If $a_0 = a_1 = a_2 = 0$, $a_3 = -4i$ and $r = -1$, then we obtain the KdV equation

$$q_t + 6qq_x + q_{xxx} = 0.$$

If $a_0 = a_1 = a_2 = 0$, $a_3 = -4i$ and $r = q$, then we obtain the mKdV equation

$$q_t - 6q^2q_x + q_{xxx} = 0.$$

[Note that if $a_0 = a_1 = a_3 = 0$, $a_2 = -2i$ and $r = -q^*$, then we obtain the nonlinear Schrödinger equation (8.42).]

We can also consider expansions of $A$, $B$ and $C$ in inverse powers of $k$.

**Example 8.3** $n = -1$. Suppose that

$$A = \frac{a(x,t)}{k}, \qquad B = \frac{b(x,t)}{k}, \qquad C = \frac{c(x,t)}{k};$$

then the compatibility conditions (8.38) are satisfied if

$$a_x = \tfrac{i}{2}(qr)_t, \qquad q_{xt} = -4iaq, \qquad r_{xt} = -4iar.$$

Special cases of these are

(i)

$$a = \tfrac{1}{4}i\cos u, \qquad b = -c = \tfrac{1}{4}i\sin u, \qquad q = -r = -\tfrac{1}{2}u_x:$$

then $u$ satisfies the sine–Gordon equation

$$u_{xt} = \sin u.$$

(ii)

$$a = \tfrac{1}{4}i\cosh u, \qquad b = -c = -\tfrac{1}{4}i\sinh u, \qquad q = r = \tfrac{1}{2}u_x,$$

where $u$ satisfies the sinh–Gordon equation:

$$u_{xt} = \sinh u.$$

These examples only show a few of the many nonlinear evolution equations that may be obtained by this procedure. We saw above that when $r = -1$, the scattering problem (8.33) reduced to the Schrödinger equation (8.35). In this case we can take an alternative associated time dependence

$$v_t = Av + Bv_x.$$

By requiring that this and ($\lambda = k^2$)

$$v_{xx} + (\lambda + q)v = 0$$

are compatible and assuming that $d\lambda/dt = 0$, yields equations for $A$ and $B$ analogous to (8.38), then by expanding in powers of $\lambda$, we obtain a general class of equations.

## 8.7 More general classes of nonlinear evolution equations

A natural question is: *what is a general class of "solvable" nonlinear evolution equations?* Here we use the $2 \times 2$ scattering problem

$$v_{1,x} = -ikv_1 + q(x,t)v_2, \tag{8.45a}$$

$$v_{2,x} = -ikv_2 + r(x,t)v_1, \tag{8.45b}$$

to investigate this issue. Ablowitz, Kaup, Newell and Segur (1974) answered this question by considering the equations (8.45) associated with the functions $A$, $B$ and $C$ in (8.34). Under certain restrictions, a relation was found that gives a class of solvable nonlinear equations. This relationship depends upon the dispersion relation of the linearized form of the nonlinear equations and a certain integro-differential operator. Suppose that $q$ and $r$ vanish sufficiently rapidly as $|x| \to \infty$: then a general evolution equation can be shown to be

$$\begin{pmatrix} r \\ -q \end{pmatrix}_t + 2A_0(L)\begin{pmatrix} r \\ q \end{pmatrix} = 0 \tag{8.46a}$$

where $A_0(k) = \lim_{|x| \to \infty} A(x,t,k)$ (we may think of $A_0(k)$ as the ratio of two entire functions), and $L$ is the integro-differential operator given by

$$L = \frac{1}{2i}\begin{pmatrix} \partial_x - 2r(I_-q) & 2r(I_-r) \\ -2q(I_-q) & -\partial_x + 2q(I_-r) \end{pmatrix}, \tag{8.46b}$$

where $\partial_x \equiv \partial/\partial x$ and

$$(I_- f)(x) \equiv \int_{-\infty}^{x} f(y)dy. \tag{8.46c}$$

Note that $L$ operates on $(r,q)$, and $I_-$ operates both on the functions immediately to its right and also on the functions to which $L$ is applied. Equation (8.46) may be written in matrix form

$$\sigma_3 \mathbf{u}_t + 2A_0(L)\mathbf{u} = 0,$$

where

$$\sigma_3 = \begin{pmatrix} 1 & 0 \\ 0 & -1 \end{pmatrix}, \quad \mathbf{u} = \begin{pmatrix} r \\ q \end{pmatrix}.$$

It is significant that $A_0(k)$ is closely related to the dispersion relation of the associated *linearized* problem. If $f(x)$ and $g(x)$ vanish sufficiently rapidly as $|x| \to \infty$, then in the limit $|x| \to -\infty$,

$$f(x)(I_-g)(x) \equiv f(x)\int_{-\infty}^{x} g(y)dy \to 0,$$

and so for infinitesimal $q$ and $r$, we have that $L$ is the diagonal differential operator

$$L = \frac{1}{2i}\frac{\partial}{\partial x}\begin{pmatrix} 1 & 0 \\ 0 & -1 \end{pmatrix}.$$

Hence in this limit (8.46) yields

$$r_t + 2A_0\left(-\frac{1}{2}i\partial_x\right)r = 0,$$

$$-q_t + 2A_0\left(\frac{1}{2}i\partial_x\right)q = 0.$$

The above equations are linear (decoupled) partial differential equations solvable by Fourier transform methods. Substituting the wave solutions $r(x,t) = \exp(i(kx - \omega_r(2k)t))$, $q(x,t) = \exp(i(kx - \omega_q(-2k)t))$, into the above equations, we obtain the relationship

$$A_0(k) = \frac{1}{2}i\omega_r(2k) = -\frac{1}{2}i\omega_q(-2k). \tag{8.47}$$

Therefore a general evolution equation is given by

$$\begin{pmatrix} r \\ -q \end{pmatrix} = -i\omega(2L)\begin{pmatrix} r \\ q \end{pmatrix} \tag{8.48}$$

with $\omega_r(k) = \omega(k)$; in matrix form this equation takes the form

$$\sigma_3\mathbf{u}_t + i\omega(2L)\mathbf{u} = 0.$$

Hence, the form of the nonlinear evolution equation is characterized by the dispersion relation of its associated linearized equations and an integro-differential operator; further details can be found in Ablowitz et al. (1974).

For the nonlinear Schrödinger equation

$$iq_t = q_{xx} \pm 2q^2q^*, \tag{8.49}$$

the associated linear equation is

$$iq_t = q_{xx}.$$

Therefore the dispersion relation is $\omega_q(k) = -k^2$, and so from (8.47), $A_0(k)$ is given by

$$A_0(k) = 2ik^2. \tag{8.50}$$

The evolution is obtained from either (8.46) or (8.48); therefore we have

$$\begin{pmatrix} r \\ -q \end{pmatrix}_t = -4iL^2\begin{pmatrix} r \\ q \end{pmatrix} = -2L\begin{pmatrix} r_x \\ q_x \end{pmatrix} = i\begin{pmatrix} r_{xx} - 2r^2q \\ q_{xx} - 2q^2r \end{pmatrix}.$$

When $r = \mp q^*$, both of these equations are equivalent to the nonlinear Schrödinger equation (8.49). Note that (8.50) is in agreement with (8.41) with $a = 2i$ (that is $\lim_{|x| \to \infty} A = 2ik^2$, since $\lim_{|x| \to \infty} r(x, t) = 0$). This explains *á posteriori* why the expansion of $A$, $B$ and $C$ in powers of $k$ are related so closely to the dispersion relation. Indeed, now in retrospect, the fact that the nonlinear Schrödinger equation is related to (8.50) implies that an expansion starting at $k^2$ will be a judicious choice. Similarly, the modified KdV equation and sine–Gordon equation can be obtained from the operator equation (8.48) using the dispersion relations $\omega(k) = -k^3$ and $\omega(k) = k^{-1}$, respectively. These dispersion relations suggest expansions commencing in powers of $k^3$ and $k^{-1}$, respectively, that indeed we saw to be the case in the earlier section.

The derivation of (8.48) required that $q$ and $r$ tend to 0 as $|x| \to 0$ and therefore we cannot simply set $r = -1$ in order to obtain the equivalent result for the Schrödinger scattering problem

$$v_{xx} + (k^2 + q)v = 0. \tag{8.51}$$

However the essential ideas are similar and Ablowitz, Kaup, Newell and Segur (1974) also showed that a general evolution equation in this case is

$$q_t + \gamma(L)q_x = 0, \tag{8.52a}$$

where

$$L \equiv -\frac{1}{4}\partial_x^2 - q + \frac{1}{2}q_x I_+, \tag{8.52b}$$

$$I_+ f(x) \equiv \int_x^\infty f(y)dy. \tag{8.52c}$$

$$\gamma(k^2) = \frac{\omega(2k)}{2k}, \tag{8.52d}$$

with $\partial_x = \partial/\partial x$ and $\omega(k)$, is the dispersion relation of the associated linear equation with $q = exp(i(kx - \omega(k)t))$.

For the KdV equation

$$q_t + 6qq_x + q_{xxx} = 0,$$

the associated linear equation is

$$q_t + q_{xxx} = 0.$$

Therefore $\omega(k) = -k^3$ and so $\gamma(k^2) = -4k^2$, thus $\gamma(L) = -4L$. Hence (8.52) yields

$$q_t - 4Lq_x = 0,$$

thus

$$q_t - 4\left(-\frac{1}{4}\partial_x^2 - q + \frac{1}{2}q_x I_+\right)q_x = 0,$$

which is the KdV equation!

Associated with the operator $L$ there is a hierarchy of equations (sometimes referred to as the Lenard hierarchy) given by

$$u_t + L^k u_x = 0, \qquad k = 1, 2, \ldots$$

The first two subsequent higher-order equations in the KdV hierarchy are given by

$$u_t - \frac{1}{4}\left(u_{5x} + 10uu_{3x} + 20u_x u_{xx} + 30u^2 u_x\right) = 0,$$

$$u_t + \frac{1}{16}(u_{7x} + 14uu_{5x} + 42u_x u_{4x} + 70u_{xx}u_{3x} + 70u^2 u_{3x}$$
$$+ 280uu_x u_{xx} + 70u_x^3 + 140u^3 n_x) = 0$$

where $u_{nx} = \partial^n u/\partial x^n$.

In the above we studied two different scattering problems, namely the classical time-independent Scrödinger equation

$$v_{xx} + (u + k^2)v = 0, \tag{8.53}$$

which can be used to linearize the KdV equation, and the $2 \times 2$ scattering problem

$$\mathbf{v}_x = k\begin{pmatrix} -i & 0 \\ 0 & i \end{pmatrix}\mathbf{v} + \begin{pmatrix} 0 & q \\ r & 0 \end{pmatrix}\mathbf{v}, \tag{8.54}$$

which can linearize the nonlinear Schrödinger, mKdV and Sine–Gordon equations. While (8.53) can be interpreted as a special case of (8.54), from the point of view of possible generalizations, however, we regard them as different scattering problems.

There have been numerous applications and generalizations of this method only a few of which we will discuss below. One generalization includes that of Wadati et al. (1979a,b) (see also Shimizu and Wadati, 1980; Ishimori, 1981, 1982; Wadati and Sogo, 1983; Konno and Jeffrey, 1983). For example, instead of (8.33), suppose we consider the scattering problem

$$\begin{aligned} v_{1,x} &= -f(k)v_1 + g(k)q(x,t)v_2, \\ v_{2,x} &= f(k)v_2 + g(k)r(x,t)v_1, \end{aligned} \tag{8.55}$$

where $f(k)$ and $g(k)$ are polynomial functions of the eigenvalue $k$, and the time dependence is given (as previously) by

$$
\begin{aligned}
v_{1,t} &= Av_1 + Bv_2, \\
v_{2,t} &= Cv_1 + Dv_2.
\end{aligned}
\tag{8.56}
$$

The compatibility of (8.55) and (8.56) requires that $A$, $B$, $C$, $D$ satisfy linear equations that generalize equations (8.36)–(8.38).

As earlier, expanding $A$, $B$, $C$, $f$ and $g$ in finite power series expansions in $k$ (where the expansions for $f$ and $g$ have constant coefficients), then one obtains a variety of physically interesting evolution equations (cf. Ablowitz and Clarkson, 1991, for further details).

Extensions to higher-order scattering problems have been considered by several authors. There are two important classes of one-dimensional linear scattering problems, associated with the solvable nonlinear evolution equations: scalar and systems of equations. Some generalizations are now mentioned.

Ablowitz and Haberman (1975) studied an $N \times N$ matrix generalization of the scattering problem (8.54):

$$
\frac{\partial v}{\partial x} = ik\mathbf{J}v + \mathbf{Q}v,
\tag{8.57a}
$$

where the matrix of potentials $\mathbf{Q}(x) \in M_N(\mathbb{C})$ (the space of $N \times N$ matrices over $\mathbb{C}$) with $Q^{ii} = 0$, $\mathbf{J} = \mathrm{diag}(J^2, J^2, \ldots, J^N)$, where $J^i \neq J^j$ for $i \neq j$, $i, j = 1, 2, \ldots, N$, and $\mathbf{v}(x, t)$ is an $N$-dimensional vector eigenfunction. In this case, to obtain evolution equations, the associated time dependence is chosen to be

$$
\frac{\partial \mathbf{v}}{\partial t} = \mathbf{Tv},
\tag{8.57b}
$$

where $\mathbf{T}$ is also an $N \times N$ matrix. The compatibility of (8.57) yields

$$
\mathbf{T}_x = \mathbf{Q}_t + ik[\mathbf{J}, \mathbf{T}] + [\mathbf{Q}, \mathbf{T}].
$$

In the same way for the $2 \times 2$ case discussed earlier in the chapter, associated nonlinear evolution equations may be found by assuming an expansion for $\mathbf{T}$ in powers or inverse powers of the eigenvalue $k$

$$
\mathbf{T} = \sum_{j=0}^{n} k^j \mathbf{T}_j.
\tag{8.58}
$$

Ablowitz and Haberman showed that this scattering problem can be used to solve several physically interesting equations such as the three-wave interaction equations in one spatial dimension (with $N = 3$ and $n = 1$) and the

Boussinesq equation (with $N = 3$ and $n = 2$). The associated recursion operators have been obtained by various authors, cf. Newell (1978) and Fokas and Anderson (1982). The inverse problem associated with the scattering problem (8.57) in the general $N \times N$ case has been rigorously studied by Beals and Coifman (1984, 1985), Caudrey (1982), Mikhailov (1979, 1981) and Zhou (1989).

The following $N$th order generalization of the scattering problem (8.53) was considered by Gel'fand and Dickii (1977)

$$\frac{d^N v}{dx^N} + \sum_{j=2}^{N} u_j(x)\frac{d^{N-j}v}{dx^{N-j}} = \lambda v, \tag{8.59}$$

where $u_j(x)$ are considered as potentials; they investigated many of the *algebraic* properties of the nonlinear evolution equations solvable through this scattering problem. The general inverse problem associated with this scattering problem is treated in detail in the monograph of Beals, Deift, and Tomei (1988).

The scattering problems (8.53) and (8.54), and their generalizations (8.59) and (8.57), are only some of the examples of scattering problems used in connection with the solution of nonlinear partial differential equations in (1+1)-dimensions; i.e., one spatial and one temporal dimension. The reader can find many more interesting scattering problems related to integrable systems in the literature (Camassa and Holm, 1993a; Konopelchenko, 1993; Rogers and Schief, 2002; Matveev and Salle, 1991; Dickey, 2003; Newell, 1985).

## Exercises

8.1 Use (8.17) to find two conserved quantities additional to (8.19); i.e., $w_5$ and $w_6$. Then also obtain the conserved fluxes.

8.2 Use the method explained in Section 8.6 associated with (8.33)–(8.34) to find the nonlinear evolution equation corresponding to

$$A = a_4 k^4 + a_3 k^3 + a_2 k^2 + a_1 k + a_0.$$

Show that this agrees with the nonlinear equations obtained via the operator approach in Section 8.7, i.e., equations (8.46a)–(8.48).

8.3 Use the pair of equations

$$v_{xx} + (\lambda + q)v = 0$$
$$v_t = Av + Bv_x$$

and the method of Section 8.6 to find an "integrable" nonlinear evolution equation of order 5 in space, i.e., of the form $u_t + u_{xxxxx} + \cdots = 0$. Show that this agrees with the the nonlinear equations obtained via the operator approach in Section 8.7, i.e., equation (8.52).

8.4 Suppose we wish to find a real integrable equation associated with (8.51) having a dispersion relation

$$\omega = \frac{1}{1 + k^2}.$$

Find this equation via (8.52).

8.5 Consider the scattering problem (8.57a) of order $3 \times 3$ with time dependence (8.57) with $T$ given by (8.58) with $n = 1$. Find the nonlinear evolution equation. Hint: see Ablowitz and Haberman (1975). Extend this to scattering problems of order $N \times N$.

8.6 Consider the KdV equation

$$u_t + 6uu_x + u_{xxx} = 0.$$

(a) Show that $u(x, t) = -2k^2 \operatorname{csech}^2[k(x - 4k^2t)]$ is a singular solution.
(b) Obtain the rational solution $u(x, t) = -2/x^2$ by letting $k \to 0$.
(c) Show that the singular solution can also be obtained from the classical soliton solution $u(x, t) = 2k^2 \operatorname{sech}^2[k(x - 4k^2t) - x_0]$ by setting $\exp(2x_0) = -1$.

8.7 Show that the concentric KdV equation

$$u_t + \frac{1}{2t}u + 6uu_x + u_{xxx} = 0$$

has the following conservation laws:

(i) $\displaystyle\int_{-\infty}^{\infty} \sqrt{t}u \, dx = \text{constant},$

(ii) $\displaystyle\int_{-\infty}^{\infty} tu^2 \, dx = \text{constant},$

(iii) $\displaystyle\int_{-\infty}^{\infty} \left( \sqrt{t}xu - 6t^{3/2}u^2 \right) dx = \text{constant}.$

8.8 Consider the KP equation

$$(u_t + 6uu_x + u_{xxx}) + 3\sigma u_{yy} = 0$$

where $\sigma = \pm 1$. Show that the KP equation has the conserved quantities:

(i) $\displaystyle I_1 = \int_{-\infty}^{\infty} \int_{-\infty}^{\infty} u \, dxdy = \text{constant};$

(ii) $\displaystyle I_2 = \int_{-\infty}^{\infty} \int_{-\infty}^{\infty} u^2 \, dxdy = \text{constant};$

(iii) $\displaystyle\int_{-\infty}^{\infty} u_{yy}\,dx = \text{constant}$

and the integral relation

(iv) $\displaystyle\frac{d}{dt}\int_{-\infty}^{\infty}\int_{-\infty}^{\infty} xu\,dxdy = \alpha I_2$

where $\alpha$ is constant.

8.9    Show that the NLS equation

$$iu_t + u_{xx} + |u|^2 u = 0$$

has the combined rational and oscillatory solution

$$u(x,t) = \left[1 - \frac{4(1 + 2it)}{1 + 2x^2 + 4t^2}\right]\exp(it).$$

8.10  (a)  Use the pair of equations

$$v_{xx} + (\lambda q)v = 0$$
$$v_t = Av + Bv_x$$

and the method of Section 8.6, with $A, B$ taken to be linear in $\lambda$ with the "isospectral" condition $\lambda_t = 0$, to find the nonlinear evolution

$$q_t = (q^{-1/2})_{xxx}.$$

This equation is usually referred to as the Harry–Dym equation, which can be shown to be integrable via "hodograph transformations" cf. Ablowitz and Clarkson (1991).

(b)  Show that the method of Section 6.8 with

$$v_{xx} + \lambda(m + \alpha)v = 0, \quad \alpha = \kappa + 1/4$$
$$v_t = Av + Bv_x, \quad A = -\frac{u_x}{2}, \quad B = u - \frac{1}{2\lambda}$$

and $m = u_{xx} - u$ results in the "isospectral" equation

$$u_t + 2\kappa u_x - u_{xxt} + 3uu_x - 2u_x u_{xx} - uu_{xxx} = 0.$$

The above equation was shown to have an infinite number of symmetries by Fokas and Fuchssteiner (1981); the inverse scattering problem, solutions and relation to shallow-water waves was considered by Camassa and Holm (1993b). When $\kappa = 0$, the equation has traveling wave "peakon" solutions, which have cusps; they have a shape of the form $e^{-|x-ct|}$. The above equation, which is usually referred to as the Camassa–Holm equation, has been studied extensively due to its many interesting properties.

8.11  Consider the Harry–Dym equation

$$u_t + 2(u^{-1/2})_{xxx}.$$

(a) Let $u = v_x^2$ and show that $v$ satisfies

$$v_t = \frac{v_{xxx}}{v_x^3} - \frac{3v_{xx}^2}{2v_x^4}.$$

(b) Show that the "hodograph" transformation $\tau = t, \xi = v, \eta = x$, yields the relations

$$\eta_\xi = \frac{1}{v_x}, \quad \eta_\tau = \frac{-v_t}{v_x}.$$

(c) Using the above relations find the equation

$$w_\tau = w_{\xi\xi\xi} - \left(\frac{3w_\xi^2}{2w}\right)_\xi$$

where $w = \eta_\xi$, which is known to be "integrable" (linearizable). Hint: see Clarkson et al. (1989).

# The inverse scattering transform for the Korteweg–de Vries (KdV) equation

We have seen in the previous chapter that the Korteweg–de Vries (KdV) equation is the result of compatibility between

$$v_{xx} + (\lambda + u(x,t))v = 0 \tag{9.1}$$

and

$$v_t = (\gamma + u_x)v + (4\lambda + 2u)v_x, \tag{9.2}$$

where $\gamma$ is constant. More precisely, the equality of the mixed derivatives $v_{xxt} = v_{txx}$ with $\lambda_t = 0$ ("isospectrality") leads to the KdV equation

$$u_t + 6uu_x + u_{xxx} = 0. \tag{9.3}$$

In the main part of this chapter we will use the above compatibility relations in order to carry out the inverse scattering transform (IST) for the KdV equation on the infinite interval. This includes a basic description of the direct and inverse problems of the time-independent Schrödinger equation and the time evolution of the scattering data. Special soliton solutions as well as the scattering data for a delta function and box potential are given. Also included are the derivation of the conserved quantities/conservation laws of the KdV equation from IST. Finally, as an extension, the IST associated with the second-order scattering system and associated time (8.33)–(8.34), discussed in the previous chapter, is carried out.

For (9.1), the eigenvalues and the behavior of the eigenfunctions as $|x| \to \infty$ determine what we call the *scattering data* $S(\lambda, t)$ at any time $t$, which depends upon the potential $u(x,t)$. The *direct scattering problem* maps the potential into the scattering data. The *inverse scattering problem* reconstructs the potential $u(x,t)$ from the scattering data $S(\lambda, t)$. The initial value problem for the KdV equation is analyzed as follows. At $t = 0$ we give initial data $u(x,0)$ which we assumes decays sufficiently rapidly at infinity. The initial data is mapped to

214

$S(\lambda, t = 0)$ via (9.1). The evolution of the scattering data $S(\lambda, t)$ is determined from (9.2). Then $u(x, t)$ is recovered from inverse scattering.

## 9.1 Direct scattering problem for the time-independent Schrödinger equation

Suppose $\lambda = -k^2$; then the time-independent Schrödinger equation (9.1) becomes

$$v_{xx} + \{u(x) + k^2\}v = 0, \tag{9.4}$$

where, for convenience, we have suppressed the time dependence in $u$. We shall further assume that $u(x)$ decays sufficiently rapidly, which for our purposes means that $u$ lies in the space of functions

$$L_n^1 : \quad \int_{-\infty}^{\infty} (1 + |x|^n)|u(x)|\, dx < \infty,$$

with $n = 2$ (Deift and Trubowitz, 1979). We also remark that the space $L_1^1$ was the original space used by Faddeev (1963); this space was subsequently used by Marchenko (1986) and Melin (1985). Associated with (9.4) are two complete sets of eigenfunctions for real $k$ that are bounded for all values of $x$, and that have appropriate analytic extensions. These eigenfunctions are defined by the equation and boundary conditions; that is, we identify four eigenfunctions defined by the following asymptotic boundary conditions:

$$\begin{aligned} \phi(x; k) \sim e^{-ikx}, \quad \overline{\phi}(x; k) \sim e^{ikx} \quad &\text{as} \quad x \to -\infty, \\ \psi(x; k) \sim e^{ikx}, \quad \overline{\psi}(x; k) \sim e^{-ikx} \quad &\text{as} \quad x \to \infty. \end{aligned} \tag{9.5}$$

Therefore $\phi(x; k)$, for example, is that solution of (9.4) that tends to $e^{-ikx}$ as $x \to -\infty$. Note that here the bar does not denote complex conjugate, for which we will use the $*$ notation. From (9.4) and the boundary conditions (9.5), we see that

$$\phi(x; k) = \overline{\phi}(x; -k) = \phi^*(x, -k), \tag{9.6a}$$

$$\psi(x; k) = \overline{\psi}(x; -k) = \psi^*(x, -k). \tag{9.6b}$$

Since (9.4) is a linear second-order ordinary differential equation, by the linear independence of its solutions we obtain the following relationships between the eigenfunctions

$$\phi(x;k) = a(k)\bar{\psi}(x;k) + b(k)\psi(x;k), \qquad (9.7a)$$

$$\bar{\phi}(x;k) = -\bar{a}(k)\psi(x;k) + \bar{b}(k)\bar{\psi}(x;k) \qquad (9.7b)$$

with $k$-dependent functions: $a(k)$, $\bar{a}(k)$, $b(k)$, $\bar{b}(k)$. These functions form the basis of the scattering data we will need.

The *Wronskian* of two functions $\psi, \phi$ is defined as

$$W(\phi, \psi) = \phi\psi_x - \phi_x\psi$$

and for (9.4), from Abel's theorem, the Wronskian is constant. Hence from $\pm\infty$:

$$W(\psi, \bar{\psi}) = -2ik = -W(\phi, \bar{\phi}).$$

The Wronskian also has the following properties

$$W(\phi(x;k), \psi(x;k)) = -W(\psi(x;k), \phi(x;k)),$$

$$W(c_1\phi(x;k), c_2\psi(x;k)) = c_1c_2\, W(\phi(x;k), \psi(x;k)),$$

with $c_1, c_2$ arbitrary constants.

From (9.7) it follows that

$$a(k) = \frac{W(\phi(x;k), \psi(x;k))}{2ik}, \qquad b(k) = -\frac{W(\phi(x;k), \bar{\psi}(x;k))}{2ik} \qquad (9.8)$$

and

$$\bar{a}(k) = \frac{W(\bar{\phi}(x;k), \bar{\psi}(x;k))}{2ik}, \qquad \bar{b}(k) = \frac{W(\bar{\phi}(x;k), \psi(x;k))}{2ik}.$$

The formula for $a(k)$ follows directly from $W(\phi(x;k), \psi(x;k))$ and using the first of equations (9.7); the others are similar.

Rather than work with the eigenfunctions $\phi(x;k)$, $\bar{\phi}(x;k)$, $\psi(x;k)$ and $\bar{\psi}(x;k)$, it is more convenient to work with the (modified) eigenfunctions $M(x;k)$, $\bar{M}(x;k)$, $N(x;k)$ and $\bar{N}(x;k)$, defined by

$$M(x;k) = \phi(x;k)\,e^{ikx}, \qquad \bar{M}(x;k) := \bar{\phi}(x;k)\,e^{ikx}, \qquad (9.9a)$$

$$N(x;k) = \psi(x;k)\,e^{ikx}, \qquad \bar{N}(x;k) = \bar{\psi}(x;k)\,e^{ikx}; \qquad (9.9b)$$

then

$$M(x;k) \sim 1, \qquad \bar{M}(x;k) \sim e^{2ikx}, \qquad \text{as} \quad x \to -\infty, \quad (9.10a)$$

$$N(x;k) \sim e^{2ikx}, \qquad \bar{N}(x;k) \sim 1, \qquad \text{as} \quad x \to \infty. \quad (9.10b)$$

Equations (9.7a) and (9.9) imply that

$$\frac{M(x;k)}{a(k)} = \bar{N}(x;k) + \rho(k)N(x;k),$$
$$\frac{\bar{M}(x;k)}{\bar{a}(k)} = -N(x;k) + \bar{\rho}(k)\bar{N}(x;k),$$

(9.11)

where

$$\rho(k) = \frac{b(k)}{a(k)},$$

(9.12a)

$$\bar{\rho}(k) = \frac{\bar{b}(k)}{\bar{a}(k)};$$

(9.12b)

$\tau(k) = 1/a(k)$ and $\rho(k)$ are called the *transmission* and *reflection* coefficients respectively. Equations (9.6b), (9.9b) imply that

$$N(x;k) = \bar{N}(x;-k)\,e^{2ikx}.$$

(9.13)

Due to this symmetry relation we will only need two eigenfunctions. Namely, from (9.11) and (9.13) we obtain

$$\frac{M(x;k)}{a(k)} = \bar{N}(x;k) + \rho(k)\,e^{2ikx}\bar{N}(x;-k),$$

(9.14)

which will be the fundamental equation. Later we will show that (9.14) is equivalent to what we call a generalized *Riemann–Hilbert boundary value problem* (RHBVP); this RHBVP is central to the inverse problem and is a consequence of the analyticity properties of $M(x;k)$, $\bar{N}(x;k)$ and $a(k)$ that are established in the following lemma.

**Lemma 9.1** (i) *$M(x;k)$ can be analytically extended to the upper half k-plane and tends to unity as $|k| \to \infty$ for $\operatorname{Im} k > 0$;*
(ii) *$\bar{N}(x;k)$ can be analytically extended to the lower half k-plane and tends to unity as $|k| \to \infty$ for $\operatorname{Im} k < 0$.*

These analyticity properties are established by studying the linear integral equations that govern $M(x;k)$ and $\bar{N}(x;k)$. We begin by transforming (9.4) via $v(x;k) = m(x;k)e^{-ikx}$. Then $m(x;k)$ satisfies

$$m_{xx}(x;k) - 2ikm_x(x;k) = -u(x)m(x;k).$$

Then assuming that $m \to 1$, either as $x \to \infty$, or as $x \to -\infty$, we find

$$m(x;k) = 1 + \int_{-\infty}^{\infty} G(x-\xi;k)\,u(\xi)\,m(\xi;k)\,d\xi,$$

where $G(x; k)$ is the Green's function that solves

$$G_{xx} - 2ikG_x = -\delta(x).$$

Using the Fourier transform method, i.e., $G(x; k) = \int_C \hat{G}(p, k)e^{ipx}\, dp$ with $\delta(x) = \int e^{ipx}\, dp$, we find

$$G(x; k) = \frac{1}{2\pi} \int_C \frac{e^{ipx}}{p(p - 2k)}\, dp,$$

where $C$ is an appropriate contour. We consider $G_\pm(x; k)$ defined by

$$G_\pm(x; k) = \frac{1}{2\pi} \int_{C_\pm} \frac{e^{ipx}}{p(p - 2k)}\, dp,$$

where $C_+$ and $C_-$ are the contours from $-\infty$ to $\infty$ that pass below and above, respectively, both of the singularities at $p = 0$ and $p = 2k$. Hence by contour integration

$$G_+(x; k) = \frac{1}{2ik}(1 - e^{2ikx})H(x) \;=\; \begin{cases} \dfrac{1}{2ik}(1 - e^{2ikx}), & \text{if } x > 0, \\[2mm] 0, & \text{if } x < 0, \end{cases}$$

$$G_-(x; k) = -\frac{1}{2ik}(1 - e^{2ikx})H(-x) = \begin{cases} 0, & \text{if } x > 0, \\[2mm] -\dfrac{1}{2ik}(1 - e^{2ikx}), & \text{if } x < 0, \end{cases}$$

where $H(x)$ is the Heaviside function given by

$$H(x) = \begin{cases} 1, & \text{if } x > 0, \\ 0, & \text{if } x < 0. \end{cases}$$

We require that $M(x; k)$ and $\bar{N}(x; k)$ satisfy the boundary conditions (9.10). Since from (9.10a), $M(x; k) \to 1$ as $x \to -\infty$, we associate $M(x; k)$ with the Green's function $G_+(x; k)$, so the integral is from $-\infty$ to $x$, and analogously since $\bar{N}(x; k) \to 1$ as $x \to \infty$, we associate $\bar{N}(x; k)$ with the Green's function $G_-(x; k)$. Therefore it follows that $M(x; k)$ and $\bar{N}(x; k)$ are the solutions of the following integral equations

$$M(x; k) = 1 + \int_{-\infty}^{\infty} G_+(x - \xi; k)\, u(\xi)\, M(\xi; k)\, d\xi$$

$$= 1 + \frac{1}{2ik} \int_{-\infty}^{x} \left\{1 - e^{2ik(x-\xi)}\right\} u(\xi)\, M(\xi; k)\, d\xi, \qquad (9.15)$$

$$\bar{N}(x; k) = 1 + \int_{-\infty}^{\infty} G_-(x - \xi; k)\, u(\xi)\, \bar{N}(\xi; k)\, d\xi$$

$$= 1 - \frac{1}{2ik} \int_{x}^{\infty} \left\{1 - e^{2ik(x-\xi)}\right\} u(\xi)\, \bar{N}(\xi; k)\, d\xi. \qquad (9.16)$$

These are Volterra integral equations that have unique solutions; their Neumann series converge uniformly with $u$ in the function space $L_2^1$, for each $k$: $M(x; k) \operatorname{Im} k \geq 0$, and for $\overline{N}(x; k) \operatorname{Im} k \leq 0$ (Deift and Trubowitz, 1979). One can show that their Neumann series converge uniformly for $u$ in this function space. We note: $G_{\pm}(x; k)$ is analytic for $\operatorname{Im} k \gtrless 0$ and vanishes as $|k| \to \infty$; $M(x; k)$ and $\overline{N}(x; k)$ are analytic for $\operatorname{Im} k > 0$ and $\operatorname{Im} k < 0$ respectively, and tend to unity as $|k| \to \infty$. In the same way it can be shown that $N(x; k)e^{-2ikx} = \psi(x; k)e^{ikx}$ is analytic for $\operatorname{Im} k > 0$ and $\overline{M}(x; k)e^{2ikx} = \overline{\psi}(x; k)e^{-ikx}$ is analytic for $\operatorname{Im} k < 0$.

## 9.2 Scattering data

A corollary to the above lemma is that $a(k)$ can be analytically extended to the upper half $k$-plane and tends to unity as $|k| \to \infty$ for $\operatorname{Im} k > 0$. We note that Deift and Trubowitz (1979) also show that $a(k)$ is continuous for $\operatorname{Im} k \geq 0$. The analytic properties of $a(k)$ may be established either from the Wronskian relationship (9.8), noting that $W(\phi(x; k), \psi(x; k)) = W(M(x; k), N(x; k))$ or by using a suitable integral representation for $a(k)$. To derive this integral representation we define

$$\Delta(x; k) = M(x; k) - a(k)\overline{N}(x; k),$$

and then from (9.15) and (9.16) it follows that

$$\Delta(x; k) = 1 - a(k) + \frac{1}{2ik} \int_{-\infty}^{\infty} \left\{ 1 - e^{2ik(x-\xi)} \right\} u(\xi) M(\xi; k) \, d\xi$$
$$- \frac{1}{2ik} \int_{x}^{\infty} \left\{ 1 - e^{2ik(x-\xi)} \right\} u(\xi) \Delta(\xi; k) \, d\xi. \quad (9.17)$$

Also, from (9.14) we have

$$\Delta(x; k) = b(k) e^{2ikx} \overline{N}(x; -k);$$

therefore from (9.16)

$$\Delta(x; k) = b(k) e^{2ikx} - \frac{1}{2ik} \int_{x}^{\infty} \left\{ 1 - e^{2ik(x-\xi)} \right\} u(\xi) \Delta(\xi; k) \, d\xi. \quad (9.18)$$

Hence by comparing (9.17) and (9.18), and equating coefficients of 1 and $e^{2ikx}$, we obtain the following integral representations for $a(k)$ and $b(k)$:

$$a(k) = 1 + \frac{1}{2ik} \int_{-\infty}^{\infty} u(\xi) M(\xi; k) \, d\xi, \quad (9.19)$$

$$b(k) = -\frac{1}{2ik} \int_{-\infty}^{\infty} u(\xi) M(\xi; k) e^{-2ik\xi} \, d\xi. \quad (9.20)$$

From these integral representations it follows that $a(k)$ is analytic for $\text{Im } k > 0$, and tends to unity as $|k| \rightarrow \infty$, while $b(k)$ cannot, in general, be continued analytically off the real $k$-axis. Furthermore, since $M(x; k) \rightarrow 1$ as $\kappa \rightarrow \infty$, then it follows from (9.19) that $a(k) \rightarrow 1$ as $\kappa$ (for $\text{Im } k > 0$). Similarly it can be shown that $\bar{a}(k)$ can be analytically extended to the lower half $k$-plane and tends to unity as $|k| \rightarrow \infty$ for $\text{Im } k < 0$.

The functions $a(k)$, $\bar{a}(k)$, $b(k)$, $\bar{b}(k)$ satisfy the condition

$$a(k)\bar{a}(k) + b(k)\bar{b}(k) = -1. \qquad (9.21)$$

This follows from the Wronskian and its relations:

$$\begin{aligned}
W(\phi(x; k), \bar{\phi}(x; k)) &= W(a(k)\bar{\psi}(x; k) + b(k)\psi(x; k), \\
&\quad - \bar{a}(k)\psi(x; k) + \bar{b}(k)\bar{\psi}(x; k)) \\
&= \left\{ a(k)\bar{a}(k) + b(k)\bar{b}(k) \right\} W(\psi(x; k), \bar{\psi}(x; k)),
\end{aligned}$$

with $W(\phi(x; k), \bar{\phi}(x; k)) = 2ik = -W(\psi(x; k), \bar{\psi}(x; k))$. From the definitions of the reflection and transmission coefficients, we see that $\tau(k)$ and $\rho(k)$ satisfy

$$|\rho(k)|^2 + |\tau(k)|^2 = 1.$$

Using (9.4)–(9.7) and their complex conjugates we have the symmetry conditions

$$\begin{aligned}
\bar{a}(k) &= -a(-k) = -a^*(k^*), \\
\bar{b}(k) &= b(-k) = b^*(k^*),
\end{aligned} \qquad (9.22)$$

where $a^*(k^*)$, $b^*(k^*)$ are the complex conjugates of $a(k)$, $b(k)$, respectively.

The zeros of $a(k)$ play an important role in the inverse problem; they are poles in the generalized Riemann–Hilbert problem obtained from (9.14). We note the following: the function $a(k)$ can have a finite number of zeros at $k_1, k_2, \ldots, k_N$, where $k_j = i\kappa_j$, $\kappa_j \in \mathbb{R}$, $j = 1, 2, \ldots, N^\#$ (i.e., they all lie on the imaginary axis), in the upper half $k$-plane.

First we see that from equations (9.21), (9.22) that

$$|a(k)|^2 - |b(k)|^2 = 1,$$

for $k \in \mathbb{R}$; hence $a(k) \neq 0$ for $k \in \mathbb{R}$. Recall that $\phi(x; k)$, $\psi(x; k)$ and $\bar{\psi}(x; k)$ are solutions of the Schrödinger equation (9.4) satisfying the boundary conditions

$$\begin{aligned}
\phi(x; k) &\sim e^{-ikx}, &\quad \text{as} &\quad x \rightarrow -\infty, \\
\psi(x; k) &\sim e^{ikx}, &\quad \text{as} &\quad x \rightarrow \infty, &\qquad (9.23) \\
\bar{\psi}(x; k) &\sim e^{-ikx}, &\quad \text{as} &\quad x \rightarrow \infty,
\end{aligned}$$

together with the relationship

$$\phi(x; k) = a(k)\bar{\psi}(x; k) + b(k)\psi(x; k). \tag{9.24}$$

From (9.8) it follows that

$$W(\phi(x; k), \psi(x; k)) = \phi(x; k)\psi_x(x; k) - \phi_x(x; k)\psi(x; k) = 2ika(k).$$

Thus at any zero $k_j$ of $a(k)$, we have that $\phi(x; k), \psi(x; k)$ are linearly independent. In order to have a rapidly decaying state; i.e., a bound state where $\phi(x; k_j)$ is in $L^2(\mathbb{R})$, we must have that $\operatorname{Im} k_j > 0$.

The eigenfunction $\phi(x; k)$ and its complex conjugate $\phi^*(x; k^*)$ satisfy the equations

$$\begin{aligned}
\phi_{xx} + \left\{ u(x) + k^2 \right\} \phi &= 0, \\
\phi_{xx}^* + \left\{ u(x) + (k^*)^2 \right\} \phi^* &= 0,
\end{aligned} \tag{9.25}$$

respectively. Hence

$$\frac{\partial}{\partial x} W(\phi, \phi^*) + \left[ (k^*)^2 - k^2 \right] \phi\phi^* = 0,$$

which implies

$$\left[ (k^*)^2 - k^2 \right] \int_{-\infty}^{\infty} |\phi(x; k)|^2 \, dx = 0 \tag{9.26}$$

since $\phi$ and its derivatives vanish as $x \to \pm\infty$. Further, calling $k_j = \xi_j + \eta_j$, we see by taking the real part of (9.26): $\left( k_j^* \right)^2 - k_j^2 = -2\xi_j \eta_j = 0$, any zero of $a$ must be purely imaginary; i.e., $\operatorname{Re} k_j = \xi_j = 0$.

To show that $a(k)$ has only a finite number of zeros, we note that $a \to 1$ as $|k| \to \infty$; furthermore, it is shown by Deift and Trubowitz (1979) that $a(k)$ is continuous for $\operatorname{Im} k \geq 0$. Hence there can be no cluster points of zeros of $a(k)$ either in the upper half $k$-plane or on the real $k$-axis and so necessarily $a(k)$ can possess only a finite number of zeros.

We can also show that the zeros of $a(k)$ are simple by studying the derivative $da/dk$ with respect to $k$: but we will not do this here (see Deift and Trubowitz, 1979).

Hence, as long as $u \in L_2^1$, $M(x; k)/a(k)$ is a meromorphic function in the upper half $k$-plane with a finite number of simple poles at $k = i\kappa_1, i\kappa_2, \ldots, i\kappa_N{}^{\#}$. Then set

$$\frac{M(x; k)}{a(k)} = \mu_+(x; k) + \sum_{j=1}^{N^{\#}} \frac{A_j(x)}{k - i\kappa_j}, \tag{9.27}$$

where $\mu_+(x;k)$ is analytic in the upper half $k$-plane. Integrating (9.27) around $i\kappa_j$ and using (9.14) and the fact that at a discrete eigenvalue $M, N$ are proportional, $M(x;i\kappa_j) = b_j N(x;i\kappa_j)$, shows that

$$A_j(x) = C_j \bar{N}(x;-i\kappa_j) \exp(-2\kappa_j x),$$

where $C_j = b_j/a'_j$. Hence from (9.14) we obtain

$$\mu_+(x;k) = \bar{N}(x;k) - \sum_{j=1}^{N^\#} \frac{C_j}{k - i\kappa_j} \exp(-2\kappa_j x)\bar{N}(x;-i\kappa_j)$$

$$+ \rho(k) \exp(2ikx)\bar{N}(x;-k). \tag{9.28}$$

Furthermore, $\mu_+(x;k) \to 1$ as $\kappa \to 1$, for $\operatorname{Im} k > 0$.

## 9.3 The inverse problem

Equation (9.28) defines a Riemann–Hilbert problem (cf. Ablowitz and Fokas, 2003) in terms of the scattering data $S(\lambda) = \left\{(\kappa_j, C_j)_{j=1}^{N^\#}, \rho(k), a(k)\right\}$. We will use the following.

Consider the $\mathcal{P}^\pm$ projection operator defined by

$$(\mathcal{P}^\pm f)(k) = \frac{1}{2\pi i} \int_{-\infty}^{\infty} \frac{f(\zeta)}{\zeta - (k \pm i0)}\, d\zeta = \lim_{\epsilon \downarrow 0} \left\{ \frac{1}{2\pi i} \int_{-\infty}^{\infty} \frac{f(\zeta)}{\zeta - (k \pm i\epsilon)}\, d\zeta \right\}. \tag{9.29}$$

Suppose that $f_\pm(k)$ is analytic in the upper/lower half $k$-plane and $f_\pm(k) \to 0$ as $|k| \to \infty$ (for $\operatorname{Im} k \gtrless 0$). Then

$$(\mathcal{P}^\pm f_\mp)(k) = 0,$$
$$(\mathcal{P}^\pm f_\pm)(k) = \pm f_\pm(k).$$

These results follow from contour integration.

To explain the ideas most easily, we first assume that there are no poles; that is, $a(k) \neq 0$. Then (9.28) reduces to (9.14), because $M(x;k)/a(k) = \mu_+$. In (9.14) unity from both $\bar{N}$ and $M/a$ and operating on (9.14) with $\mathcal{P}^-$ and using the analytic properties of $M, \bar{N}$ and $a$ yields

$$\mathcal{P}^- \left[ \left( \frac{M(x;k)}{a(k)} - 1 \right) \right] = \mathcal{P}^- \left[ (\bar{N}(x;k) - 1) + \rho(k)e^{2ikx}\bar{N}(x;-k) \right].$$

This implies

$$0 = (\bar{N}(x;k) - 1) + \frac{1}{2\pi i} \int_{-\infty}^{\infty} \frac{\rho(\zeta)\, e^{2i\zeta x}\, \bar{N}(x;-\zeta)}{\zeta - (k - i0)}\, d\zeta.$$

Then employing the symmetry: $N(x;k) = e^{2ikx}\bar{N}(x;-k)$ and $N(x;-k) = e^{-2ikx}\bar{N}(x;k)$, it follows that

$$\bar{N}(x;k) = 1 + \frac{1}{2\pi i} \int_{-\infty}^{\infty} \frac{\rho(\zeta)\,N(x;\zeta)}{\zeta - (k-i0)}\,d\zeta,$$

$$N(x;k) = e^{2ikx}\left\{1 + \frac{1}{2\pi i} \int_{-\infty}^{\infty} \frac{\rho(\zeta)\,N(x;\zeta)}{\zeta + k + i0}\,d\zeta\right\}.$$

$$(9.30)$$

As $k \to \infty$, equation (9.30a) shows that

$$\bar{N}(x;k) \sim 1 - \frac{1}{2\pi i}\frac{1}{k}\int_{-\infty}^{\infty} \rho(\zeta)\,N(x;\zeta)\,d\zeta, \qquad (9.31)$$

From the integral equation for $\bar{N}(x;k)$, equation (9.16), using the Riemann–Lebesgue lemma, we also see that as $\kappa \to \infty$

$$\bar{N}(x;k) \sim 1 - \frac{1}{2ik}\int_{x}^{\infty} u(\xi)\,d\xi. \qquad (9.32)$$

Therefore, by comparing (9.31) and (9.32), it follows that the potential can be reconstructed from

$$u(x) = -\frac{\partial}{\partial x}\left\{\frac{1}{\pi}\int_{-\infty}^{\infty} \rho(\zeta)\,N(x;\zeta)\,d\zeta\right\}. \qquad (9.33)$$

For the case when $a(k)$ has zeros, one can extend the above result. Suppose that

$$a(i\kappa_j) = 0, \qquad \kappa_j \in \mathbb{R}, \qquad j = 1, 2, \ldots, N^{\#};$$

then define

$$M_j(x) = M(x;i\kappa_j),$$
$$N_j(x) = \exp(-2\kappa_j x)\bar{N}(x;-i\kappa_j),$$

for $j = 1, 2, \ldots, N^{\#}$. In (9.28) subtracting unity from both $\mu_+(x,k)$ and $\bar{N}(x;k)$ and applying the $\mathcal{P}^-$ projection operator (9.29) to (9.28), recalling that $N(x;k) = e^{2ikx}\bar{N}(x;-k)$ and carrying out the contour integration of simple poles, we find the following integral equations

$$N(x;k) = e^{2ikx}\left\{1 - \sum_{j=1}^{N^{\#}} \frac{C_j N_j(x)}{k + i\kappa_j} + \frac{1}{2\pi i}\int_{-\infty}^{\infty} \frac{\rho(\zeta)\,N(x;\zeta)}{\zeta + k + i0}\,d\zeta\right\}, \qquad (9.34)$$

$$N_p(x) = \exp(-2\kappa_p x)\left\{1 + i\sum_{j=1}^{N^{\#}} \frac{C_j N_j(x)}{\kappa_p + \kappa_j} + \frac{1}{2\pi i}\int_{-\infty}^{\infty} \frac{\rho(\zeta)\,N(x;\zeta)}{\zeta + i\kappa_p}\,d\zeta\right\},$$

$$(9.35)$$

for $p = 1, 2, \ldots N^{\#}$, with the potential reconstructed from [following the same method that led to (9.33)]

$$u(x) = \frac{\partial}{\partial x} \left\{ 2i \sum_{j=1}^{N^{\#}} C_j N_j(x) - \frac{1}{\pi} \int_{-\infty}^{\infty} \rho(k) N(x; k) \, dk \right\}. \tag{9.36}$$

Existence and uniqueness of solutions to equations such as (9.34)–(9.36) is given by Beals et al. (1988). We also note that the above integral equations (9.34) and (9.35) can be transformed to Gel'fand–Levitan–Marchenko integral equations, see Section 9.5 below, from which one can also deduce the existence and uniqueness of solutions (cf. Marchenko, 1986).

## 9.4 The time dependence of the scattering data

In this section we will find out how the scattering data evolves. The time evolution of the scattering data may be obtained by analyzing the asymptotic behavior of the associated time evolution operator, which as we have seen earlier for the KdV equation is

$$v_t = (u_x + \gamma)v + (4k^2 - 2u)v_x,$$

with $\gamma$ a constant. If we let $v = \phi(x; k)$ and make the transformation

$$\phi(x, t; k) = M(x, t; k) e^{-ikx},$$

$M$ then satisfies

$$M_t = (\gamma - 4ik^3 + u_x + 2iku)M + (4k^2 - 2u)M_x. \tag{9.37}$$

Recall that from (9.11)

$$M(x, t; k) = a(k, t)\bar{N}(x, t; k) + b(k, t)N(x, t; -k),$$

where $\rho(k, t) = b(k, t)/a(k, t)$. From (9.10) the asymptotic behavior of $M(x, t; k)$ is given by

$$\begin{aligned} M(x, t; k) &\to 1, & \text{as} \quad x \to -\infty, \\ M(x, t; k) &\to a(k, t) + b(k, t) e^{2ikx}, & \text{as} \quad x \to \infty. \end{aligned}$$

By using using the fact that $u \to 0$ rapidly as $x \to \pm\infty$ and the above equation it follows from (9.37) that

$$\begin{aligned} \gamma - 4ik^3 &= 0, & x \to -\infty \\ a_t + b_t \, e^{2ikx} &= 8ik^3 b \, e^{2ikx}, & x \to +\infty, \end{aligned}$$

and by equating coefficients of $e^0$, $e^{2ikx}$ we find

$$a_t = 0, \qquad b_t = 8ik^3 b. \tag{9.38}$$

Solving (9.38) yields

$$a(k,t) = a(k,0), \qquad b(k,t) = b(k,0)\exp(8ik^3t), \tag{9.39}$$

and so

$$\rho(k,t) \equiv \frac{b(k,t)}{a(k,t)} = \rho(k,0)\exp(8ik^3t).$$

Since $a(k,t)$ does not evolve in time, the discrete eigenvalues, which are the zeros of $a(k)$ that are finite in number, simple, and lie on the imaginary axis, satisfy

$$k_j = i\kappa_j = \text{constant}, \qquad j = 1, 2, \ldots, N^{\#}.$$

Since the eigenvalues are constant in time, we say this is an "isospectral flow". Similarly we find that the time dependence of the $C_j(t)$ is given by

$$C_j(t) = C_j(0) \, \exp\left(8ik_j^3 t\right) = C_j(0) \, \exp\left(8\kappa_j^3 t\right). \tag{9.40}$$

## 9.5 The Gel'fand–Levitan–Marchenko integral equation

The Gel'fand–Levitan–Marchenko integral equation may be derived from the above formulation. We assume a "triangular" kernel $K(x, s; t)$ (see also below) such that

$$N(x,t;k) = e^{2ikx}\left\{1 + \int_x^{\infty} K(x, s; t)\, e^{ik(s-x)}\, ds\right\}, \tag{9.41}$$

where $K(x, s; t)$ is assumed to decay rapidly as $x$ and $s \to \infty$. Substituting (9.41) into (9.34)–(9.35) and taking the Fourier transform, i.e., operating with

$$\frac{1}{2\pi}\int_{-\infty}^{\infty} dk\, e^{ik(x-y)}, \qquad \text{for} \quad y > x$$

we find that

$$K(x, y; t) + F(x + y; t) + \int_x^{\infty} K(x, s; t)\, F(s + y; t)\, ds = 0, \qquad y > x, \tag{9.42}$$

where

$$F(x; t) = \sum_{j=1}^{N^{\#}} -iC_j(0)\exp\left(8\kappa_j^3 t - \kappa_j x\right) + \frac{1}{2\pi}\int_{-\infty}^{\infty} \rho(k, t)\, e^{ikx}\, dk. \tag{9.43}$$

Substituting (9.41) into (9.36) we have

$$u(x, t) = 2 \frac{\partial}{\partial x} K(x, x; t). \qquad (9.44)$$

Thus we have shown that the Gel'fand–Levitan–Marchenko equation (9.42) arises from the integral equations that result from the Riemann–Hilbert boundary value problem.

We can also show that $K(x, s, t)$ satisfies a PDE in $x$, $s$; i.e., it is independent of $k$. Using (9.41), then from

$$N = e^{ikx} \psi$$

we have

$$\psi = e^{ikx} + \int_x^\infty K(x, s, t) e^{iks} \, ds.$$

We wish to use this in

$$\psi_{xx} + (u(x, t) + k^2)\psi = 0 \qquad (9.45)$$

so we calculate

$$\psi_x = ike^{ikx} - K(x, x, t)e^{ikx} + \int_x^\infty K_x(x, s, t) e^{iks} \, ds$$

and

$$\psi_{xx} = -k^2 e^{ikx} - \frac{d}{dx} K(x, x, t) e^{ikx}$$
$$- ikK(x, x, t)e^{ikx} + \int_x^\infty K_{xx}(x, s, t) e^{iks} \, ds - K_x(x, x, t)e^{ikx}.$$

Part of the third term in (9.45) becomes after integrating by parts

$$k^2 \int_x^\infty K(x, s, t) e^{iks} \, ds = -\int_x^\infty K(x, s, t) \frac{\partial^2}{\partial s^2} e^{iks} \, ds$$
$$= K(x, x, t) ike^{ikx} + \int_x^\infty \frac{\partial K}{\partial s} \frac{\partial}{\partial s} e^{iks} \, ds$$
$$= -ikK(x, x, t)e^{ikx} - \frac{\partial K}{\partial s}(x, x, t)e^{ikx}$$
$$- \int_x^\infty \frac{\partial^2 K}{\partial s^2} e^{iks} \, ds.$$

Then using all of the above in (9.45) gives

$$-\frac{d}{dx}K(x,x,t)e^{ikx} + \int_x^\infty \frac{\partial^2}{\partial x^2}K(x,s,t)e^{iks}\,ds - \frac{\partial}{\partial x}K(x,s=x,t)e^{ikx}$$

$$+ ue^{ikx} + \int_x^\infty u(x)K(x,s,t)e^{iks}\,ds - \frac{\partial K}{\partial s}(x,x,t)e^{ikx}$$

$$- \int_x^\infty \frac{\partial^2 K}{\partial s^2}(x,s,t)e^{iks}\,ds = 0$$

which in turn leads to

$$\int_x^\infty [(K_{xx} - K_{ss}) - uK(x,s,t)]e^{ikx}\,ds$$

$$+ \left[ \underbrace{-\frac{d}{dx}K(x,x,t) - \frac{\partial}{\partial x}K(x,s=x,t) - \frac{\partial}{\partial s}K(x,x,t)}_{-\frac{d}{dx}K(x,x,t)} + u \right]e^{ikx} = 0$$

which for arbitrary kernels leads to

$$(K_{xx} - K_{ss}) - u(x)K(x,s,t) = 0, \quad u(x,t) = 2\frac{d}{dx}K(x,x,t), \quad s > x$$

where $t$ is a parameter and $K(x,s;t)$ decays rapidly as $x$ and $s \to \infty$. Since $K$ satisfies the above system, which is a well-posed Goursat problem (Garabedian, 1984), this indicates the solution exists and is unique. Hence, this shows that $K(x,y,t)$ satisfies a wave-like PDE independent of $k$.

## 9.6 Outline of the inverse scattering transform for the KdV equation

We have seen that associated with the KdV equation (9.3) is a linear scattering problem (9.1) and time-dependent equation (9.2). These equations are compatible if the eigenvalues $\lambda \to i\kappa_j, j = 1,\ldots N^\#$, are time-independent, or isospectral, i.e. $\kappa_{j,t} = 0$.

The direct scattering problem maps a decaying potential $(u \in L_2^1)$ into the scattering data. The inverse scattering problem reconstructs the potential from the scattering data. The time evolution equation tells us how the scattering data evolves.

The solution of the KdV equation (9.3) proceeds as follows: at time $t = 0$, given $u(x,0)$ we solve the direct scattering problem. This involves solving the integral equations (9.15)–(9.16) and using these results to obtain the scattering data (9.19)–(9.20). The scattering data needed are:

$$S(k,0) = \left( \{ \kappa_j, C_j(0) \}_{j=1}^{N^{\#}}, \rho(k,0) \equiv \frac{b(k,0)}{a(k,0)} \right),$$

where $\kappa_j$, $j = 1, 2, \ldots, N^{\#}$, are the locations of the zeros of $a(k,0)$ and $C_j(0)$, $j = 1, 2, \ldots, N^{\#}$, are often called the associated "normalization constants" that were also defined above.

Next we determine the time evolution of the scattering data using equations (9.39)–(9.40). This yields the scattering data at any time $t$:

$$S(k,t) = \left( \{ \kappa_j, C_j(t) \}_{j=1}^{N^{\#}}, \rho(k,t) \equiv \frac{b(k,t)}{a(k,t)} \right).$$

For the inverse problem, at time $t$, we solve either the integral equations for $N(x,t)$ and reconstruct $u(x,t)$ via (9.34)–(9.36) or solve the Gel'fand–Levitan–Marchenko equations for $K(x,y,t)$ and reconstruct $u(x,t)$ via (9.42)–(9.44).

This method is conceptually analogous to the Fourier transform method for solving linear equations (discussed in Chapter 2), except however the step of solving the inverse scattering problem requires one to solve a linear integral equation (i.e., a Riemann–Hilbert boundary value problem) that adds complications. Schematically the solution is given by

$$u(x,0) \xrightarrow{\text{direct scattering}} S(k,0) = \left( \{ \kappa_n, C_j(0) \}_{j=1}^{N^{\#}}, \rho(k,0), a(k,0) \right)$$

$$\downarrow t: \text{time evolution}$$

$$u(x,t) \xleftarrow{\text{inverse scattering}} S(k,t) = \left( \{ \kappa_n, C_j(t) \}_{j=1}^{N^{\#}}, \rho(k,t), a(k,t) \right)$$

In the analogy of IST to the Fourier transform method for linear partial differential equations, the direct scattering problem and the scattering data play the role of the Fourier transform and the inverse scattering problem the inverse Fourier transform.

## 9.7 Soliton solutions of the KdV equation

In this section we will describe how the inverse scattering method can be used to find soliton solutions of the Korteweg–de Vries equation (9.3).

### 9.7.1 The Riemann–Hilbert method

Pure soliton solutions may be obtained from (9.34)–(9.36) in the case when $\rho(k,t) = 0$ on the real $k$-axis. This is usually called a reflectionless potential. We assume a special form for $\rho(k;t)$, namely

$$\rho(k;t) = \begin{cases} 0, & \text{if } k \text{ real,} \\ \displaystyle\sum_{j=1}^{N^{\#}} \frac{C_j(0)}{k - i\kappa_j} \exp\left(8\kappa_j^3 t\right), & \text{if Im } k > 0. \end{cases}$$

Then we obtain a linear algebraic system for the $N_j(x,t)$

$$N_j(x,t) - \sum_{m=1}^{N^{\#}} \frac{iC_m(0)}{\kappa_m + \kappa_j} \exp\left(-2\kappa_m x + 8\kappa_m^3 t\right) N_m(x,t) = \exp(-2\kappa_j x), \quad (9.46)$$

where $j = 1, 2, \ldots, N$, $N_j(x,t) := N(x; k = i\kappa_j; t)$, and the potential $u(x,t)$ is given by

$$u(x,t) = 2\frac{\partial}{\partial x}\left\{\sum_{j=1}^{N^{\#}} \exp\left(8\kappa_j^3 t\right) iC_j(0) N_j(x,t)\right\}.$$

In the case of the one-soliton solution, the number of eigenvalues is $N^{\#} = 1$ and (9.46) reduces to

$$N_1(x,t) - \frac{iC_1(0)}{2\kappa_1} \exp\left(-2\kappa_1 x + 8\kappa_1^3 t\right) N_1(x,t) = \exp(-2\kappa_1 x),$$

and so

$$N_1(x,t) = \frac{2\kappa_1 \exp(-2\kappa_1 x)}{2\kappa_1 - iC_1(0)\exp\left(-2\kappa_1 x + 8\kappa_1^3 t\right)}.$$

Therefore

$$
\begin{aligned}
u(x,t) &= 2\frac{\partial}{\partial x}\left\{\exp\left(8\kappa_1^3 t\right) iC_1(0) N_1(x,t)\right\} \\
&= 2\frac{\partial}{\partial x}\left\{\frac{2\kappa_1 iC_1(0)\exp\left(-2\kappa_1 x + 8\kappa_1^3 t\right)}{2\kappa_1 - iC_1(0)\exp\left(-2\kappa_1 x + 8\kappa_1^3 t\right)}\right\} \\
&= 2\frac{\partial}{\partial x}\left\{\frac{2\kappa_1 iC_1(0)}{2\kappa_1 \exp\left(2\kappa_1 x - 8\kappa_1^3 t\right) - iC_1(0)}\right\} \\
&= \frac{8\kappa_1^2}{\left[\exp\left(\kappa_1 x - \kappa_1^3 t - \kappa_1 x_1\right) + \exp\left(-\kappa_1 x + \kappa_1^3 t + \kappa_1 x_1\right)\right]^2}
\end{aligned}
$$

where we define $-iC_1(0) = 2\kappa_1 \exp(2\kappa_1 x_1)$; hence we obtain the one-soliton solution

$$u(x,t) = 2\kappa_1^2 \operatorname{sech}^2 \left\{ \kappa_1 \left( x - 4\kappa_1^2 t - x_1 \right) \right\}.$$

In the case of the two-solution solution, we have to first solve a linear system

$$N_1(x,t) + \sum_{j=1}^{2} \frac{C_j(0)}{\kappa_j + \kappa_1} \exp\left( -2\kappa_j x + 8\kappa_j^3 t \right) N_j(x,t) = \exp(-2\kappa_1 x),$$

$$N_2(x,t) + \sum_{j=1}^{2} \frac{C_j(0)}{\kappa_j + \kappa_2} \exp\left( -2\kappa_j x + 8\kappa_j^3 t \right) N_j(x,t) = \exp(-2\kappa_2 x),$$

for $N_1(x,t)$ and $N_2(x,t)$. After some algebra, the two-soliton solution of the KdV equation is found to be

$$u(x,t) = \frac{4\left( \kappa_2^2 - \kappa_1^2 \right) \left[ \left( \kappa_2^2 - \kappa_1^2 \right) + \kappa_1^2 \cosh(2\kappa_2 \xi_2) + \kappa_2^2 \cosh(2\kappa_1 \xi_1) \right]}{\left[ (\kappa_2 - \kappa_1) \cosh(\kappa_1 \xi_1 + \kappa_2 \xi_2) + (\kappa_2 + \kappa_1) \cosh(\kappa_2 \xi_2 - \kappa_1 \xi_1) \right]^2},$$

where $\xi_i = x - 4\kappa_i^2 t - x_i$, and $C_i(0) = 2\kappa_i \exp(2\kappa_i x_i)$, for $i = 1, 2$.

### 9.7.2 Gel'fand–Levitan–Marchenko approach

Again we suppose that the potential $u(x,0)$ is reflectionless, $\rho(k,0) = 0$, and the time-independent Schrödinger equation (9.4) has $N^\#$ discrete eigenvalues $k_n = i\kappa_n$, $n = 1, 2, \ldots, N^\#$, such that $0 < \kappa_1 < \kappa_2 < \cdots < \kappa_{N^\#}$. To connect with the standard approach, in the kernel $F(x,t)$ in the Gel'fand–Levitan–Marchenko equation (9.42)–(9.43) we take $-iC_n = c_n^2$. In this way the associated eigenfunctions $\psi_n(x)$, $n = 1, 2, \ldots, N^\#$, satisfy the normalization condition

$$\int_{-\infty}^{\infty} \psi_n^2(x)\, dx = 1,$$

and the asymptotic limit as $x \to +\infty$, $\lim_{x\to+\infty} \psi_n(x) \exp(\kappa_n x) = c_n(0)$, where $c_j(0)$ are called the Gel'fand–Levitan–Marchenko normalization constants; and $c_n(t) = c_n(0) \exp\left( 4\kappa_n^3 t \right)$.

Then this is solvable by a separation of variables (cf. Kay and Moses, 1956; Gardner et al., 1974). Here we take

$$F(x;t) = \sum_{n=1}^{N^\#} c_n^2(0) \exp\left( 8\kappa_n^3 t - \kappa_n x \right),$$

and so the Gel'fand–Levitan–Marchenko equation becomes

$$
K(x, y; t) + \sum_{n=1}^{N^\#} c_n^2(t) \exp\{-\kappa_n(x + y)\}
$$

$$
+ \int_x^\infty K(x, s; t) \sum_{n=1}^{N} c_n^2(t) \exp\{-\kappa_n(s + y)\} \, ds = 0. \tag{9.47}
$$

Now we seek a solution of this equation in the form

$$
K(x, y; t) = \sum_{n=1}^{N^\#} L_n(x, t) \exp(-\kappa_n y),
$$

and then the integral equation (9.47) reduces to a linear algebraic system

$$
L_n(x, t) + \sum_{m=1}^{N^\#} \frac{c_n^2(t)}{\kappa_m + \kappa_n} \exp\{-(\kappa_m + \kappa_n)x\} L_m(x, t) + c_n^2(t) \exp(-\kappa_n x) = 0,
$$

for $n = 1, 2, \ldots, N^\#$. We can, in principle, solve this linear system. We will go a little further.

If we let $L_n(x, t) = -c_n(t)w_n(x, t)$, for $n = 1, 2, \ldots, N^\#$, this yields a system that can be written in the convenient form

$$
(\mathsf{I} + \mathsf{B})\mathbf{w} = \mathbf{f}, \tag{9.48}
$$

where $\mathbf{w} = (w_1, w_2, \ldots, w_{N^\#})$, $\mathbf{f} = (f_1, f_2, \ldots, f_{N^\#})$ with $f_n(x, t) = c_n(t) \exp(-\kappa_n x)$ for $n = 1, 2, \ldots, N^\#$, $\mathsf{I}$ is the $N^\# \times N^\#$ identity matrix and $\mathsf{B}$ is a symmetric, $N^\# \times N^\#$ matrix with entries

$$
B_{mn}(x, t) = \frac{c_m(t)c_n(t)}{\kappa_m + \kappa_n} \exp\{-(\kappa_m + \kappa_n)x\}, \qquad m, n = 1, 2, \ldots, N^\#.
$$

The system (9.48) has a unique solution when $\mathsf{B}$ is positive definite. If we consider the quadratic form

$$
\mathbf{v}^{\mathsf{T}} \mathsf{B} \mathbf{v} = \sum_{m=1}^{N^\#} \sum_{n=1}^{N^\#} \frac{c_m(t)c_n(t)v_m v_n}{\kappa_m + \kappa_n} \exp\{-(\kappa_m + \kappa_n)x\}
$$

$$
= \int_x^\infty \left[ \sum_{n=1}^{N^\#} c_n(t)v_n \exp(-\kappa_n y) \right]^2 dy,
$$

where $\mathbf{v} = (v_1, v_2, \ldots, v_{N^\#})$, which is positive and vanishes only if $\mathbf{v} = \mathbf{0}$; then $\mathsf{B}$ is positive definite.

If in the kernel, $F(x,t)$, of the Gel'fand–Levitan–Marchenko equation (9.47), we take $N^\# = 1$ and define $c_1^2(0) = 2\kappa_1 \exp(2\kappa_1 x_1)$, then we find the one-soliton solution

$$u(x,t) = 2\kappa_1^2 \operatorname{sech}^2\left\{\kappa_1\left(x - 4\kappa_1^2 t - x_1\right)\right\},$$

which is the same as derived from the RHBVP approach. The general $N^\#$-soliton case was analyzed in detail by Gardner et al. (1974).

## 9.8 Special initial potentials

In the previous sections we obtained pure soliton solutions for the KdV equation corresponding to reflectionless potentials. However, more general choices of the initial data $u(x,0)$ give rise to non-zero reflection coefficients; unfortunately, in this case it generally is not possible to solve the inverse equations in explicit form. However, long-time asymptotic information can be obtained (Ablowitz and Segur, 1979; Deift and Zhou, 1992; Deift et al., 1994).

### 9.8.1 Delta function initial profile

As an example consider the delta function initial profile given by

$$u(x;0) = Q\delta(x),$$

where $Q$ is constant and $\delta(x)$ is the Dirac delta function. We use the Schrödinger equation

$$v_{xx} + (k^2 + u(x,0))v = 0$$

with the scattering definitions given by (9.23)–(9.24). So

$$\phi(x;k) = e^{-ikx}, \qquad \text{for} \qquad x < 0,$$
$$\psi(x;k) = e^{ikx}, \qquad \text{for} \qquad x > 0, \qquad \text{and}$$
$$\bar{\psi}(x;k) = e^{-ikx}, \qquad \text{for} \qquad x > 0.$$

Then requiring the function $v$ to be continuous and satisfy the jump condition

$$[v_x]_{0^-}^{0^+} + Qv(0) = 0,$$

and letting $v(x,k) = \phi(x;k)$ yields the following equations

$$a(k,0) + b(k,0) = 1$$
$$a(k,0)(-ik) + b(k,0)(ik) = -Q - ik.$$

Hence

$$a(k,0) = \frac{2ik + Q}{2ik} = a(k,t) = \frac{1}{\tau(k,t)}, \quad b(k,0) = \frac{-Q}{2ik}, \quad \rho(k,t) = \frac{-Q}{2ik + Q}.$$

Thus in this case, when $Q > 0$, there is a single discrete eigenvalue $\lambda = -\kappa_1^2$, where $\kappa_1 = Q/2$. The reflection coefficient is non-zero and given by

$$\rho(k,0) = \frac{-\kappa_1}{\kappa_1 + ik},$$

and similarly we find $C_1(0) = i\kappa_1$. The time evolution follows as above; it is given by

$$\kappa_1 = \text{constant}, \quad C_1(t) = i\kappa_1 \exp\left(8\kappa_1^3 t\right), \quad \rho(k,t) = \frac{-\kappa_1 \exp(8ik^3 t)}{\kappa_1 + ik}.$$

### 9.8.2 Box initial profile

Here we outline how to find the data associated with a box profile at $t = 0$:

$$u(x,0) = \begin{cases} H^2, & |x| < L/2 \\ 0, & |x| > L/2 \end{cases}$$

where $H, L > 0$ are constant. Again, we use the Schrödinger equation

$$v_{xx} + (k^2 + u(x,0))v = 0$$

with the scattering functions/coefficients satisfying (9.23)–(9.24).

Since this is a linear, constant coefficient, second-order differential equation we have

$$v(x) = \begin{cases} Ae^{ikx} + Be^{-ikx}, & x < -L/2 \\ Ce^{i\lambda x} + De^{-i\lambda x}, & |x| < L/2 \\ Ee^{ikx} + Fe^{-ikx}, & x > L/2 \end{cases}$$

where $\lambda \equiv \sqrt{H^2 + k^2}$ and so

$$v'(x) = \begin{cases} Aike^{ikx} - Bike^{-ikx}, & x < -L/2 \\ Ci\lambda e^{i\lambda x} - Di\lambda e^{-i\lambda x}, & |x| < L/2 \\ Eike^{ikx} - Fike^{-ikx}, & x > L/2. \end{cases}$$

Let us first consider the case when $A = 0$ and $B = 1$; then, requiring $v$ and $v_x$ to be continuous at $x = \pm L/2$ implies

$$e^{ikL/2} = Ce^{-i\lambda L/2} + De^{i\lambda L/2},$$
$$-ike^{ikL/2} = Ci\lambda e^{-i\lambda L/2} - Di\lambda e^{i\lambda L/2},$$
$$Ee^{ikL/2} + Fe^{-ikL/2} = Ce^{i\lambda L/2} + De^{-i\lambda L/2},$$
$$Eike^{ikL/2} - Fike^{-ikL/2} = Ci\lambda e^{i\lambda L/2} - Di\lambda e^{-i\lambda L/2}.$$

Solving for the constants $C, D, E, F$ gives

$$C = \frac{\lambda - k}{2\lambda} e^{i(k+\lambda)L/2},$$

$$D = \frac{\lambda + k}{2\lambda} e^{i(k-\lambda)L/2},$$

$$E = -\frac{k^2 - \lambda^2}{4k\lambda} \left( e^{iL\lambda} - e^{-iL\lambda} \right),$$

$$F = -\frac{e^{iL(k-\lambda)}}{4k\lambda} \left\{ (k-\lambda)^2 e^{2iL\lambda} - (k+\lambda)^2 \right\}.$$

We define the function $\phi \sim e^{-ikx}$ as $x \to -\infty$, so

$$\phi(x) \equiv \begin{cases} e^{-ikx}, \ x < -L/2 \\ e^{ikL/2} \left\{ \cos\left[ \lambda(L/2 + x) \right] - \frac{ik}{\lambda} \sin\left[ \lambda(L/2 + x) \right] \right\}, \ |x| < L/2 \\ e^{-ik(x-L)} \cos(\lambda L) \\ \quad - \frac{i \sin(\lambda L)}{2k\lambda} \left[ e^{-ik(x-L)}(\lambda^2 + k^2) - e^{ikx}(\lambda^2 - k^2) \right], \ x > L/2. \end{cases}$$

Now, let us consider the case when $A = 1$ and $B = 0$; then solving for $C, D,$ $E, F$ by enforcing continuity of $v$ and $v_x$ at $x = \pm L/2$ gives

$$C = \frac{\lambda + k}{2\lambda} e^{i(k-\lambda)L/2},$$

$$D = \frac{\lambda - k}{2\lambda} e^{i(k+\lambda)L/2},$$

$$E = e^{-ikL} \left\{ \cos(\lambda L) + i\frac{\lambda^2 + k^2}{2k\lambda} \sin(\lambda L) \right\},$$

$$F = i\frac{\lambda^2 - k^2}{2k\lambda} \sin(\lambda L).$$

We then define $\overline{\phi} \sim e^{-ikx}$ as $x \to -\infty$ and find

$$\overline{\phi}(x) \equiv \begin{cases} e^{ikx}, \ x < -L/2 \\ e^{-ikL/2} \left\{ \cos\left[ \lambda(L/2 + x) \right] + \frac{ik}{\lambda} \sin\left[ \lambda(L/2 + x) \right] \right\}, \ |x| < L/2 \\ e^{ik(x-L)} \cos(\lambda L) \\ \quad + \frac{i \sin(\lambda L)}{2k\lambda} \left[ e^{ik(x-L)}(\lambda^2 + k^2) - e^{-ikx}(\lambda^2 - k^2) \right], \ x > L/2. \end{cases}$$

Recall the eigenfunctions $\psi, \overline{\psi}$ have the asymptotic boundary conditions: $\psi \sim$ $e^{ikx}$ as $x \to \infty$ and $\overline{\psi} \sim e^{-ikx}$ as $x \to \infty$; we write:

$$\phi(x) = a(k)\overline{\psi}(x) + b(k)\psi(x) \sim a(k)e^{-ikx} + b(k)e^{ikx},$$

as $x \to \infty$. By comparing coefficients we find that

$$a(k) = e^{ikL} \left\{ \cos\left(L \sqrt{H^2 + k^2}\right) - i\frac{H^2 + 2k^2}{2k \sqrt{H^2 + k^2}} \sin\left(L \sqrt{H^2 + k^2}\right) \right\},$$

$$b(k) = \frac{iH^2}{2k \sqrt{H^2 + k^2}} \sin\left(L \sqrt{H^2 + k^2}\right).$$

Since the zeros of $a(k)$ are purely imaginary, let $k = i\kappa$ where $\kappa \in \mathbb{R}$. Then

$$a(i\kappa) = e^{-\kappa L} \left\{ \cos\left(L \sqrt{H^2 - \kappa^2}\right) - \frac{H^2 - 2\kappa^2}{2\kappa \sqrt{H^2 - \kappa^2}} \sin\left(L \sqrt{H^2 - \kappa^2}\right) \right\}.$$

Thus, the zeros of $a(i\kappa)$ occur when

$$\tan\left(L \sqrt{H^2 - \kappa^2}\right) = \frac{2\kappa \sqrt{H^2 - \kappa^2}}{H^2 - 2\kappa^2}.$$

Note that the right-hand side is a monotonically increasing function from $[0, \infty)$ on $0 \le \kappa < H/\sqrt{2}$ and from $(-\infty, 0]$ on $H/\sqrt{2} < \kappa \le H$, and that $\tan\left(LH \sqrt{1 - \kappa^2/H^2}\right)$ has $LH/\pi$ periods for $\kappa$ between $0$ and $H$. Thus, the number of zeros of $a(i\kappa)$ depends on the size of $LH$, such that there are $n \in \mathbb{N}$ zeros of $a(i\kappa)$ for $0 < \kappa < H$ when

$$(n - 1)\pi < LH \le n\pi.$$

Additionally, using the same reasoning as above, there are up to $n$ zeros for $-H < \kappa < 0$ when $LH < n\pi$. For $|\kappa| > H$, we can write

$$\tanh\left(L \sqrt{\kappa^2 - H^2}\right) = \frac{2\kappa \sqrt{\kappa^2 - H^2}}{H^2 - 2\kappa^2},$$

and it is seen that there are no zeros for $\kappa > H$. As earlier, the time dependence of the scattering data can be determined from $a(k, t) = a(k, 0), b(k, t) = b(k, 0) \exp(8ik^3 t)$, etc.

## 9.9 Conserved quantities and conservation laws

As mentioned in Chapter 8, one of the major developments in the study of integrable systems was the recognition that special equations, such as the KdV equation, have an infinite number of conserved quantities/conservation laws. These were derived by Gardner et al. (1967) by making use of the Miura transformation. That derivation can be found in Section 8.3. In this section we use IST to derive these results.

Recall from Lemma 9.1 that $a(k)$ and $M(x; k) = \phi(x; k) \exp(-ikx)$ are analytic in the upper half $k$-plane and that $a(k)$ has a finite number of simple zeros on the imaginary axis: $\{k_m = iK_m\}_{m=1}^{N}$, in the upper half $k$-plane and $a(k) \to 1$ as $k \to \infty$. Moreover we have shown in Section 9.4 that $a(k)$ is constant in time. We will use this fact to obtain the conserved densities.

We can relate $\log a(k)$ to the potential $u$ and its derivatives. From (9.5) and (9.8) we have

$$a(k) = \frac{1}{2ik} W(\phi, \psi)$$

$$= \frac{1}{2ik}(\phi\psi_x - \phi_x\psi) = \lim_{x \to +\infty} \frac{1}{2ik}\left(\phi i k e^{ikx} - \phi_x e^{ikx}\right).$$

Letting

$$\phi = e^{p - ikx}, \quad p \to 0 \text{ as } x \to -\infty \tag{9.49}$$

yields

$$a(k) = \lim_{x \to +\infty} e^p \left(1 - \frac{p_x}{2ik}\right)$$

or

$$\log a = \lim_{x \to +\infty} p + \lim_{x \to +\infty} \log\left(1 - \frac{p_x}{2ik}\right). \tag{9.50}$$

Substituting (9.49) into

$$\phi_{xx} + (k^2 + u)\phi = 0$$

and calling $\mu = p_x = ik + \dfrac{\phi_x}{\phi}$ gives

$$\mu_x + \mu^2 - 2ik\mu + u = 0 \tag{9.51}$$

which is a Ricatti equation for $\mu$ in terms of $u$. This equation has a series solution, in terms of inverse powers of $k$, of the form

$$\mu = \sum_{n=1}^{\infty} \frac{\mu_n}{(2ik)^n} = \frac{\mu_1}{2ik} + \frac{\mu_2}{(2ik)^2} + \frac{\mu_3}{(2ik)^3} + \cdots, \tag{9.52}$$

which yields a recursion relation for the $\mu_n$. From (9.51) the first three terms are

$$\mu_1 = u, \quad \mu_2 = u_x, \quad \mu_3 = u^2 + u_{xx} \tag{9.53}$$

and in general $\mu_n$ satisfies

$$\mu_{n,x} + \sum_{q=1}^{n-1} \mu_q \mu_{n-q} - \mu_{n+1} = 0 \tag{9.54}$$

which further yields $\mu_4 = (2u^2 + u_{xx})_x$, $\mu_5 = \mu_{4,x} + (u^2)_{xx} + 2u^3 - u_x^2$, and so on. From (9.53)–(9.54) we see that all of $\mu_n$ are polynomial functions of $u$ and its derivatives, i.e., $\mu_n = \mu_n(u, u_x, u_{xx}, \ldots)$ and $\mu_n \to 0$ as $x \to \infty$. Hence,

$$\lim_{x \to +\infty} \log\left(1 - \frac{p_x}{2ik}\right) = 0.$$

Then from (9.50) and (9.52),

$$\log a = \lim_{x \to +\infty} \int_{-\infty}^x p_x \, dx = \int_{-\infty}^\infty \mu \, dx$$

$$= \sum_{n=1}^\infty \frac{1}{(2ik)^n} \int_{-\infty}^\infty \mu_n(u, u_x, u_{xx}, \ldots) \, dx = \sum_{n=1}^\infty \frac{C_n}{(2ik)^2}, \qquad (9.55)$$

which implies we have an infinite number of conserved quantities

$$C_n = \int_{-\infty}^\infty \mu_n(u, u_x, u_{xx}, \ldots) \, dx.$$

The first three non-trivial conserved quantities are, from (9.53)–(9.54) ($C_{2n}$ are trivial)

$$C_1 = \int_{-\infty}^\infty u \, dx, \quad C_3 = \int_{-\infty}^\infty u^2 \, dx, \quad C_5 = \int_{-\infty}^\infty \left(2u^3 - u_x^2\right) dx.$$

We can also use the large $k$ expansion to retrieve the infinite number of conservation laws

$$\frac{\partial T_n}{\partial t} + \frac{\partial F_n}{\partial x} = 0, \quad n = 1, 2, 3, \ldots$$

associated with the KdV equation. As discussed earlier there is an associated linear dependence with the Schrödinger scattering problem; that is,

$$\phi_t = (u_x + 4ik^3)\phi + (4k^2 - 2u)\phi_x. \qquad (9.56)$$

We note that the fixed boundary condition $\phi \sim \exp(-ikx)$ as $x \to -\infty$ is consistent with (9.56). Substituting $\phi = \exp(p - ikx)$, $\mu = p_x$ into (9.56) and taking a derivative with respect to $x$ yields after some manipulation

$$\mu_t = \frac{\partial}{\partial x}[2iku + u_x + (4k^2 - 2u)\mu].$$

Then the expansion (9.52) gives

$$\frac{\partial \mu_n}{\partial t} + \frac{\partial}{\partial x}(\mu_{n+2} + 2u\mu_n) = 0, \quad n \geq 1$$

with the $\mu$ obtained from (9.53)–(9.54). Substitution of $\mu_n$ yields the infinite number of conservation laws. The first two non-trivial conservation laws are

$$\frac{\partial}{\partial t}u + \frac{\partial}{\partial x}(u_{xx} + 3u^2) = 0$$

$$\frac{\partial}{\partial t}(u^2 + u_{xx}) + \frac{\partial}{\partial x}\left[u_{xxxx} + 4(u^2)_{xx} + 4u^3 - 3u_x^2\right] = 0.$$

Finally, we note that the conserved densities can also be computed in terms of the scattering data. This requires some complex analysis.

Consider the function

$$\alpha(k) = a(k)\prod_{m=1}^{N^\#} \frac{k - k_m^*}{k - k_m}, \tag{9.57}$$

where $k_m^*$ is the complex conjugate of $k_m$. This function is analytic in the upper half $k$-plane with no zeros and $\alpha(k) \to 1$ as $k \to \infty$. By the Schwarz reflection principle (see Ablowitz and Fokas, 2003) the complex conjugate of (9.57)

$$\alpha(k)^* = a(k)^*\prod_{m=1}^{N^\#} \frac{k - k_m}{k - k_m^*} \tag{9.58}$$

is analytic in the lower half $k$-plane with no zeros. Then by Cauchy's integral theorem for Im$\{k\} > 0$

$$\log \alpha(k) = \frac{1}{2\pi i}\int_{-\infty}^{\infty} \frac{\log \alpha(z)}{z - k}\, dz$$

$$0 = \frac{1}{2\pi i}\int_{-\infty}^{\infty} \frac{\log \alpha(z)^*}{z - k}\, dz.$$

Adding the above equations for Im$\{k\} > 0$ and using (9.57) and (9.58) gives

$$\log a(k) = -\frac{1}{2\pi i k}\int_{-\infty}^{\infty} \frac{\log aa^*}{1 - z/k}\, dz + \sum_{m=1}^{N^\#} \log\left(\frac{1 - k_m/k}{1 - k_m^*/k}\right).$$

Then expanding for large $k$ gives

$$\log a(k) = -\frac{1}{2\pi i}\sum_{n=1}^{\infty}\frac{1}{k^n}\int_{-\infty}^{\infty} z^{n-1}\log(aa^*(z))\, dz$$

$$+ \sum_{n=1}^{\infty}\frac{1}{k^n}\sum_{m=1}^{N^\#}\frac{((k_m^*)^n - k_m^n)}{n} = \sum_{n=1}^{\infty}\frac{C_n}{(2ik)^n}.$$

The right-hand side of the above equation is also obtained from (9.55); thus $C_n$ are the conserved densities now written in terms of scattering data. Recall,

$$1 - \left|\frac{b}{a}(k)\right|^2 = 1 - |\rho(k)|^2 = \frac{1}{|a(k)|^2}.$$

Thus we can write the non-trivial conserved densities in terms of the data obtained from $\log a(k)$:

$$C_{2n+1} = (2i)^{2n+1}\left[\frac{1}{2\pi i}\int_{-\infty}^{\infty} z^{2n}\log(1-|\rho|^2)\,dz + \sum_{m=1}^{N^\#}\frac{(k_m^*)^{2n+1}-k_m^{2n+1}}{2n+1}\right],$$

for $n \geq 1$, or noting that $k_m = i\kappa_m$, $\kappa_m > 0$

$$C_{2n+1} = \frac{1}{\pi}\int_{-\infty}^{\infty}(-1)^n(2z)^{2n}\log(1-|\rho(z)|^2)\,dz + \frac{2}{2n+1}\sum_{m=1}^{N^\#}(2\kappa_m)^{2n+1},$$

which gives the conserved densities in terms of the scattering data $\{\rho(z), \kappa_m|_{m=1}^N\}$.

## 9.10 Outline of the IST for a general evolution system – including the nonlinear Schrödinger equation with vanishing boundary conditions

In the previous sections of this chapter we concentrated on the IST for the KdV equation with vanishing boundary values. In this section we will discuss the main steps in the IST associated with the following compatible linear $2 \times 2$ system (Lax pair):

$$v_x = \begin{pmatrix} -ik & q \\ r & ik \end{pmatrix} v, \tag{9.59a}$$

and

$$v_t = \begin{pmatrix} A & B \\ C & -A \end{pmatrix} v, \tag{9.59b}$$

where $v$ is a two-component vector, $v(x,t) = \left(v^{(1)}(x,t), v^{(2)}(x,t)\right)^{\mathrm{T}}$.[1] In Chapter 8 we described how the compatibility of the above general linear system yielded a class of nonlinear evolution equations. This class includes as special cases the nonlinear Schrödinger (NLS), modified Korteweg–de Vries (mKdV)

---

[1] Unless otherwise specified, superscripts $^{(\ell)}$ with $\ell = 1, 2$ denote the $\ell$th component of the corresponding two-component vector and $^{\mathrm{T}}$ denotes matrix transpose.

and sine–, sinh–Gordon equations. The equality of the mixed derivatives of $v$, i.e., $v_{xt} = v_{tx}$, is equivalent to the statement that $q$ and $r$ satisfy these evolution equations, if $k$, the scattering parameter or eigenvalue, is independent of $x$ and $t$. As before, we refer to the equation with the $x$ derivative, (9.59a), as the scattering problem: its solutions are termed eigenfunctions (with respect to the eigenvalue $k$) and the equation with the $t$ derivative, (9.59b), is called the time equation.

For example, the particular case of the NLS equation results when we take the time-dependent system to be

$$v_t = \begin{pmatrix} A & B \\ C & -A \end{pmatrix} v = \begin{pmatrix} 2ik^2 + iqr & 2kq + iq_x \\ 2kr + ir_x & -2ik^2 - iqr \end{pmatrix} v, \qquad (9.60)$$

which by compatibility of (9.59a)–(9.60) yields

$$iq_t = q_{xx} - 2rq^2 \qquad (9.61a)$$
$$-ir_t = r_{xx} - 2qr^2. \qquad (9.61b)$$

Then the reduction $r = \sigma q^*$ with $\sigma = \mp 1$, gives, as a special case, the NLS equation

$$iq_t = q_{xx} \pm 2|q|^2 q. \qquad (9.62)$$

We say the NLS equation is focusing or defocusing corresponding to $\sigma = \mp 1$.

In what follows we will assume that the potentials, i.e., the solutions of the nonlinear evolution equations $q(x, t), r(x, t)$, decay sufficiently rapidly as $x \to \pm\infty$; at least $q, r \in L^1$. The reader can also see the references mentioned earlier in this chapter and in Chapter 8. Here we will closely follow the methods and notation employed by Ablowitz et al. (2004b).

We can rewrite the scattering problem (9.59a) as

$$v_x = (ik\mathbf{J} + \mathbf{Q}) v$$

where

$$\mathbf{J} = \begin{pmatrix} -1 & 0 \\ 0 & 1 \end{pmatrix}, \quad \mathbf{Q} = \begin{pmatrix} 0 & q \\ r & 0 \end{pmatrix}.$$

and $\mathbf{I}$ is the $2 \times 2$ identity matrix.

## 9.10.1 Direct scattering problem

### Eigenfunctions and integral equations

When the potentials $q, r \to 0$ sufficiently rapidly as $x \to \pm\infty$ the eigenfunctions are asymptotic to the solutions of

$$v_x = \begin{pmatrix} -ik & 0 \\ 0 & ik \end{pmatrix} v$$

when $|x|$ is large. We single out the solutions of (9.59a) (eigenfunctions) defined by the following boundary conditions:

$$\phi(x; k) \sim \begin{pmatrix} 1 \\ 0 \end{pmatrix} e^{-ikx}, \qquad \overline{\phi}(x; k) \sim \begin{pmatrix} 0 \\ 1 \end{pmatrix} e^{ikx} \qquad \text{as } x \to -\infty \qquad (9.63a)$$

$$\psi(x; k) \sim \begin{pmatrix} 0 \\ 1 \end{pmatrix} e^{ikx}, \qquad \overline{\psi}(x; k) \sim \begin{pmatrix} 1 \\ 0 \end{pmatrix} e^{-ikx} \qquad \text{as } x \to +\infty. \qquad (9.63b)$$

Note that here and in the following the bar does not denote complex conjugate, for which we will use the $*$ notation. It is convenient to introduce eigenfunctions with constant boundary conditions by defining

$$M(x; k) = e^{ikx}\phi(x; k), \qquad \overline{M}(x; k) = e^{-ikx}\overline{\phi}(x; k)$$
$$N(x; k) = e^{-ikx}\psi(x; k), \qquad \overline{N}(x; k) = e^{ikx}\overline{\psi}(x; k).$$

In terms of the above notation, the eigenfunctions, also called *Jost solutions*, $M(x; k)$ and $\overline{N}(x; k)$, are solutions of the differential equation

$$\partial_x \chi(x; k) = ik\,(\mathbf{J} + \mathbf{I})\,\chi(x; k) + (\mathbf{Q}\chi)\,(x; k), \qquad (9.64)$$

while the eigenfunctions $N(x; k)$ and $\overline{M}(x; k)$ satisfy

$$\partial_x \overline{\chi}(x; k) = ik\,(\mathbf{J} - \mathbf{I})\,\overline{\chi}(x; k) + (\mathbf{Q}\overline{\chi})\,(x; k), \qquad (9.65)$$

with the constant boundary conditions

$$M(x; k) \to \begin{pmatrix} 1 \\ 0 \end{pmatrix}, \qquad \overline{M}(x; k) \to \begin{pmatrix} 0 \\ 1 \end{pmatrix} \qquad \text{as } x \to -\infty \qquad (9.66a)$$

$$N(x; k) \to \begin{pmatrix} 0 \\ 1 \end{pmatrix}, \qquad \overline{N}(x; k) \to \begin{pmatrix} 1 \\ 0 \end{pmatrix} \qquad \text{as } x \to +\infty. \qquad (9.66b)$$

Solutions of the differential equations (9.64) and (9.65) can be represented by means of the following integral equations

$$\chi(x;k) = w + \int_{-\infty}^{+\infty} \mathbf{G}(x - x';k)\mathbf{Q}(x')\chi(x';k)\,dx'$$

$$\overline{\chi}(x;k) = \overline{w} + \int_{-\infty}^{+\infty} \overline{\mathbf{G}}(x - x';k)\mathbf{Q}(x')\overline{\chi}(x';k)\,dx'$$

where $w = (1, 0)^{\mathrm{T}}$, $\overline{w} = (0, 1)^{\mathrm{T}}$ and the (matrix) Green's functions $\mathbf{G}(x;k)$ and $\overline{\mathbf{G}}(x;k)$ satisfy the differential equations

$$[\mathbf{I}\partial_x - ik\,(\mathbf{J} + \mathbf{I})]\,\mathbf{G}(x;k) = \delta(x),$$
$$[\mathbf{I}\partial_x - ik\,(\mathbf{J} - \mathbf{I})]\,\overline{\mathbf{G}}(x;k) = \delta(x),$$

where $\delta(x)$ is the Dirac delta (generalized) function. The Green's functions are not unique, and, as we will show below, the choice of the Green's functions and the choice of the inhomogeneous terms $w$, $\overline{w}$ together uniquely determine the eigenfunctions and their analytic properties in $k$.

By using the Fourier transform method, one can represent the Green's functions in the form

$$\mathbf{G}(x;k) = \frac{1}{2\pi i} \int_C \begin{pmatrix} 1/p & 0 \\ 0 & 1/(p - 2k) \end{pmatrix} e^{ipx}\,dp$$

$$\overline{\mathbf{G}}(x;k) = \frac{1}{2\pi i} \int_{\overline{C}} \begin{pmatrix} 1/(p + 2k) & 0 \\ 0 & 1/p \end{pmatrix} e^{ipx}\,dp$$

where $C$ and $\overline{C}$ will be chosen as appropriate contour deformations of the real $p$-axis. It is natural to consider $\mathbf{G}_\pm(x;k)$ and $\overline{\mathbf{G}}_\pm(x;k)$ defined by

$$\mathbf{G}_\pm(x;k) = \frac{1}{2\pi i} \int_{C_\pm} \begin{pmatrix} 1/p & 0 \\ 0 & 1/(p - 2k) \end{pmatrix} e^{ipx}\,dp$$

$$\overline{\mathbf{G}}_\pm(x;k) = \frac{1}{2\pi i} \int_{\overline{C}_\pm} \begin{pmatrix} 1/(p + 2k) & 0 \\ 0 & 1/p \end{pmatrix} e^{ipx}\,dp$$

where $C_\pm$ and $\overline{C}_\pm$ are the contours from $-\infty$ to $+\infty$ that pass below (+ functions) and above (− functions) both singularities at $p = 0$ and $p = 2k$, or respectively, $p = 0$ and $p = -2k$. Contour integration then gives

$$\mathbf{G}_\pm(x;k) = \pm\theta(\pm x) \begin{pmatrix} 1 & 0 \\ 0 & e^{2ikx} \end{pmatrix},$$

$$\overline{\mathbf{G}}_\pm(x;k) = \mp\theta(\mp x) \begin{pmatrix} e^{-2ikx} & 0 \\ 0 & 1 \end{pmatrix},$$

where $\theta(x)$ is the Heaviside function ($\theta(x) = 1$ if $x > 0$ and $\theta(x) = 0$ if $x < 0$). The "+" functions are analytic and bounded in the upper half-plane of $k$ and the "−" functions are analytic in the lower half-plane. By taking into account the boundary conditions (9.66), we obtain the following integral equations for the eigenfunctions:

$$M(x;k) = \begin{pmatrix} 1 \\ 0 \end{pmatrix} + \int_{-\infty}^{+\infty} \mathbf{G}_+(x - x';k)\mathbf{Q}(x')M(x';k)dx' \qquad (9.67a)$$

$$N(x;k) = \begin{pmatrix} 0 \\ 1 \end{pmatrix} + \int_{-\infty}^{+\infty} \overline{\mathbf{G}}_+(x - x';k)\mathbf{Q}(x')N(x';k)dx' \qquad (9.67b)$$

$$\overline{M}(x;k) = \begin{pmatrix} 0 \\ 1 \end{pmatrix} + \int_{-\infty}^{+\infty} \overline{\mathbf{G}}_-(x - x';k)\mathbf{Q}(x')\overline{M}(x';k)dx' \qquad (9.67c)$$

$$\overline{N}(x;k) = \begin{pmatrix} 1 \\ 0 \end{pmatrix} + \int_{-\infty}^{+\infty} \mathbf{G}_-(x - x';k)\mathbf{Q}(x')\overline{N}(x';k)dx'. \qquad (9.67d)$$

Equations (9.67) are Volterra integral equations, whose solutions can be sought in the form of Neumann series iterates. In the following lemma we will show that if $q, r \in L^1(\mathbb{R})$, the Neumann series associated to the integral equations for $M$ and $N$ converge absolutely and uniformly (in $x$ and $k$) in the upper $k$-plane, while the Neumann series of the integral equations for $\overline{M}$ and $\overline{N}$ converge absolutely and uniformly (in $x$ and $k$) in the lower $k$-plane. This implies that the eigenfunctions $M(x;k)$ and $N(x;k)$ are analytic functions of $k$ for $\text{Im}\,k > 0$ and continuous for $\text{Im}\,k \geq 0$, while $\overline{M}(x;k)$ and $\overline{N}(x;k)$ are analytic functions of $k$ for $\text{Im}\,k < 0$ and continuous for $\text{Im}\,k \leq 0$.

**Lemma 9.2** *If $q, r \in L^1(\mathbb{R})$, then $M(x;k)$, $N(x;k)$ defined by (9.67a) and (9.67b) are analytic functions of $k$ for $\text{Im}\,k > 0$ and continuous for $\text{Im}\,k \geq 0$, while $\overline{M}(x;k)$ and $\overline{N}(x;k)$ defined by (9.67c) and (9.67d) are analytic functions of $k$ for $\text{Im}\,k < 0$ and continuous for $\text{Im}\,k \leq 0$. Moreover, the solutions of the corresponding integral equations are unique in the space of continuous functions.*

The proof can be found in Ablowitz et al. (2004b).

Note that simply requiring $q, r \in L^1(\mathbb{R})$ does not yield analyticity of the eigenfunctions on the real $k$-axis, for which more stringent conditions must be imposed. Having $q, r$ vanishing faster than any exponential as $|x| \to \infty$ implies that all four eigenfunctions are entire functions of $k \in \mathbb{C}$.

From the integral equations (9.67) we can compute the asymptotic expansion at large $k$ (in the proper half-plane) of the eigenfunction. Integration by parts yields

$$M(x;k) = \begin{pmatrix} 1 - \frac{1}{2ik}\int_{-\infty}^{x} q(x')r(x')dx' \\ -\frac{1}{2ik}r(x) \end{pmatrix} + O(1/k^2) \tag{9.68a}$$

$$\overline{N}(x;k) = \begin{pmatrix} 1 + \frac{1}{2ik}\int_{x}^{+\infty} q(x')r(x')dx' \\ -\frac{1}{2ik}r(x) \end{pmatrix} + O(1/k^2) \tag{9.68b}$$

$$N(x;k) = \begin{pmatrix} \frac{1}{2ik}q(x) \\ 1 - \frac{1}{2ik}\int_{x}^{+\infty} q(x')r(x')dx' \end{pmatrix} + O(1/k^2) \tag{9.68c}$$

$$\overline{M}(x;k) = \begin{pmatrix} \frac{1}{2ik}q(x) \\ 1 + \frac{1}{2ik}\int_{-\infty}^{x} q(x')r(x')dx' \end{pmatrix} + O(1/k^2). \tag{9.68d}$$

## 9.10.2 Scattering data

The two eigenfunctions with fixed boundary conditions as $x \to -\infty$ are linearly independent, and the same holds for the two eigenfunctions with fixed boundary conditions as $x \to +\infty$. Indeed, let $u(x,k) = \left(u^{(1)}(x,k), u^{(2)}(x,k)\right)^{\mathrm{T}}$ and $v(x,k) = \left(v^{(1)}(x,k), v^{(2)}(x,k)\right)^{\mathrm{T}}$ be any two solutions of (9.59a) and let us define the Wronskian of $u$ and $v$, $W(u,v)$, as

$$W(u,v) = u^{(1)}v^{(2)} - u^{(2)}v^{(1)}.$$

Using (9.59a) it can be verified that

$$\frac{d}{dx}W(u,v) = 0$$

where $W(u,v)$ is a constant. Therefore, from the prescribed asymptotic behavior for the eigenfunctions, it follows that

$$W\left(\phi,\overline{\phi}\right) = \lim_{x\to-\infty} W\left(\phi(x;k),\overline{\phi}(x;k)\right) = 1$$
$$W\left(\psi,\overline{\psi}\right) = \lim_{x\to+\infty} W\left(\psi(x;k),\overline{\psi}(x;k)\right) = -1,$$

which shows that the solutions $\phi$ and $\overline{\phi}$ are linearly independent, as are $\psi$ and $\overline{\psi}$. As a consequence, we can express $\phi(x;k)$ and $\overline{\phi}(x;k)$ as linear combinations of $\psi(x;k)$ and $\overline{\psi}(x;k)$, or vice versa, with the coefficients of these linear combinations depending on $k$ only. Hence, the relations

$$\phi(x;k) = b(k)\psi(x;k) + a(k)\overline{\psi}(x;k) \tag{9.69a}$$
$$\overline{\phi}(x;k) = \overline{a}(k)\psi(x;k) + \overline{b}(k)\overline{\psi}(x;k) \tag{9.69b}$$

hold for any $k \in \mathbb{C}$ such that the four eigenfunctions exist. In particular, (9.69) hold for Im $k = 0$ and define the scattering coefficients $a(k)$, $\overline{a}(k)$, $b(k)$, $\overline{b}(k)$ for

$k \in \mathbb{R}$. Comparing the asymptotics of $W(\phi, \overline{\phi})$ as $x \rightarrow \pm\infty$ with (9.69) shows that the scattering data satisfy the following equation:

$$a(k)\overline{a}(k) - b(k)\overline{b}(k) = 1 \qquad \forall k \in \mathbb{R}.$$

The scattering coefficients can, in turn, be represented as Wronskians of the eigenfunctions. Indeed, from (9.69) it follows that

$$a(k) = W(\phi, \psi), \qquad \overline{a}(k) = W(\overline{\psi}, \overline{\phi}) \tag{9.70a}$$

$$b(k) = W(\overline{\psi}, \phi), \qquad \overline{b}(k) = W\left(\overline{\phi}, \psi\right). \tag{9.70b}$$

Therefore, if $q, r \in L^1(\mathbb{R})$, Lemma 9.2 and equations (9.70) imply that $a(k)$ admit analytic continuation in the upper half $k$-plane, while $\overline{a}(k)$ can be analytically continued in the lower half $k$-plane. In general, $b(k)$ and $\overline{b}(k)$ cannot be extended off the real $k$-axis. It is also possible to obtain integral representations for the scattering coefficients in terms of the eigenfunctions, but we will not do so here.

Equation (9.69) can be written as

$$\mu(x, k) = \overline{N}(x; k) + \rho(k)e^{2ikx}N(x; k) \tag{9.71a}$$

$$\overline{\mu}(x, k) = N(x; k) + \overline{\rho}(k)e^{-2ikx}\overline{N}(x; k), \tag{9.71b}$$

where we introduced

$$\mu(x, k) = M(x; k)/a(k), \qquad \overline{\mu}(x, k) = \overline{M}(x; k)/\overline{a}(k) \tag{9.72}$$

and the *reflection coefficients*

$$\rho(k) = b(k)/a(k), \qquad \overline{\rho}(k) = \overline{b}(k)/\overline{a}(k). \tag{9.73}$$

Note that from the representation of the scattering data as Wronskians of the eigenfunctions and from the asymptotic expansions (9.68), it follows that

$$a(k) = 1 + O(1/k), \qquad \overline{a}(k) = 1 + O(1/k)$$

in the proper half-plane, while $b(k), \overline{b}(k)$ are $O(1/k)$ as $|k| \rightarrow \infty$ on the real axis. We further assume that $a(k), \overline{a}(k)$ are continuous for real $k$. Since $a(k)$ is analytic for $\text{Im } k > 0$, continuous for $\text{Im } k = 0$ and $a(k) \rightarrow 1$ for $|k| \rightarrow \infty$, there cannot be a cluster point of zeros for $\text{Im } k \geq 0$. Hence the number of zeros is finite. Similarly for $\overline{a}(k)$.

### Proper eigenvalues and norming constants

A *proper* (or *discrete*) *eigenvalue* of the scattering problem (9.59a) is a (complex) value of $k$ corresponding to a bounded solution $v(x, k)$ such that $v(x, k) \rightarrow 0$ as $x \rightarrow \pm\infty$; usually one requires $v \in L^2(\mathbb{R})$ with respect to $x$. We also call such solutions *bound states*. When $q = r^*$ in (9.59a), the scattering problem is self-adjoint and there are no eigenvalues/bound states. So the only possibility for eigenvalues is when $q = -r^*$. We will see that these eigenvalues correspond to solitons that decay rapidly at infinity; sometimes called "bright" solitons. The so-called "dark" solitons, which tend to constant states at infinity, are contained in an extended theory (cf. Zakharov and Shabat, 1973; Prinari et al., 2006).

Suppose that $k_j = \xi_j + i\eta_j$, with $\eta_j > 0$, is such that $a(k_j) = 0$. Then from (9.70a) it follows that $W(\phi(x; k_j), \psi(x; k_j)) = 0$ and therefore $\phi_j(x) := \phi(x; k_j)$ and $\psi_j(x) = \psi(x; k_j)$ are linearly dependent; that is, there exists a complex constant $c_j$ such that

$$\phi_j(x) = c_j \psi_j(x).$$

Hence, by (9.63a) and (9.63b) it follows that

$$\phi_j(x) \sim \begin{pmatrix} 1 \\ 0 \end{pmatrix} e^{\eta_j x - i\xi_j x} \qquad \qquad \text{as } x \to -\infty$$

$$\phi_j(x) = c_j \psi_j(x) \sim \begin{pmatrix} 0 \\ 1 \end{pmatrix} e^{-\eta_j x + i\xi_j x} \qquad \qquad \text{as } x \to +\infty.$$

Since we have strong decay for large $|x|$, it follows that $k_j$ is a proper eigenvalue. On the other hand, if $a(k) \neq 0$ for $\text{Im}\, k > 0$, then solutions of the scattering problem blow up in one or both directions. We conclude that the proper eigenvalues in the region $\text{Im}\, k > 0$ are precisely the zeros of the scattering coefficient $a(k)$. Similarly, the eigenvalues in the region $\text{Im}\, k < 0$ are given by the zeros of $\overline{a}(k)$, and these zeros $\overline{k}_j = \overline{\xi}_j + i\overline{\eta}_j$ are such that $\overline{\phi}(x; \overline{k}_j) = \overline{c}_j \overline{\psi}(x; \overline{k}_j)$ for some complex constant $\overline{c}_j$. The coefficients $\{c_j\}_{j=1}^{J}$ and $\{\overline{c}_j\}_{j=1}^{\overline{J}}$ are often called *norming constants*. In terms of the eigenfunctions, the norming constants are defined by

$$M_j(x) = c_j\, e^{2ik_j x} N_j(x), \qquad \overline{M}_j(x) = \overline{c}_j\, e^{-2i\overline{k}_j x} \overline{N}_j(x), \qquad (9.74)$$

where $M_j(x) = M(x; k_j)$, $\overline{M}_j(x) = \overline{M}(x; \overline{k}_j)$ and similarly for $N_j(x)$ and $\overline{N}_j(x)$. As we will see in the following, discrete eigenvalues and associated norming constants are part of the scattering data (i.e., the data necessary to uniquely solve the inverse problem and reconstruct $q(x, t)$ and $r(x, t)$).

Note that if the potentials $q, r$ are rapidly decaying, such that (9.69) can be extended off the real axis, then

$$c_j = b(k_j) \qquad \text{for } j = 1, \ldots, J$$
$$\overline{c}_j = \overline{b}(\overline{k}_j) \qquad \text{for } j = 1, \ldots, \overline{J}.$$

## Symmetry reductions

The NLS equation is a special case of the system (9.61) under the symmetry reduction $r = \pm q^*$. This symmetry in the potentials induces a symmetry between the eigenfunctions analytic in the upper $k$-plane and the ones analytic in the lower $k$-plane. In turn, this symmetry of the eigenfunctions induces symmetries in the scattering data.

Indeed, if $v(x, k) = \left(v^{(1)}(x, k), v^{(2)}(x, k)\right)^{\mathrm{T}}$ satisfies (9.59a) and $r = \mp q^*$, then $\hat{v}(x, k) = \left(v^{(2)}(x, k^*), \mp v^{(1)}(x, k^*)\right)^{\dagger}$ (where $^{\dagger}$ denotes conjugate transpose) also satisfies the same (9.59a). Taking into account the boundary conditions (9.63), we get

$$\overline{\psi}(x; k) = \left( \begin{array}{c} \psi^{(2)}(x; k^*) \\ \mp \psi^{(1)}(x; k^*) \end{array} \right)^*, \qquad \overline{\phi}(x; k) = \left( \begin{array}{c} \mp \phi^{(2)}(x; k^*) \\ \phi^{(1)}(x; k^*) \end{array} \right)^*,$$

$$\overline{N}(x; k) = \left( \begin{array}{c} N^{(2)}(x; k^*) \\ \mp N^{(1)}(x; k^*) \end{array} \right)^*, \qquad \overline{M}(x; k) = \left( \begin{array}{c} \mp M^{(2)}(x; k^*) \\ M^{(1)}(x; k^*) \end{array} \right)^*.$$

Then, from the Wronskian representations for the scattering data (9.70) there follows

$$\overline{a}(k) = a^*(k^*), \qquad \overline{b}(k) = \mp b^*(k^*),$$

which implies that $k_j$ is a zero of $a(k)$ in the upper half $k$-plane if and only if $k_j^*$ is a zero of $\overline{a}(k)$ in the lower $k$-plane and vice versa. This means that the zeros of $a(k)$ and $\overline{a}(k)$ come in pairs and the number of zeros of each is the same, i.e., $\overline{J} = J$. Hence when there are eigenvalues associated with the potential $q$, with $q = -r^*$, we have that

$$\overline{k}_j = k_j^*, \qquad \overline{c}_j = \mp c_j^* \qquad j = 1, \ldots, J.$$

Note: when $r = q^*$ the equation (9.59a) is Hermitian and in this case the spectrum lies on the real axes. Moreover in this case $|a^2|(k) = 1 + |b^2|(k)$ so $|a|(k) > 0$.

On the other hand, if we have the symmetry $r = \mp q$, with $q$ real, then $\hat{v}(x, k) = \left(v^{(2)}(x, -k), \mp v^{(1)}(x, -k)\right)^{\mathrm{T}}$ also satisfies the same equation (9.59a). Taking into account the boundary conditions (9.63), we get

$$\overline{\psi}(x; k) = \left( \begin{array}{c} \psi^{(2)}(x; -k) \\ \mp \psi^{(1)}(x; -k) \end{array} \right), \qquad \overline{\phi}(x; k) = \left( \begin{array}{c} \mp \phi^{(2)}(x; -k) \\ \phi^{(1)}(x; -k) \end{array} \right),$$

$$\overline{N}(x; k) = \left( \begin{array}{c} N^{(2)}(x; -k) \\ \mp N^{(1)}(x; -k) \end{array} \right), \qquad \overline{M}(x; k) = \left( \begin{array}{c} \mp M^{(2)}(x; -k) \\ M^{(1)}(x; -k) \end{array} \right).$$

Then, from the Wronskian representations for the scattering data (9.70), there follows

$$\overline{a}(k) = a(-k), \qquad \overline{b}(k) = \mp b(-k), \qquad (9.75)$$

which implies that $k_j$ is a zero of $a(k)$ in the upper half $k$-plane if and only if $-k_j$ is a zero of $\bar{a}(k)$ in the lower $k$-plane, and vice versa. As a consequence, when $r = -q$, with $q$ real, the zeros of $a(k)$ and $\bar{a}(k)$ are paired, their number is the same: $\bar{J} = J$ and

$$\bar{k}_j = -k_j, \qquad \bar{c}_j = -c_j \qquad j = 1, \ldots, J. \qquad (9.76)$$

Thus if $r = -q$, with $q$ real, both of the above symmetry conditions must hold and when $k_j$ is an eigenvalue so is $-k_j^*$; i.e., either the eigenvalues come in pairs, $\{k_j, -k_j^*\}$, or they lie on the imaginary axis.

### 9.10.3  Inverse scattering problem

The inverse problem consists of constructing a map from the scattering data, that is:

(i)  the reflection coefficients $\rho(k)$ and $\bar{\rho}(k)$ for $k \in \mathbb{R}$, defined by (9.73);

(ii)  the discrete eigenvalues $\left\{k_j\right\}_{j=1}^J$ (zeros of the scattering coefficient $a(k)$ in the upper half plane of $k$) and $\left\{\bar{k}_j\right\}_{j=1}^{\bar{J}}$ (zeros of the scattering coefficient $\bar{a}(k)$ in the lower half plane of $k$);

(iii)  the norming constants $\left\{c_j\right\}_{j=1}^J$ and $\left\{\bar{c}_j\right\}_{j=1}^{\bar{J}}$, cf. (9.74); back to the potentials $q$ and $r$.

First, we use these data to reconstruct the eigenfunctions (for instance, $N(x; k)$ and $\bar{N}(x; k)$), and then we recover the potentials from the large $k$ asymptotics of the eigenfunctions, cf. equations (9.68). Note that the inverse problem is solved at fixed $t$, and therefore the explicit time dependence is omitted. In fact, in the inverse problem both $x$ and $t$ are treated as parameters.

### Riemann–Hilbert approach

In the previous section, we showed that the eigenfunctions $N(x; k)$ and $\bar{N}(x; k)$ exist and are analytic in the regions $\operatorname{Im} k > 0$ and $\operatorname{Im} k < 0$, respectively, if $q, r \in L^1(\mathbb{R})$. Similarly, under the same conditions on the potentials, the functions $\mu(x, k)$ and $\bar{\mu}(x, k)$ introduced in (9.72) are meromorphic in the regions $\operatorname{Im} k > 0$ and $\operatorname{Im} k < 0$, respectively, with poles at the zeros of $a(k)$ and $\bar{a}(k)$. Therefore, in the inverse problem we assume these analyticity properties for the unknown eigenfunctions ($N(x; k)$ and $\bar{N}(x; k)$) or modified eigenfunctions ($\mu(x, k)$ and $\bar{\mu}(x, k)$). With these assumptions, (9.71) can be considered as the "jump" conditions of a Riemann–Hilbert problem. To recover the sectionally meromorphic functions from the scattering data, we will convert the Riemann–Hilbert problem into a system of linear integral equations.

Suppose that $a(k)$ and $\bar{a}(k)$ have a finite number of simple zeros in the regions $\operatorname{Im} k > 0$ and $\operatorname{Im} k < 0$, respectively, which we denote as $\left\{k_j, \operatorname{Im} k_j > 0\right\}_{j=1}^J$ and

$\left\{\overline{k}_j, \ \mathrm{Im}\,\overline{k}_j < 0\right\}_{j=1}^{\overline{J}}$. We will also assume that $a(\xi) \neq 0$ and $\overline{a}(\xi) \neq 0$ for $\xi \in \mathbb{R}$ (i.e., $a$ and $\overline{a}$ have no zeros on the real axis) and $a(k)$ is continuous for $\mathrm{Im}\,k \geq 0$.

Let $f(\zeta)$, $\zeta \in \mathbb{R}$, be an integrable function and consider the projection operators

$$P^{\pm}[f](k) = \frac{1}{2\pi i} \int_{-\infty}^{+\infty} \frac{f(\zeta)}{\zeta - (k \pm i0)} d\zeta.$$

If $f_+$ (resp. $f_-$) is analytic in the upper (resp. lower) $k$-plane and $f_{\pm}(k) \to 0$ as $|k| \to \infty$ for $\mathrm{Im}\,k \gtrless 0$, then

$$P^{\pm}\left[f_{\pm}\right] = \pm f_{\pm}, \qquad P^{\pm}\left[f_{\mp}\right] = 0$$

$[P^{\pm}$ are referred to as projection operators into the upper/lower half $k$-planes]. Let us now apply the projector $P^-$ to both sides of (9.71a) and $P^+$ to both sides of (9.71b). Note that $\mu(x,k) = M(x;k)/a(k)$ in (9.71a) does not decay for large $k$; rather it tends to $(1,0)^{\mathrm{T}}$. In addition it has poles at the zeros of $a(k)$. We subtract these contributions from both sides of (9.71a) and then take the projector; similar statements apply to $\overline{\mu}(x,k)$ in (9.71b). So, taking into account the analyticity properties of $N, \overline{N}, \mu, \overline{\mu}$ and the asymptotics (9.68) and using (9.74), we obtain

$$\overline{N}(x;k) = \begin{pmatrix} 1 \\ 0 \end{pmatrix} + \sum_{j=1}^{J} \frac{C_j e^{2ik_j x}}{k - k_j} N_j(x) + \frac{1}{2\pi i} \int_{-\infty}^{+\infty} \frac{\rho(\zeta) e^{2i\zeta x}}{\zeta - (k - i0)} N(x;\zeta)\, d\zeta \qquad (9.77a)$$

$$N(x;k) = \begin{pmatrix} 0 \\ 1 \end{pmatrix} + \sum_{j=1}^{\overline{J}} \frac{\overline{C}_j e^{-2i\overline{k}_j x}}{k - \overline{k}_j} \overline{N}_j(x) - \frac{1}{2\pi i} \int_{-\infty}^{+\infty} \frac{\overline{\rho}(\zeta) e^{-2i\zeta x}}{\zeta - (k + i0)} \overline{N}(x;\zeta)\, d\zeta, \quad (9.77b)$$

where $N_j(x) = N(x;k_j)$, $\overline{N}_j(x) = \overline{N}(x;\overline{k}_j)$ and we introduced

$$C_j = \frac{c_j}{a'(k_j)} \qquad \text{for } j = 1, \ldots, J$$

$$\overline{C}_j = \frac{\overline{c}_j}{\overline{a}'(\overline{k}_j)} \qquad \text{for } j = 1, \ldots, \overline{J}$$

with $'$ denoting the derivative of $a(k)$ and $\overline{a}(k)$ with respect to $k$. We see that the equations defining the inverse problem for $N(x;k)$ and $\overline{N}(x;k)$ depend on the extra terms $\left\{N_j(x)\right\}_{j=1}^{J}$ and $\left\{\overline{N}_j(x)\right\}_{j=1}^{\overline{J}}$. In order to close the system, we evaluate (9.77a) at $k = k_j$ for $j = 1, \ldots, J$ and (9.77b) at $k = \overline{k}_j$ for $j = 1, \ldots, \overline{J}$, thus obtaining

$$\overline{N}_\ell(x) = \begin{pmatrix} 1 \\ 0 \end{pmatrix} + \sum_{j=1}^{J} \frac{C_j e^{2ik_jx}}{\overline{k}_\ell - k_j} N_j(x) + \frac{1}{2\pi i} \int_{-\infty}^{+\infty} \frac{\rho(\zeta) e^{2i\zeta x}}{\zeta - \overline{k}_\ell} N(x;\zeta)\,d\zeta \qquad (9.78a)$$

$$N_j(x) = \begin{pmatrix} 0 \\ 1 \end{pmatrix} + \sum_{m=1}^{\overline{J}} \frac{\overline{C}_m e^{-2i\overline{k}_m x}}{k_j - \overline{k}_m} \overline{N}_j(x) - \frac{1}{2\pi i} \int_{-\infty}^{+\infty} \frac{\overline{\rho}(\zeta) e^{-2i\zeta x}}{\zeta - k_j} \overline{N}(x;\zeta)\,d\zeta. \qquad (9.78b)$$

Equations (9.77) and (9.78) together constitute a linear algebraic – integral system of equations that, in principle, solve the inverse problem for the eigenfunctions $N(x;k)$ and $\overline{N}(x;k)$.

By comparing the asymptotic expansions at large $k$ of the right-hand sides of (9.77) with the expansions (9.68), we obtain

$$r(x) = -2i \sum_{j=1}^{J} e^{2ik_jx} C_j N_j^{(2)}(x) + \frac{1}{\pi} \int_{-\infty}^{+\infty} \rho(\zeta) e^{2i\zeta x} N^{(2)}(x;\zeta)\,d\zeta \qquad (9.79a)$$

$$q(x) = 2i \sum_{j=1}^{\overline{J}} e^{-2i\overline{k}_jx} \overline{C}_j \overline{N}_j^{(1)}(x) + \frac{1}{\pi} \int_{-\infty}^{+\infty} \overline{\rho}(\zeta) e^{-2i\zeta x} \overline{N}^{(1)}(x;\zeta)\,d\zeta \qquad (9.79b)$$

which reconstruct the potentials in terms of the scattering data and thus complete the formulation of the inverse problem (as before, the superscript $^{(\ell)}$ denotes the $\ell$-component of the corresponding vector).

We mention that the issue of establishing existence and uniqueness of solutions for the equations of the inverse problem is usually carried out by converting the inverse problem into a set of Gelfand–Levitan–Marchenko integral equations – which is given next.

### Gel'fand–Levitan–Marchenko integral equations

As an alternative inverse procedure we provide a reconstruction for the potentials by developing the Gel'fand–Levitan–Marchenko (GLM) integral equations, instead of using the projection operators (cf. Zakharov and Shabat, 1972; Ablowitz and Segur, 1981). To do this we represent the eigenfunctions in terms of triangular kernels

$$N(x;k) = \begin{pmatrix} 0 \\ 1 \end{pmatrix} + \int_{x}^{+\infty} K(x,s) e^{-ik(x-s)}\,ds, \qquad s > x, \quad \text{Im}\,k > 0 \qquad (9.80)$$

$$\overline{N}(x;k) = \begin{pmatrix} 1 \\ 0 \end{pmatrix} + \int_{x}^{+\infty} \overline{K}(x,s) e^{ik(x-s)}\,ds, \qquad s > x, \quad \text{Im}\,k < 0. \qquad (9.81)$$

Applying the operator $\frac{1}{2\pi}\int_{-\infty}^{+\infty} dk\, e^{-ik(x-y)}$ for $y > x$ to (9.77a), we find

$$\overline{K}(x,y) + \begin{pmatrix} 0 \\ 1 \end{pmatrix} F(x+y) + \int_x^{+\infty} K(x,s)F(s+y)\,ds = 0 \qquad (9.82)$$

where

$$F(x) = \frac{1}{2\pi}\int_{-\infty}^{+\infty}\rho(\xi)e^{i\xi x}\,d\xi - i\sum_{j=1}^{J} C_j e^{ik_j x}.$$

Analogously, operating on (9.77b) with $\frac{1}{2\pi}\int_{-\infty}^{+\infty} dk\, e^{ik(x-y)}$ for $y > x$ gives

$$K(x,y) + \begin{pmatrix} 1 \\ 0 \end{pmatrix} \overline{F}(x+y) + \int_x^{+\infty} \overline{K}(x,s)\overline{F}(s+y)\,ds = 0 \qquad (9.83)$$

where

$$\overline{F}(x) = \frac{1}{2\pi}\int_{-\infty}^{+\infty}\overline{\rho}(\xi)e^{-i\xi x}\,d\xi + i\sum_{j=1}^{\overline{J}} \overline{C}_j e^{-i\overline{k}_j x}.$$

Equations (9.82) and (9.83) constitute the Gel'fand–Levitan–Marchenko equations.

Inserting the representations (9.80)–(9.81) for the eigenfunctions into (9.79) we obtain the reconstruction of the potentials in terms of the kernels of the GLM equations, i.e.,

$$q(x) = -2K^{(1)}(x,x), \qquad r(x) = -2\overline{K}^{(2)}(x,x) \qquad (9.84)$$

where, as usual, $K^{(j)}$ and $\overline{K}^{(j)}$ for $j = 1,2$ denote the $j$th component of the vectors $K$ and $\overline{K}$ respectively.

If the symmetry $r = \mp q^*$ holds, then, taking into account (9.75)–(9.76), one can verify that

$$\overline{F}(x) = \mp F^*(x)$$

and consequently

$$\overline{K}(x,y) = \begin{pmatrix} K^{(2)}(x,y) \\ \mp K^{(1)}(x,y) \end{pmatrix}^*.$$

In this case (9.82)–(9.83) solving the inverse problem reduce to

$$K^{(1)}(x,y) = \pm F^*(x+y) \mp \int_x^{+\infty} ds \int_x^{+\infty} ds'\, K^{(1)}(x,s')F(s+s')F^*(y+s)$$

and the potentials are reconstructed by means of the first of (9.84). We also note that when $r = \mp q$ with $q$ real, then $F(x)$ and $K^{(1)}(x,y)$ are real.

## 9.10.4 Time evolution

We will now show how to determine the time dependence of the scattering data. Then, by the inverse transform we establish the solution $q(x, t)$ and $r(x, t)$.

The operator equation (9.59b) determines the evolution of the eigenfunctions, which can be written as

$$\partial_t v = \begin{pmatrix} A & B \\ C & -A \end{pmatrix} v \qquad (9.85)$$

where $B, C \to 0$ as $x \to \pm\infty$ (since $q, r \in L^1(\mathbb{R})$). Then the time-dependent eigenfunctions must asymptotically satisfy the differential equation

$$\partial_t v = \begin{pmatrix} A_0 & 0 \\ 0 & -A_0 \end{pmatrix} v \qquad \text{as } x \to \pm\infty \qquad (9.86)$$

with

$$A_0 = \lim_{|x| \to \infty} A(x, k).$$

The system (9.86) has solutions that are linear combinations of

$$v^+ = \begin{pmatrix} e^{A_0 t} \\ 0 \end{pmatrix}, \qquad v^- = \begin{pmatrix} 0 \\ e^{-A_0 t} \end{pmatrix}.$$

However, such solutions are not compatible with the fixed boundary conditions of the eigenfunctions, i.e., equations (9.63a)–(9.63b). Therefore, we define time-dependent functions

$$\Phi(x, t; k) = e^{A_0 t} \phi(x, t; k), \qquad \overline{\Phi}(x, t; k) = e^{-A_0 t} \overline{\phi}(x, t; k)$$
$$\Psi(x, t; k) = e^{-A_0 t} \psi(x, t; k), \qquad \overline{\Psi}(x, t; k) = e^{A_0 t} \overline{\psi}(x, t; k)$$

to be solutions of the differential equation (9.85). Then the evolution for $\phi$ and $\overline{\phi}$ becomes

$$\partial_t \phi = \begin{pmatrix} A - A_0 & B \\ C & -A - A_0 \end{pmatrix} \phi, \quad \partial_t \overline{\phi} = \begin{pmatrix} A + A_0 & B \\ C & -A + A_0 \end{pmatrix} \overline{\phi}, \quad (9.87)$$

so that, taking into account (9.69) and evaluating (9.87) as $x \to +\infty$, we obtain

$$\partial_t a = 0, \qquad \partial_t \overline{a} = 0$$
$$\partial_t b = -2A_0 b, \qquad \partial_t \overline{b} = 2A_0 \overline{b}$$

or, explicitly,

$$a(k, t) = a(k, 0), \qquad \overline{a}(k, t) = \overline{a}(k, 0) \qquad (9.88a)$$
$$b(k, t) = b(k, 0)e^{-2A_0(k)t}, \qquad \overline{b}(k, t) = \overline{b}(k, 0)e^{2A_0(k)t}. \qquad (9.88b)$$

From (9.88a) it follows that the discrete eigenvalues (i.e., the zeros of $a$ and $\bar{a}$) are constant as the solution evolves. Not only the number of eigenvalues, but also their locations are fixed. Thus, the eigenvalues are time-independent discrete states of the evolution. In fact, this time invariance is the underlying mechanism of the elastic soliton interaction for the integrable soliton equations. On the other hand, the evolution of the reflection coefficients (9.73) is given by

$$\rho(k,t) = \rho(k,0)e^{-2A_0(k)t}, \qquad \bar{\rho}(k,t) = \bar{\rho}(k,0)e^{2A_0(k)t}$$

and this also gives the evolution of the norming constants:

$$C_j(t) = C_j(0)e^{-2A_0(k_j)t}, \qquad \overline{C}_j(t) = \overline{C}_j(0)e^{2A_0(\bar{k}_j)t}. \qquad (9.89)$$

The expressions for the evolution of the scattering data allow one to solve the initial value problem for *all* of the solutions in the class associated with the scattering problem (9.59a). Namely we can solve (see the previous chapter for details) all the equations in the class given by the general evolution operator

$$\begin{pmatrix} r \\ -q \end{pmatrix}_t + 2A_0(L)\begin{pmatrix} r \\ q \end{pmatrix} = 0, \qquad (9.90)$$

where $A_0(k) = \lim_{|x|\to\infty} A(x,t,k)$ (here $A_0(k)$ may be the ratio of two entire functions), and $L$ is the integro-differential operator given by

$$L = \frac{1}{2i}\begin{pmatrix} \partial_x - 2r(I_-q) & 2r(I_-r) \\ -2q(I_-q) & -\partial_x + 2q(I_-r) \end{pmatrix},$$

where $\partial_x \equiv \partial/\partial x$ and

$$(I_-f)(x) \equiv \int_{-\infty}^{x} f(y)\,dy. \qquad (9.91)$$

Note that $L$ operates on $(r,q)$, and $I_-$ operates both on the functions immediately to its right and also on the functions to which $L$ is applied.

Special cases are listed below:

- When $A_0 = 2ik^2, r = \mp q^*$ we obtain the solution of the NLS equation

$$iq_t = q_{xx} \pm 2|q|^2 q$$

i.e., (9.62).
- If $A_0 = -4ik^3$ when $r = \mp q$, $q$ real we find the solution of the modified KdV (mKdV) equation,

$$q_t \pm 6q^2 q_x + q_{xxx} = 0. \qquad (9.92)$$

- If on the other hand $r = \mp q^*$ the solution of the complex mKdV equation results:

$$q_t \pm 6|q|^2 q_x + q_{xxx} = 0.$$

- Finally, if $A_0 = \frac{i}{4k}$ and
  (i) $q = -r = -\frac{1}{2} u_x$, we obtain the solution of the sine – Gordon equation

$$u_{xt} = \sin u; \tag{9.93}$$

  (ii) or, if $q = r = \frac{1}{2} u_x$, then we can find the solution of the sinh–Gordon equation

$$u_{xt} = \sinh u.$$

In summary the solution procedure is as follows:
  (i) The scattering data are calculated from the initial data $q(x) = q(x,0)$ and $r(x) = r(x,0)$ according to the direct scattering method described in Section 9.10.1.
 (ii) The scattering data at a later time $t > 0$ are determined from (9.88)–(9.89).
(iii) The solutions $q(x,t)$ and $r(x,t)$ are recovered from the scattering data using inverse scattering using the reconstruction formulas via the Riemann–Hilbert formulation (9.77)–(9.79) or Gel-fand–Levitan–Marchenko equations (9.82)–(9.83).
If we further require symmetry, $r = \pm q^*$, as discussed in the above section on symmetry reductions, we obtain the solution of the reduced evolution equation.

### 9.10.5 Soliton solutions

In the case where the scattering data comprise proper eigenvalues but $\rho(k) = \bar{\rho}(k) \equiv 0$ for all $k \in \mathbb{R}$ (corresponding to the so-called reflectionless solutions), the algebraic-integral system (9.77) and (9.78) reduces to a linear algebraic system, namely

$$\overline{N}_l(x) = \begin{pmatrix} 1 \\ 0 \end{pmatrix} + \sum_{j=1}^{J} \frac{C_j e^{2ik_j x} N_j(x)}{\overline{k}_l - k_j}$$

$$N_j(x) = \begin{pmatrix} 0 \\ 1 \end{pmatrix} + \sum_{m=1}^{\overline{J}} \frac{\overline{C}_m e^{-2i\overline{k}_m x} \overline{N}_m(x)}{k_j - \overline{k}_m},$$

that can be solved in closed form. The one-soliton solution, in particular, is obtained for $J = \overline{J} = 1$ (i.e., one single discrete eigenvalue $k_1 = \xi + i\eta$ and corresponding norming constant $C_1$). In the relevant physical case, when the symmetry $r = -q^*$ holds, using (9.75) and (9.76) in the above system, we get

$$N_1^{(1)}(x) = -\frac{C_1}{k_1 - k_1^*} e^{-2ik_1^* x} \left[ 1 - \frac{|C_1|^2 \, e^{2i(k_1 - k_1^*)x}}{\left(k_1 - k_1^*\right)^2} \right]^{-1},$$

$$N_1^{(2)}(x) = \left[ 1 - \frac{|C_1|^2 \, e^{2i(k_1 - k_1^*)x}}{\left(k_1 - k_1^*\right)^2} \right]^{-1},$$

where, as before, $N_1(x) = \left( N_1^{(1)}(x), N_1^{(2)}(x) \right)^{\mathrm{T}}$. Then, if we set

$$k_1 = \xi + i\eta, \qquad e^{2\delta} = \frac{|C_1|}{2\eta},$$

it follows from (9.79) that

$$q(x) = -2i\eta \frac{C_1^*}{|C_1|} e^{-2i\xi x} \operatorname{sech}(2\eta x - 2\delta).$$

Taking into account the time dependence of $C_1$ as given by (9.89), we find

$$q(x) = 2\eta e^{-2i\xi x + 2i\,\mathrm{Im}\,A_0(k_1)t - i\psi_0} \operatorname{sech}\left[2\left(\eta(x - x_0) + \mathrm{Re}\,A_0(k_1)t\right)\right],$$

where $C_1(0) = 2\eta e^{2\eta x_0 + i(\psi_0 + \pi/2)}$.

Thus when we take

(a) $A_0(k_1) = 2ik_1^2 = -4\xi\eta + 2i(\xi^2 - \eta^2)$ where $k_1 = \xi + i\eta, r = -q^*$ we get the well-known bright soliton solution of the NLS equation

$$q(x, t) = 2\eta e^{-2i\xi x + 4i(\xi^2 - \eta^2)t - i\psi_0} \operatorname{sech}\left[2\eta(x - 4\xi t - x_0)\right].$$

(b) $A_0(k_1) = -4ik_1^3 = -4\eta^3$ where $k_1 = i\eta, r = -q$, real; recall that due to symmetry all eigenvalues must come in pairs $\{k_j, -k_j^*\}$ hence with only one eigenvalue $\mathrm{Re}\,k_1 = \xi = 0$. Then we get the bright soliton solution of the mKdV equation (9.92):

$$q(x, t) = 2\eta \operatorname{sech}\left[2\eta\left(x - 4\eta^2 t - x_0\right)\right].$$

(c) $A_0(k_1) = \frac{i}{4k} = \frac{1}{4\eta}$ where $k_1 = i\eta, r = -q = \frac{u_x}{2}$, real; again due to symmetry $\mathrm{Re}\,k_1 = \xi = 0$. Then we get the soliton-kink solution of the sine–Gordon equation (9.93):

$$q(x, t) = -\frac{u_x}{2} = -2\eta \operatorname{sech}\left[2\eta\left(x + \frac{1}{4\eta}t - x_0\right)\right],$$

or in terms of $u$

$$u(x, t) = 4 \tan^{-1} \exp\left[2\eta\left(x + \frac{1}{4\eta}t - x_0\right)\right].$$

## Exercises

9.1   Use (9.15) and (9.16) to find the Neumann series for $M(x;k)$ and $\overline{N}(x;k)$. Show these series converge uniformly when $u(x)$ decays appropriately (e.g., $u \in L_2^1$).

9.2   Using the results in Exercise 9.1 establish that $a(k)$ is analytic for $\operatorname{Im} k > 0$ and continuous for $\operatorname{Im} k = 0$.

9.3   Find a one-soliton solution to an equation associated with the time-independent Schrödinger equation (9.4) when

$$b(k,t) = b(k,0)e^{-32ik^5t}, \qquad a(k,t) = a(k,0),$$

$$c_j(t) = c_j(0)e^{32k_j^5t}, \qquad k_j = i\kappa_j.$$

Use the concepts of Chapter 8 to determine the nonlinear evolution equation this soliton solution solves. (Hint: use (8.51)–(8.52).)

9.4   Use (8.51)–(8.52) to deduce the time-dependence of the scattering data for a nonlinear evolution equation associated with (9.4) whose linear dispersion relation is $\omega(k)$.

9.5   From (9.53)–(9.54) find the next two conserved quantities $C_7$, $C_9$ associated with the KdV equation. Determine the conservation laws.

9.6   Use (9.67) to establish that $M(x;k)$, $N(x;k)$ are analytic for $\operatorname{Im} k > 0$, $\overline{M}(x;k)$, $\overline{N}(x;k)$ for $\operatorname{Im} k < 0$ where $q, r \in L^1(\mathbb{R})$.

9.7   Suppose $A_0(k) = 8ik^4$, associated with (9.85) and (9.90) with $r = -q^*$. Find the time dependence of the scattering data and the one-soliton solution. Use the results of Chapter 8 to find the nonlinear evolution equation that this soliton solution satisfies.

9.8   Consider the equation

$$(u_t + 6uu_x + u_{xxx})_x - 3u_{yy} = 0.$$

(a) Show that the equation has the rational solution

$$u(x,t) = 4\frac{p^2y^2 - X^2 + 1/p^2}{p^2y^2 + X^2 + 1/p^2}$$

where $X = x + 1/p - 3p^2t$ and $p$ is a real constant.

(b) Make the transformation $y \to iy$ to formally derive the following KP equation

$$(u_t + 6uu_x + u_{xxx})_x + 3u_{yy} = 0$$

and thus show that the above solution is a singular solution of the KP.

9.9 Suppose $K(x, z; t)$ satisfies the Marchenko equation

$$K(x, z; t) + F(x, z; t) + \int_x^\infty K(x, y; t) F(y, z; t) \, dy = 0,$$

where $F$ is a solution of the pair of equations

$$F_{xx} - F_{zz} = (x - z)F$$
$$3t F_t - F + F_{xxx} + F_{zzz} = x F_x + z F_z.$$

(a) Show that

$$u_t + \frac{u}{2t} + 6u u_x + u_{xxx} = 0,$$

where $u(x, t) = \frac{2}{(12t)^{2/3}} \frac{\partial}{\partial X} K(X, X; t)$ with $X = \frac{x}{(12t)^{1/3}}$.

(b) Show that a solution for $F$ is

$$F(x, z; t) = \int_{-\infty}^\infty f(y t^{1/3}) \, \text{Ai}(x + y) \, \text{Ai}(y + z) \, dy$$

where $f$ is an arbitrary function and $\text{Ai}(x)$ is the Airy function.
Hint: See Ablowitz and Segur (1981) on how to operate on $K$.

9.10 Show that the KP equation

$$(u_t + 6u u_x + u_{xxx})_x + 3\sigma^2 u_{yy} = 0$$

with $\sigma = \pm 1$ can be derived from the compatibility of the system

$$\frac{\partial^2 v}{\partial x^2} + \sigma \frac{\partial v}{\partial y} + u v = 0$$

$$v_t + 4 \frac{\partial^3 v}{\partial x^3} + 6u \frac{\partial v}{\partial x} + \left( 3u_x - 3\sigma \int_\infty^\infty u_y \, dx + \alpha \right) v = 0$$

where $\alpha$ is constant. Note: there is no "eigenvalue" in this equation. The scattering parameter is inserted when carrying out the inverse scattering (see Ablowitz and Clarkson, 1991).

9.11 Consider the following time-independent Schrödinger equation

$$v_{xx} + (k^2 + Q \, \text{sech}^2 \, x) v = 0$$

where $Q$ is constant.
(a) Make the transformation $v(x) = \Psi(\xi)$, with $\xi = \tanh x$ (hence $-1 < \xi < 1$ corresponds to $-\infty < x < \infty$) to find

$$(1 - \xi^2) \frac{d^2 \Psi}{d\xi^2} - 2\xi \frac{d\Psi}{d\xi} + \left( Q + \frac{k^2}{1 - \xi^2} \right) \Psi = 0$$

which is the *associated Legendre equation*, cf. Abramowitz and Stegun (1972).

(b) Show that there are eigenvalues $k = i\kappa$ when

$$\kappa = \left[\left(Q + \frac{1}{4}\right)^{1/2} - m - \frac{1}{2}\right] > 0,$$

where $m = 0, 1, 2, \ldots$ and therefore a necessary condition for eigenvalues is $Q + \frac{1}{4} > \frac{1}{4}$.

(c) Show that when $Q = N(N + 1), N = 1, 2, 3, \ldots$, the reflection coefficient vanishes, the discrete eigenvalues are given by $\kappa_n = n, n = 1, 2, 3, \ldots, N$ and the discrete eigenfunctions are proportional to the associated Legendre function $P_N^n(\tanh x)$.

9.12 Show that the solution to the Gel'fand–Levitan–Marchenko (GLM) equation (9.42)–(9.43) with pure continuous spectra is unique. Hint: one method is to show the homogeneous equation has only the zero solution, hence by the Fredholm alternative, the solution to the GLM equation is unique. Then in the homogeneous equation let $K(x, y) = 0, y < x$ and take the Fourier transform.

# PART III

APPLICATIONS OF NONLINEAR WAVES IN
OPTICS

# 10

## *Communications*

Nonlinear optics is the branch of optics that describes the behavior of light in nonlinear media; such as, media in which the induced dielectric polarization responds nonlinearly to the electric field of the light. This nonlinearity is typically observed at very high light intensities such as those provided by pulsed lasers. In this chapter, we focus on the application of high bit-rate communications. We will see that the nonlinear Schrödinger (NLS) equation and the dispersion-managed nonlinear Schrödinger (DMNLS) equation play a central role.

## 10.1 Communications

In 1973 Hasegawa and Tappert (Hasegawa and Tappert, 1973a; Hasegawa and Kodama, 1995) showed that the nonlinear Schrödinger equation derived in Chapter 7 [see (7.26), and the subsequent discussion] described the propagation of quasi-monochromatic pulses in optical fibers. Motivated by the fact that the NLS equation supports special stable, localized, soliton solutions, Mollenauer et al. (1980) demonstrated experimentally that solitons can propagate in a real fiber. However, it was soon apparent that due to unavoidable damping in optical fibers, solitons lose most of their energy over relatively short distances. In the mid-1980s all-optical amplifiers (called erbium doped fiber amplifiers: EDFAs) were developed. However with such amplifiers there is always some additional small amount of noise. Gordon and Haus (1986) (see also Elgin, 1985) showed that solitons suffered seriously from these noise effects. The frequency and temporal position of the soliton was significantly shifted over long distances, thereby limiting the available transmission distance and speed of soliton-based systems. Subsequently researchers began to seriously consider so-called wavelength division multiplexed (WDM) communication systems in which many optical pulses are transmitted simultaneously. The inevitable pulse interactions caused other serious problems, called four-wave mixing instabilities and collision-induced time shifts. In order to alleviate these penalties, researchers began to study dispersion-managed (DM) transmission systems. A DM transmission line consists of fibers with different types of dispersion fused together. The success of DM technology has already led to its use in commercial systems. In this chapter we discuss some of the important issues in the context of nonlinear waves and asymptotic analysis. There are a number of texts that cover this topic, which the interested reader can consult. These include Agrawal (2002, 2001), Hasegawa and Kodama (1995), Hasegawa and Matsumoto (2002), and Molleneauer and Gordon (2006).

### 10.1.1 The normalized NLS equation

In Chapter 7 the NLS equation was derived for weakly nonlinear, quasi-monochromatic electromagnetic waves. It was also discussed how the NLS equation should be modified to apply to optical fibers. Using the assumption for electromagnetic fields in the $x$-direction

$$E_x = \varepsilon(A(Z, T)e^{i(k_3 - \omega t)} + cc) + O(\varepsilon^2),$$

where $Z = \epsilon z$, $T = \epsilon t$, the NLS equation in optical fibers was found, in Chapter 7 [see (7.27) and the equations following], to satisfy

$$i\frac{\partial A}{\partial \tilde{z}} + \left(-\frac{k''(\omega)}{2}\right)\frac{\partial^2 A}{\partial T^2} + v|A|^2 A = 0, \tag{10.1}$$

with the nonlinear coefficient $v = v_{\text{eff}} = \dfrac{n_2 \omega}{4 c A_{\text{eff}}}$ where the retarded coordinate frame is given by $T = \epsilon t - \dfrac{Z}{v_g} = \epsilon t - k'Z$, $\tilde{z} = \epsilon Z$ with $k = \dfrac{\omega}{c}n_0$, and the nonlinear index of refraction is given by $n = n_0 + n_2|E_x|^2 = n_0 + 4n_2|A|^2$; $v_g = 1/k'(\omega)$ is the group velocity and $A_{\text{eff}}$ is the effective area of the fiber. It is standard to introduce the following non-dimensional coordinates: $\tilde{z} = z'z_*$, $T = t't_*$, and $A = \sqrt{P_*}u$, where a subscript $*$ denotes a characteristic scale. Usually, $z_*$ is taken as being proportional to the nonlinear length, the length over which a nonlinear phase change of one radian occurs, $t_*$ is taken as being proportional to the pulse full width at half-maximum (FWHM), and $P_*$ is taken as the peak pulse power. With the use of this non-dimensionalization, (10.1) becomes

$$i\frac{\partial u}{\partial z'} + \left(\frac{z_*}{t_*^2}\right)(-k'')\frac{1}{2}\frac{\partial^2 u}{\partial t'^2} + z_* P_* v|u|^2 u = 0.$$

For solitons, $z_* P_* v = 1$ gives the nonlinear length as $z_* = 1/vP_*$. On the other hand the linear dispersive length is given by $z_{*L} = t_*^2/|k''|$. Further assuming the nonlinear length to be equal to the linear dispersive length, yields the relation $P_* = |k''|/\left(vt_*^2\right)$, and the normalized NLS equation follows:

$$i\frac{\partial u}{\partial z'} + \frac{1}{2}\frac{\partial^2 u}{\partial t'^2} + |u|^2 u = 0. \tag{10.2}$$

An exact solution of (10.2), called here the classical soliton solution, is given by $u = \eta \operatorname{sech}(\eta t') e^{i\eta^2 z'/2}$ (see also Chapter 6). As mentioned earlier, solitons in optical fibers were predicted theoretically in 1973 (Hasegawa and Tappert, 1973a) and then demonstrated experimentally a few years later (Mollenauer et al., 1980; see also Hasegawa and Kodama, 1995, for additional references and historical background).

### 10.1.2 FWHM: The full width at half-maximum

The FWHM corresponds to the temporal distance between the two points where the pulse is at half its peak power. This is schematically shown in Figure 10.1. For a sech pulse this gives

$$|u|^2 = \eta^2 \operatorname{sech}^2\left(\frac{\eta T}{t_*}\right) = \frac{\eta^2}{2}.$$

Thus $T$ is such that $\operatorname{sech}^2(\eta T/t_*) = 1/2$. If we assume $\eta = 1$, then $T/t_* \approx 1.763/2 = 0.8815$. The FWHM for classical solitons is notationally taken to be $\tau$ with $\tau = 2T \approx 1.763t_*$. Recall that $t_*$ is the normalizing value of

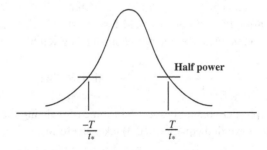

Figure 10.1 The full width at half-maximum.

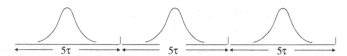

Figure 10.2 Pulses are frequently spaced five FWHM apart in a single-channel optical fiber communication system.

$t$. If $\tau = 20$ ps (ps denotes picoseconds) with pulse spacing of $5\tau = 100$ ps, this implies one bit is transmitted every 100 ps, as indicated in Figure 10.2. The bit-rate is therefore found to be:

$$\text{Bit-Rate} = \frac{1}{100 \times 10^{-12}} = 10 \times 10^9 \text{ bits/sec} = 10 \text{ gb/s}$$

where gb/s denotes gigabits per second. Typical numbers for classical solitons are: $\tau = 20$ ps which gives $t_* = 20/1.763 = 11.3$ ps.

On the other hand if the pulse were of Gaussian shape (which more closely approximates dispersion-managed solitons, described later in this chapter), then

$$u = u_0 e^{-t^2/2t_*^2},$$

$$|u|^2 = |u_0|^2 e^{-t^2/t_*^2} = |u_0|^2/2,$$

and

$$t = (\log 2)^{1/2} t_*,$$

hence in this case the FWHM is given by $\tau = 2\sqrt{\log 2}\, t_* = 1.665 t_*$.

### 10.1.3 Loss

In reality there is a damping effect, or loss, in an optical fiber. Otherwise, using typical fibers, after 100 km pulses would lose most of their energy. Usually linear damping is assumed, and the NLS equation is given by

$$i\frac{\partial A}{\partial \bar{z}} + \left(-\frac{k''(\omega)}{2}\right)\frac{\partial^2 A}{\partial T^2} + \nu|A|^2 A = -i\gamma A, \tag{10.3}$$

where $\gamma$ is the physical damping rate. Non-dimensionalizing the equation as before, see (10.2), we find equation (10.3) takes the form

$$i\frac{\partial u}{\partial z} + \frac{1}{2}\frac{\partial^2 u}{\partial t^2} + |u|^2 u = -i\Gamma u, \tag{10.4}$$

where $\Gamma = \gamma z_*$ and, for convenience, we drop the prime from the variables. We can deduce the rate at which the soliton amplitude decays. Multiplying the NLS equation, (10.4), by $u^*$ and its complex conjugate by $u$, we find:

$$iu^* \frac{\partial u}{\partial z} + u^* \frac{1}{2} \frac{\partial^2 u}{\partial t^2} + |u|^4 = -i\Gamma |u|^2$$

$$-iu \frac{\partial u^*}{\partial z} + u \frac{1}{2} \frac{\partial^2 u^*}{\partial t^2} + |u|^4 = i\Gamma |u|^2.$$

Noting

$$\frac{\partial(uu^*)}{\partial z} = u^* \frac{\partial u}{\partial z} + u \frac{\partial u^*}{\partial z}$$

and

$$\frac{\partial^2 u}{\partial t^2} u^* - \frac{\partial^2 u^*}{\partial t^2} u = \frac{\partial}{\partial t} \left( u^* \frac{\partial u}{\partial t} - u \frac{\partial u_*}{\partial t} \right),$$

we get, by substituting the above equations,

$$i \frac{\partial |u|^2}{\partial z} u^* + \frac{1}{2} \frac{\partial}{\partial t} \left( u^* \frac{\partial u}{\partial t} - u \frac{\partial u_*}{\partial t} \right) = -2\Gamma |u|^2,$$

and hence integrating over the (infinite) domain of the soliton pulse,

$$\frac{\partial}{\partial z} \int_{-\infty}^{\infty} |u|^2 \, dt = -2\Gamma \int_{-\infty}^{\infty} |u|^2 \, dt. \tag{10.5}$$

Substituting into (10.5) a classical soliton given by $u = \eta \operatorname{sech}(\eta t) \, e^{i\eta^2 z/2}$, we find

$$\frac{\partial}{\partial z} \int_{-\infty}^{\infty} \eta^2 \operatorname{sech}^2(\eta t) \, dt = -2\Gamma \int_{-\infty}^{\infty} \eta^2 \operatorname{sech}^2(\eta t) \, dt.$$

This yields the equation governing the change in the soliton parameter,

$$\frac{\partial \eta}{\partial z} = -2\Gamma \eta.$$

Thus $u = u_0 e^{-2\Gamma z}$ and we say the soliton damping rate is $2\Gamma$. This is very different from the rate one would get if one had only a linear equation

$$i \frac{\partial u}{\partial z} + \frac{1}{2} \frac{\partial^2 u}{\partial t^2} = -i\Gamma u,$$

i.e., if we neglect the nonlinear term in (10.4). In this case, if we take $u = Ae^{i\omega t} + cc$, i.e., we assume a linear periodic wave (sometimes called "CW" or continuous wave), we find by the same procedure as above, integrating over a period of the linear wave $T = 2\pi/\omega$:

$$\frac{\partial}{\partial z} \int_0^{\frac{2\pi}{\omega}} |A|^2 \cos^2(\omega t) \, dt = -2\Gamma \int_0^{\frac{2\pi}{\omega}} |A|^2 \cos^2(\omega t) \, dt$$

$$\frac{\partial}{\partial z}|A|^2 = -2\Gamma|A|^2$$

$$\frac{\partial|A|}{\partial z} = -\Gamma|A|.$$

Thus for a linear wave (CW): $|A| = |A(0)|e^{-\Gamma z}$ and therefore the amplitude of a soliton decays at "twice the linear rate"; i.e., a soliton decays at twice the damping rate as that of a linear (CW) wave.

### 10.1.4 Amplification

In practice, for sending signals over a long distance, in order to counteract the damping, the optical field must be amplified. A method, first developed in the mid-1980s, uses all-optical fibers (typically: "EDFAs": erbium doped fiber amplifiers)

In order to derive the relevant equation we assume the amplifiers occur at distinct locations $z_m = mz_a$, $m = 1, 2, \ldots$, down the fiber and are modeled as Dirac delta functions $(G - 1)\delta(z - z_m)$, where $G$ is the normalized gain. Here $z_a$ denotes the normalized amplifier distance: $z_a = \ell_a/z_*$, where $\ell_a$ is the distance between the amplifiers in physical units. The NLS equation takes the form:

$$i\frac{\partial u}{\partial z} + \frac{1}{2}\frac{\partial^2 u}{\partial t^2} + |u|^2 u = i\left(-\Gamma + \sum_m (G - 1)\delta(z - z_m)\right)u.$$

This is usually termed a "lumped" model. Integrating from $z_{n-} = z_n - \delta$ to $z_{n+} = z_n + \delta$, where $0 < \delta \ll 1$ and assuming continuity of $u$ and its temporal derivatives

$$i\int_{z_{n-}}^{z_{n+}} \frac{\partial u}{\partial z} \, dz = i\int_{z_{n-}}^{z_{n+}} \left(-\Gamma + \sum_m (G - 1)\delta(z - z_m)\right)u \, dz$$

we find

$$u(z_{n+}, t) - u(z_{n-}, t) = (G - 1)u(z_{n-}, t)$$

$$u(z_{n+}, t) = Gu(z_{n-}, t).$$

This gives a jump condition between the locations just in front of and just beyond an amplifier. On the other hand, if we look inside $z_{n-1+} < z < z_{n-}$ and assume $u = A(z)\tilde{u}$ inside $z_{n-1+} < z < z_{n-}$, we have

$$iA_z\tilde{u} + iA\frac{\partial\tilde{u}}{\partial z} + \frac{1}{2}A\frac{\partial^2\tilde{u}}{\partial t^2} + |A|^2A|\tilde{u}|^2\tilde{u} = i\left[-\Gamma + (G-1)\delta(z-z_{n-})\right]A\tilde{u}.$$

We take $A(z)$ to satisfy

$$\frac{\partial A}{\partial z} = \left[-\Gamma + (G-1)\delta(z-z_{n-})\right]A. \tag{10.6}$$

If we assume

$$A(z) = A_0e^{-\Gamma(z-z_{n-1})}, \quad z_{n-1+} < z < z_{n-} \tag{10.7}$$

then $A(z_{n-}) = A_0e^{-\Gamma z_a}$ and then integration of (10.6) using (10.7) yields,

$$A(z_{n+}) - A(z_{n-}) = (G-1)A(z_{n-}),$$

which implies $A(z_{n+}) = GA(z_{n-})$ consistent with the above result for $u$. If we further assume $A(z_{n+}) = A_0$ (i.e., $A$ returns to its original value) then we find that the gain is $G = e^{\Gamma z_a}$ in the lumped model.

With gain-compensating loss we have the following model equation:

$$i\frac{\partial\tilde{u}}{\partial z} + \frac{1}{2}\frac{\partial^2\tilde{u}}{\partial t^2} + |A(z)|^2|\tilde{u}|^2\tilde{u} = 0,$$

where $A(z)$ is a decaying exponential in $(nz_a, (n+1)z_a)$ given by $|A|^2 = |A_0|^2e^{-2\Gamma(z-nz_a)}$, $nz_a < z < (n+1)z_a$, periodically extended (see Figure 10.3).

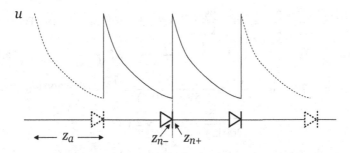

Figure 10.3 The loss of energy of the soliton $u$ and periodic amplification.

In this model $\tilde{u}$ (unlike $u$) does not have delta functions in the equation and does not change discontinuously over one period. We choose $|A|^2$ such that $\frac{1}{z_a}\int_0^{z_a} |A|^2\, dz = 1$; hence

$$\frac{1}{z_a}|A_0|^2 \int_0^{z_a} e^{-2\Gamma z}dz = \frac{|A_0|^2}{-2\Gamma z_a}(e^{-2\Gamma z_a} - 1) = 1,$$

$$|A_0|^2 = \frac{2\Gamma z_a}{1 - e^{2\Gamma z_a}}.$$

Note that as $z_a \to 0$ we have $\dfrac{2\Gamma z_a}{1 - e^{2\Gamma z_a}} \to 1$.

For notational convenience, we now drop the tilde on $u$ – but stress that the "true" normalized field is $u = A(z)\tilde{u}$, with $A = A_0 e^{-\Gamma(z-z_n)}$ where $n z_a < z < (n + 1)z_a$ periodically extended. Thus, in summary, our basic gain/loss model is the "distributed" equation

$$i\frac{\partial u}{\partial z} + \frac{1}{2}\frac{\partial^2 u}{\partial t^2} + g(z)|u|^2 u = 0, \tag{10.8}$$

where $g(z) = |A_0|^2 e^{-2\Gamma(z-n z_a)}$, $n z_a < z < (n + 1)z_a$, periodically extended. Note again that amplification occurs between $z_{n-} < z < z_{n+}$. We also see that the average of $g$ is given by $\langle g \rangle = \dfrac{1}{z_a}\int_0^{z_a} g(z)dz = 1$. This is important regarding multi-scale theory; on average we will see that we get the unperturbed NLS equation, $iu_z + \frac{1}{2}u_{tt} + |u|^2 u = 0$.

### 10.1.5 Some typical units: Classical solitons

Below we list some typical classical soliton parameter values and useful relationships, where $k = 2\pi/\lambda$ and $\omega/k = c$ (here, as is standard, $k$ is wave number, $\omega$ is frequency etc.):

$$\lambda \sim 1550\,\text{nm}; \qquad c = 3 \times 10^8\,\text{m/s}; \qquad D \sim 0.5\,\frac{\text{ps}}{\text{nm}\cdot\text{km}};$$

$$k'' = -\frac{\lambda^2 D}{2\pi c}\frac{\text{ps}^2}{\text{km}} = -0.65\frac{\text{ps}^2}{\text{km}}; \quad \omega = 1.2 \times 10^{15}\,\text{rad/s} = 193\,\text{THz};$$

$$\tau_{\text{FWHM}} \simeq 20\,\text{ps}; \quad t_* = 11.3\,\text{ps}; \quad \nu = \frac{n_2 \omega}{4 c A_{\text{eff}}} \sim 2.5\frac{1}{\text{W}\cdot\text{km}}; \quad P_* \sim 2\,\text{mW};$$

$$z_* \sim 200\,\text{km} = \frac{1}{\nu P_*} = \frac{1}{(2.5)(2 \times 10^{-3})} = \frac{1}{5} \times 10^3 = 200\,\text{km};$$

$$z_a \sim \frac{20\,\text{km}}{200\,\text{km}} \sim 0.1; \quad \gamma = 0.05\,\frac{1}{\text{km}};$$

$$\gamma_{\text{decibels}} = 4.343\gamma = 0.22\,\frac{\text{db}}{\text{km}}; \qquad \Gamma = \gamma z_* = 10.$$

## 10.2 Multiple-scale analysis of the NLS equation

We begin with (10.8),

$$i\frac{\partial u}{\partial z} + \frac{1}{2}\frac{\partial^2 u}{\partial t^2} + g(z)|u|^2 u = 0.$$

We assume that $g(z)$ is rapidly varying and denote it by $g(z) = g(z/z_a)$, where $0 < z_a \ll 1$ is the "fast" scale and $z_* \sim O(1)$ is the "slow" scale. Recall that $g(z) = A_0^2 e^{-2\Gamma(z - nz_a)}$, where $nz_a < z < (n+1)z_a$. Note that in terms of the "fast" scale $z_a$, this can be written

$$g(z) = A_0^2 e^{-2\Gamma z_a\left(\frac{z}{z_a} - n\right)}, \qquad n < \frac{z}{z_a} < (n+1).$$

Typically $z_a \sim 0.1$, $\Gamma = 10$ and we say that $g(z)$ is rapidly varying on a scale of length $z_a$. Next we assume a solution of the form $u = u(\zeta, Z, t; \epsilon)$ where $\zeta = z/z_a, Z = z$ and $\epsilon = z_a$ as our small parameter. Using

$$\frac{\partial}{\partial z} \rightarrow \frac{\partial}{\partial Z} + \frac{1}{\epsilon}\frac{\partial}{\partial \zeta},$$

we can write the NLS equation as

$$i\left(\frac{\partial u}{\partial Z} + \frac{1}{\epsilon}\frac{\partial u}{\partial \zeta}\right) + \frac{1}{2}\frac{\partial^2 u}{\partial t^2} + g(\zeta)|u|^2 u = 0.$$

Assuming the standard multiple-scale expansion

$$u = u_0 + \epsilon u_1 + \epsilon^2 u_2 + \cdots,$$

we find, to leading order, $O(1/\epsilon)$,

$$i\frac{\partial u_0}{\partial \zeta} = 0,$$

which implies $u_0$ is independent of the fast scale $\zeta = z/\epsilon$; we write $u_0 = U(z, t)$. At the next order, $O(1)$, we have

$$i\frac{\partial u_1}{\partial \zeta} + i\frac{\partial u_0}{\partial Z} + \frac{1}{2}\frac{\partial^2 u_0}{\partial t^2} + g(\zeta)|u_0|^2 u_0 = 0,$$

or

$$i\frac{\partial u_1}{\partial \zeta} = -\left(i\frac{\partial U}{\partial Z} + \frac{1}{2}\frac{\partial^2 U}{\partial t^2} + g(\zeta)|U|^2 U\right).$$

Therefore

$$u_1 = i\left(i\frac{\partial U}{\partial Z} + \frac{1}{2}\frac{\partial^2 U}{\partial t^2} + \langle g\rangle|U|^2 U\right)\zeta + \int_0^\zeta (g(\zeta) - \langle g\rangle)\, d\zeta\, |U|^2 U.$$

Since $g(\zeta)$ is periodic with unit period, $g(\zeta) = \langle g \rangle + \sum_{n \neq 0} g_n e^{in\zeta}$. We note that $g(\zeta) - \langle g \rangle$ has zero mean and hence is non-secular (i.e., it does not grow with $\zeta$). The first term, however, will grow without bound, i.e., this term is secular. To remove this secular term, we take, using $\langle g \rangle = 1$,

$$i\frac{\partial U}{\partial Z} + \frac{1}{2}\frac{\partial^2 U}{\partial t^2} + |U|^2 U = 0, \tag{10.9}$$

with $u \sim U$ to first order, and we see that we regain the lossless NLS equation. This approximation is sometimes called the "guiding center" in soliton theory – see Hasegawa and Kodama (1991a,b, 1995). An alternative multi-scale approach for classical solitons was employed later (Yang and Kath, 1997). Equation (10.9) was obtained at the beginning of the classical soliton era. However soon it was understood that noise limited the distance of propagation. Explicitly, for typical amplifier models, noise-induced amplitude jitter (Gordon and Haus, 1986) can reduce propagation distance (in units described earlier) from 10,000 km to about 4000–5000 km. Researchers developed tools to deal with this problem (see Yang and Kath, 1997), examples being soliton transmission control and the use of filters (see, e.g., Molleneauer and Gordon, 2006). But another serious problem was soon encountered. This is described in the following section.

### 10.2.1 Multichannel communications: Wavelength division multiplexing

In the mid-1990s communications systems were moving towards multichannel communications or "WDM", standing for wavelength division multiplexing. WDM allows signals to be sent simultaneously in different frequency channels. In terms of the solution of the NLS equation we assume $u$, for a two-channel system, to be initially composed of two soliton solutions in two different channels, $u = u_1 + u_2$ that is valid before any interaction occurs. Using classical solitons,

$$u_j = \eta_j \operatorname{sech}\left[\eta_j(t - \Omega_j z)\right] e^{i\Omega_j t + i\left(\eta^2 - \Omega_j^2\right)z/2 + i\phi_j},$$

with $\Omega_1 \neq \Omega_2$; usually we take

$$u_2 = \operatorname{sech}(t - \Omega z)\, e^{i\Omega t + i(1 - \Omega^2)z/2},$$
$$u_1 = \operatorname{sech}(t + \Omega z)\, e^{-i\Omega t + i(1 - \Omega^2)z/2},$$

where for simplicity $\eta_1 = \eta_2 = 1$, $\Omega_2 = -\Omega_1 = \Omega$ and we assume $\Omega \gg 1$, because in practice the different frequencies are well separated. Once interaction begins we assume a solution of the form

$$u = u_1 + u_2 + u_f, \tag{10.10}$$

where $u_f$ is due to "four-wave mixing" and we assume $|u_f| \ll 1$ because such four-wave mixing terms generated by the interaction are small. Substituting (10.10) into (10.8) we get

$$i\frac{\partial u_1}{\partial z} + i\frac{\partial u_2}{\partial z} + i\frac{\partial u_f}{\partial z} + \frac{1}{2}\left(\frac{\partial^2 u_1}{\partial t^2} + \frac{\partial^2 u_2}{\partial t^2} + \frac{\partial^2 u_f}{\partial t^2}\right) +$$

$$+g(z)(u_1 + u_2 + u_f)^2(u_1^* + u_2^* + u_f^*) = 0.$$

Since $u_1, u_2 \sim O(1)$ and $|u_f| \ll 1$, we can neglect the terms containing $u_f$. Expanding we arrive at

$$(u_1 + u_2 + u_f)^2\left(u_1^* + u_2^* + u_f^*\right) =$$

$$= \underbrace{|u_1|^2 u_1 + |u_2|^2 u_2}_{\text{SPM}} + \underbrace{2|u_1|^2 u_2 + 2|u_2|^2 u_1}_{\text{XPM}} + \underbrace{u_1^2 u_2^* + u_2^2 u_1^*}_{\text{FWM}} + O(u_f),$$

where SPM means "self-phase modulation", XPM stands for "cross-phase modulation" and FWM: "four-wave mixing". Recall that the frequencies (or frequency channels) of the solitons $u_j$, $j = 1, 2$, are given by $u_1 \sim e^{-i\Omega t}$ and $u_2 \sim e^{i\Omega t}$, giving

$$u_1^2 u_2^* \sim e^{-3i\Omega t}$$

$$u_1^* u_2^2 \sim e^{+3i\Omega t}.$$

The equation for the above FWM components (with frequencies $\pm 3i\Omega t$) is

$$i\frac{\partial u_f}{\partial z} + \frac{1}{2}\frac{\partial^2 u_f}{\partial t^2} + g(z)\left(u_1^2 u_2^* + u_2^2 u_1^*\right) = 0. \tag{10.11}$$

On the other hand, the equation for $u_1$ (sometimes called channel 1), which is forced by SPM and XPM terms, is

$$i\frac{\partial u_1}{\partial z} + \frac{1}{2}\frac{\partial^2 u_1}{\partial t^2} + g(z)|u_1|^2 u_1 = -2g(z)|u_2|^2 u_1. \tag{10.12}$$

The symmetric analog for channel 2 ($u_2$) is

$$i\frac{\partial u_2}{\partial z} + \frac{1}{2}\frac{\partial^2 u_2}{\partial t^2} + g(z)|u_2|^2 u_2 = -2g(z)|u_1|^2 u_2. \qquad (10.13)$$

Since $|\Omega_2 - \Omega_1| \gg 1$, the equations separate naturally from one another. This is most easily understood as separation of energy in Fourier space. Direct numerical simulations support this model.

### 10.2.2 Four-wave mixing

If we say $u_f = q + r$, where $q \sim e^{3i\Omega t}$ and $r \sim e^{-3i\Omega t}$, we can separate (10.11) into two parts:

$$iq_z + \frac{1}{2}q_{tt} + g(z)u_2^2 u_1^* = 0$$

$$ir_z + \frac{1}{2}r_{tt} + g(z)u_1^2 u_2^* = 0.$$

Let us look at the equation for $q$:

$$iq_z + \frac{1}{2}q_{tt} = -g(z)u_2^2 u_1^*. \qquad (10.14)$$

A similar analysis can be done for $r$. Note that the right-hand side of (10.14) is

$$\text{RHS} = u_2^2 u_1^* = e^{+3i\Omega t - \frac{1}{2}\Omega^2 z}\,\text{sech}^2(t - \Omega z)\,\text{sech}(t + \Omega z)e^{iz/2}.$$

It is natural to assume $q = F(z,t)e^{3i\Omega t - i\Omega^2 z/2}$, where the rapidly varying phase is separated out leaving the slowly varying function $F(z,t)$. The equation for $F(z,t)$ then satisfies a linear equation

$$i\left(F_z - \frac{i}{2}\Omega^2 F\right) + \frac{1}{2}\left(F_{tt} + 6i\Omega F_t - 9\Omega^2 F\right) = -g(z)R(z,t),$$

where

$$R(z,t) = \text{sech}^2(t - \Omega z)\,\text{sech}(t + \Omega z)e^{iz/2}.$$

After simplification, $F(z,t)$ now satisfies

$$iF_z + \frac{1}{2}F_{tt} + 3i\Omega F_t - 4\Omega^2 F = -g(z)R. \qquad (10.15)$$

(We remark that if we put $\omega_f = 3\Omega$ and $k_f = \dfrac{2\Omega_2^2 - \Omega_1^2}{2} = \dfrac{\Omega^2}{2}$, then the phase of $u_2^2 u_1^*$ is $e^{i(\omega_f t - k_f z)}$ and $\omega_f^2/2 - k_f = 9\Omega^2/2 - \Omega^2/2 = 4\Omega^2$.)

We may now solve (10.15) using Fourier transforms (see also Ablowitz et al., 1996). However, a simple and useful model is obtained by assuming $|F_{tt}| \ll \Omega^2 F$ and $|\Omega F_t| \ll \Omega^2 F$. In this case we have the following reduced model:

$$iF_z - 4\Omega^2 F = -g(z)R.$$

Letting $R = \tilde{R} e^{iz/2}$ leads to

$$i\frac{\partial}{\partial z}\left(F e^{4i\Omega^2 z}\right) = -g(z)\tilde{R} e^{4i\Omega^2 z + iz/2}. \tag{10.16}$$

Near the point of collision of the solitons, i.e., at $z = 0$, there is a non-trivial contribution to $F$. We consider $\tilde{R}$ as constant in this reduced model. Since $g(z)$ is periodic with period $z_a$ it can be expanded in a Fourier series, $g(z) = \sum_{-\infty}^{\infty} g_n e^{-2\pi i n z/z_a}$, in which case the right-hand side of (10.16) is proportional to

$$g(z)e^{4i\Omega^2 z + iz/2} = \sum_{-\infty}^{\infty} g_n e^{iz[4\Omega^2 - 2\pi n/z_a + \frac{1}{2}]}.$$

Then from (10.16), we see that there is an FWM *resonance* when

$$\frac{2\pi n}{z_a} = 4\Omega^2 + \frac{1}{2}, \tag{10.17}$$

where $n$ is an integer. We therefore have the growth of FWM terms whenever (10.17) is (nearly) satisfied. Otherwise $F$ can be expected to remain small. Thus, given $\Omega$ and $z_a$, the nearest integer $n$ is given by $n = (4\Omega^2 + 1/2)z_a/2\pi$. Using the typical numbers $z_a = 0.1$, $\Omega = 4$, $n \approx (0.1)\left(\dfrac{64.5}{6.28}\right) \simeq 1.03$ gives $n = 1$ as the dominant FWM contribution. See Figure 10.4 where numerical simulation of (10.1) (with typical parameters) demonstrates the significant FWM growth for classical solitons. For more information see Mamyshev and Mollenauer (1996) and Ablowitz et al. (1996).

To avoid the FWM growth/interactions (and noise effects) that created significant penalties, researchers began to use "dispersion-management" described in the following section. Dispersion-management (DM) is characterized by large varying local dispersion that changes sign. With such large local dispersion variation one expects strongly reduced FWM contribution, which has been analyzed in detail (see Ablowitz et al., 2003a).

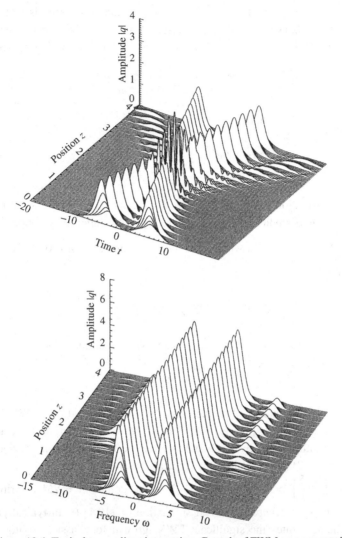

Figure 10.4 Typical two-soliton interaction. Growth of FWM components in both the physical ($t$) and the Fourier domain ($\omega$) for classical solitons can be observed.

## 10.3 Dispersion-management

We begin with the dimensional NLS equation in the form:

$$iA_z + \frac{(-k''(z))}{2}A_{tt} + v|A|^2 A = 0, \qquad (10.18)$$

where we note that now $k''$ is a function of $z$ and for simplicity we omit gain and loss for now. For classical solitons the dispersive coefficient, $k''$, is constant. However, with dispersion-management the dispersion varies with $z$ and thus $k'' = k''(z)$. To normalize this equation (see also Section 10.1.1) we take $A = \sqrt{P_*}u$, $z = z_* z'$ and $t = t_* t'$; recall the nonlinear distance is given by $z_* = 1/\nu P_*$ and $t_*$ is determined by the FWHM of the pulse, $t_{\text{FWHM}}$. Then (10.18) becomes

$$ iu_z + \frac{(-k''(z))}{2}\frac{z_*}{t_*^2}u_{tt} + \nu P_* z_* |u|^2 u = 0. $$

Taking $k''_* = t_*^2/z_*$ we get

$$ iu_z + \frac{(-k''(z)/k''_*)}{2}u_{tt} + |u|^2 u = 0. $$

The dispersion $d(z) = -k''(z)/k''_*$ is dependent on $z$ and can be written as an average plus a varying part, $k'' = \langle k'' \rangle + \delta k''(z)$, where $\langle k'' \rangle$ represents the average and is given by $\langle k'' \rangle = \left( k''_1 \ell_1 + k''_2 \ell_2 \right)/\ell$ where $\ell_1$ and $\ell_2$ are the lengths of the anomalous and normal dispersion segments, respectively, and $\ell = \ell_1 + \ell_2$, and usually $l = l_a$ ($l_a$: the length between amplifiers). The non-dimensionalized dispersion is given by $k''/k''_* = \langle d \rangle + \tilde{\Delta}(z) = d(z)$, with the average dispersion denoted by $\langle d \rangle = \langle k'' \rangle/k''_*$ and the varying part around the average: $\tilde{\Delta}(z) = \delta k''/k''_*$. Typically it is taken to be a piecewise constant function as illustrated in Figure 10.5.

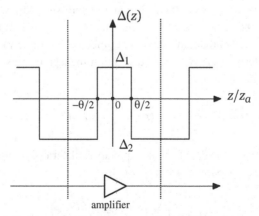

Figure 10.5 A schematic diagram of a two-fiber dispersion-managed cell. Typically the periodicity of the dispersion-management is equal to the amplifier spacing, i.e., $\ell = \ell_a$ or in normalized form, $z_a = l_a/z_*$.

The non-dimensionalized NLS equation is now

$$iu_z + \frac{d(z)}{2}u_{tt} + |u|^2 u = 0.$$

Damping and amplification can be added, as before, which leads to our fundamental model:

$$iu_z + \frac{d(z)}{2}u_{tt} + g(z)|u|^2 u = 0. \qquad (10.19)$$

If $d(z)$ does not change sign, we can simplify (10.19) by letting

$$\tilde{z}(z) = \int_0^z h(z')\,dz'$$

so that $\partial_z = h(z)\partial_{\tilde{z}}$, where $h(z)$ is to be determined. Equation (10.19) then becomes

$$iu_{\tilde{z}} + \frac{1}{2}\frac{d(z)}{h(z)}u_{tt} + \frac{g(z)}{h(z)}|u|^2 u = 0.$$

If fibers could be constructed so that $h(z) = d(z) = g(z)$, then we would obtain the classical NLS equation:

$$iu_{\tilde{z}} + \frac{1}{2}u_{tt} + |u|^2 u = 0.$$

This type of fiber is called "dispersion following the loss profile".

Unfortunately, it is very difficult to manufacture such fibers. An idea first suggested in 1980 (Lin et al., 1980) that has become standard technology is to fuse together fibers that have large and different, but (nearly) constant, dispersion characteristics. A two-step "dispersion-managed" transmission line is modeled as follows:

$$d(z) = \langle d \rangle + \frac{\Delta(z/z_a)}{z_a}, \quad \langle \Delta \rangle = \int_0^1 \Delta(\zeta)\,d\zeta = 0, \quad \zeta = \frac{z}{z_a}.$$

We quantify the parameters of a two-step map as illustrated in Figure 10.5. The variation of the dispersion is given by $\Delta(\zeta)$,

$$\Delta(\zeta) = \begin{cases} \Delta_1, & 0 \le |\zeta| < \theta/2, \\ \Delta_2, & \theta/2 < |\zeta| < 1/2. \end{cases}$$

In this two-step map we usually parameterize $\Delta_j$, $j = 1, 2$, by

$$\Delta_1 = \frac{2s}{\theta}, \qquad \Delta_2 = \frac{-2s}{1 - \theta},$$

where $\langle \Delta \rangle = 0$ and $s$ is termed the map strength parameter, defined as

$$s = \frac{\theta \Delta_1 - (1 - \theta)\Delta_2}{4} = \frac{\text{area enclosed by} \Delta}{4}. \tag{10.20}$$

Note that $\langle \Delta \rangle = \theta \Delta_1 + (1 - \theta)\Delta_2 = 0$, as it should be by its definition.

Remarkably dispersion-management is also an important technology that is used to produce ultra-short pulses in mode-locked lasers, such as in Ti:sapphire lasers (Ablowitz et al., 2004a; Quraishi et al., 2005; Ablowitz et al., 2008) discussed in the next chapter.

## 10.4 Multiple-scale analysis of DM

In this section we apply the method of multiple scales to (10.19) (see Ablowitz and Biondini, 1998):

$$iu_z + \frac{1}{2}\left(\langle d \rangle + \frac{\Delta(\zeta)}{z_a}\right)u_{tt} + g(\zeta)|u|^2 u = 0. \tag{10.21}$$

In (10.21), let us call our small parameter $z_a = \varepsilon$, assume $z_a \equiv \epsilon \ll 1$, and define $u = u(t, \zeta, Z; \epsilon)$, $\zeta = z/\epsilon$, and $Z = z$; hence, $\dfrac{\partial}{\partial z} = \dfrac{1}{\epsilon}\dfrac{\partial}{\partial \zeta} + \dfrac{\partial}{\partial Z}$. Thus,

$$\frac{1}{\epsilon}\left(i\frac{\partial u}{\partial \zeta} + \frac{1}{2}\Delta(\zeta)\frac{\partial^2 u}{\partial t^2}\right) + i\frac{\partial u}{\partial Z} + \frac{1}{2}\langle d \rangle \frac{\partial^2 u}{\partial t^2} + g(\zeta)|u|^2 u = 0.$$

With a standard multiple-scales expansion $u = u_0 + \epsilon u_1 + \epsilon^2 u_2 + \cdots$, we have at leading order, $O(1/\epsilon)$,

$$i\frac{\partial u_0}{\partial \zeta} + \frac{\Delta(\zeta)}{2}\frac{\partial^2 u_0}{\partial t^2} = 0. \tag{10.22}$$

We solve (10.22) using Fourier transforms, defined by (inverse transform)

$$u_0(t, \zeta, Z) \equiv \frac{1}{2\pi}\int_{-\infty}^{\infty} \hat{u}_0(\omega, \zeta, Z)e^{i\omega t}\,d\omega,$$

and (direct transform)

$$\hat{u}_0(\omega, \zeta, Z) = \mathcal{F}\{u_0\} \equiv \int_{-\infty}^{\infty} u_0(t, \zeta, Z)e^{-i\omega t}\,dt.$$

Taking the Fourier transform of (10.22) and solving the resulting ODE, we find

$$\frac{\partial \hat{u}_0}{\partial \zeta} - \frac{\omega^2}{2}\Delta(\zeta)\hat{u}_0 = 0, \tag{10.23a}$$

$$\hat{u}_0(\omega, \zeta, Z) = \hat{U}(\omega, Z)e^{-i\omega^2 C(\zeta)/2}, \tag{10.23b}$$

where $C(\zeta) = \int_0^\zeta \Delta(\zeta')\, d\zeta'$ is the integrated dispersion.

At the next order, $O(1)$, we have

$$i\frac{\partial u_1}{\partial \zeta} + \frac{\Delta(\zeta)}{2}\frac{\partial^2 u_1}{\partial t^2} = F_1, \tag{10.24}$$

where

$$-F_1 \equiv i\frac{\partial u_0}{\partial Z} + \frac{1}{2}\langle d \rangle\frac{\partial^2 u_0}{\partial t^2} + g(\zeta)|u_0|^2 u_0.$$

Again using the Fourier transform, (10.24) becomes

$$i\hat{u}_{1\zeta} - \frac{\omega^2}{2}\Delta(\zeta)\hat{u}_1 = \widehat{F}_1,$$

or

$$i\frac{\partial}{\partial \zeta}\left(\hat{u}_1 e^{i\omega^2 C(\zeta)/2}\right) = \widehat{F}_1 e^{i\omega^2 C(\zeta)/2}.$$

Hence

$$\left[i\hat{u}_1 e^{i\omega^2 C(\zeta)/2}\right]_{\zeta'=0}^{\zeta'=\zeta} = \int_0^\zeta \widehat{F}_1 e^{i\omega^2 C(\zeta)/2}\, d\zeta.$$

Since $C(\zeta)$ is periodic in $\zeta$ and $\Delta$ has zero mean, $\hat{u}_0$ is also periodic in $\zeta$. Therefore to remove secular terms we must have $\left\langle \widehat{F}_1 e^{i\omega^2 C(\zeta)/2}\right\rangle = 0$. Thus we require that

$$\left\langle \widehat{F}_1 e^{i\omega^2 C(\zeta)/2}\right\rangle = \int_0^1 \widehat{F}_1 e^{i\omega^2 C(\zeta)/2}\, d\zeta = 0,$$

or

$$\int_0^1 \left(i\frac{\partial \hat{u}_0}{\partial Z} - \frac{\langle d \rangle}{2}\omega^2 \hat{u}_0\right) e^{i\omega^2 C(\zeta)/2}\, d\zeta$$

$$+ \int_0^1 g(\zeta)\mathcal{F}\left\{|u_0|^2 u_0\right\} e^{i\omega^2 C(\zeta)/2}\, d\zeta = 0.$$

Using (10.23b) we get the dispersion-managed NLS (DMNLS) equation:

$$i\frac{\partial \hat{U}}{\partial Z} - \frac{\langle d \rangle}{2}\omega^2 \hat{U} + \left\langle g(\zeta)\mathcal{F}\left\{|u_0|^2 u_0\right\} e^{i\omega^2 C(\zeta)/2}\right\rangle = 0. \tag{10.25}$$

### 10.4.1 The DMNLS equation in convolution form

Equation (10.25) is useful numerically, but analytically it is often better to transform the nonlinear term to Fourier space. Using the Fourier transform

$$u_0(t, \zeta, Z) = \frac{1}{2\pi} \int_{-\infty}^{\infty} \hat{U}(\omega, Z) e^{-i\omega^2 C(\zeta)/2} e^{i\omega t} \, d\omega \tag{10.26}$$

in the nonlinear term in (10.25), we find, after interchanging integrals (see Section 10.4.2 for a detailed discussion):

$$\left\langle g(\zeta) \mathcal{F}[|u_0|^2 \, u_0] e^{i\omega^2 C(\zeta)/2} \right\rangle = \int_{-\infty}^{\infty} \int_{-\infty}^{\infty} r(\omega_1 \omega_2) \hat{U}(\omega + \omega_1, z)$$
$$\times \hat{U}(\omega + \omega_2, z) \hat{U}^*(\omega + \omega_1 + \omega_2, z) \, d\omega_1 d\omega_2, \tag{10.27}$$

where

$$r(\omega_1 \omega_2) = \frac{1}{(2\pi)^2} \int_0^1 g(\zeta) \exp\left[i\omega_1 \omega_2 C(\zeta)\right] \, d\zeta. \tag{10.28}$$

If $g(z) = 1$, i.e., the lossless case, we find for the two-step map, depicted in Figure 10.5, that

$$r(x) = \frac{\sin sx}{(2\pi)^2 sx}, \tag{10.29}$$

where $s$ is called the DM map strength [see (10.20)]. This leads to another representation of the DMNLS equation in "convolution form"

$$i\hat{U}_Z - \frac{\langle d \rangle}{2} \omega^2 \hat{U} + \int_{-\infty}^{\infty} \int_{-\infty}^{\infty} r(\omega_1 \omega_2) \hat{U}(\omega + \omega_1, z) \hat{U}(\omega + \omega_2, z)$$
$$\hat{U}^*(\omega + \omega_1 + \omega_2, z) \, d\omega_1 \, d\omega_2 = 0, \tag{10.30}$$

where $r(\omega_1 \omega_2)$ is given in (10.28) (Ablowitz and Biondini, 1998; Gabitov and Turitsyn, 1996).

The above DMNLS equation has a natural dual in the time domain. Taking the inverse Fourier transform of (10.30) yields (see Section 10.4.2 for details)

$$iU_Z + \frac{\langle d \rangle}{2} U_{tt} + \int_{-\infty}^{\infty} \int_{-\infty}^{\infty} R(t_1, t_2) U(t_1) U(t_2) U^*(t_1 + t_2 - t) \, dt_1 dt_2 = 0,$$
$$\tag{10.31}$$

where

$$R(t_1, t_2) = \int_{-\infty}^{\infty} \int_{-\infty}^{\infty} r(\omega_1 \omega_2) e^{i\omega_1 t_2} e^{i\omega_2 t_1} \, d\omega_1 \, d\omega_2.$$

Note that if $\Delta(z) \to 0$, we recover the classical NLS equation since it can be shown that as $s \to 0$, $r \to 1/(2\pi)^2$, $R(t_1, t_2) \to \delta(t_1)\delta(t_2)$, recalling that $\delta(t) = \frac{1}{2\pi} \int e^{i\omega t} \, dt$. When $g(z) = 1$, the lossless case, we also find

$$R(t_1, t_2) = \frac{1}{2\pi s} \mathrm{Ci} \left( \frac{|t_1 t_2|}{s} \right)$$

with

$$\mathrm{Ci}(x) = \int_x^{\infty} \frac{\cos u}{u} \, du.$$

### 10.4.2 Detailed derivation

In this subsection we provide the details underlying the derivation of (10.27) and (10.31) as well as the special cases of (10.29) when $g = 1$. We start by writing (10.27) explicitly:

$$I(\omega) = I \equiv \left\langle g(\zeta) \mathcal{F} \left\{ |u_0|^2 u_0 \right\} e^{i\omega^2 C(\zeta)/2} \right\rangle$$

$$= \int_0^1 g(\zeta) \int_{-\infty}^{\infty} dt \left[ \frac{1}{2\pi} \int_{-\infty}^{\infty} \hat{U}(\omega_1, Z) e^{-i\omega_1^2 C(\zeta)/2} e^{i\omega_1 t} \, d\omega_1 \right]$$

$$\times \left[ \frac{1}{2\pi} \int_{-\infty}^{\infty} \hat{U}(\omega_2, Z) e^{-i\omega_2^2 C(\zeta)/2} e^{i\omega_2 t} \, d\omega_2 \right]$$

$$\times \left[ \frac{1}{2\pi} \int_{-\infty}^{\infty} \hat{U}^*(\omega_3, Z) e^{i\omega_3^2 C(\zeta)/2} e^{-i\omega_3 t} \, d\omega_3 \right]$$

$$\times e^{-i\omega t} e^{i\omega^2 C(\zeta)/2} \, d\zeta.$$

Interchanging the order of integration and grouping the exponentials together,

$$I = \int_0^1 g(\zeta) \int_{-\infty}^{\infty} \int_{-\infty}^{\infty} \int_{-\infty}^{\infty} \left( \frac{1}{2\pi} \right)^2 \hat{U}(\omega_1, Z) \hat{U}(\omega_2, Z)$$

$$\times \hat{U}^*(\omega_3, Z) e^{-i(\omega_1^2 + \omega_2^2 - \omega_3^2 - \omega^2)C(\zeta)/2}$$

$$\times \frac{1}{2\pi} \int_{-\infty}^{\infty} e^{i(\omega_1 + \omega_2 - \omega_3 - \omega)t} \, dt \, d\omega_1 \, d\omega_2 d\omega_3 \, d\zeta.$$

Recall that the Dirac delta function can be written as

$$\delta(\omega' - \omega) = \frac{1}{2\pi} \int_{-\infty}^{\infty} e^{i(\omega' - \omega)t} \, dt,$$

where $\omega' = \omega_1 + \omega_2 - \omega_3$. Thus, with $\omega_3 = \omega_1 + \omega_2 - \omega$,

$$I = \int_0^1 g(\zeta) \int_{-\infty}^{\infty} \int_{-\infty}^{\infty} \left(\frac{1}{2\pi}\right)^2 \hat{U}(\omega_1, Z) \hat{U}(\omega_2, Z) \hat{U}^*(\omega_1 + \omega_2 - \omega, Z)$$

$$\times \exp\left[-i\left((\omega_1 + \omega_2)\omega - \omega_1\omega_2 - \omega^2\right) C(\zeta)\right] d\omega_1 \, d\omega_2 \, d\zeta. \quad (10.32)$$

A more symmetric form is obtained when we use

$$\left.\begin{array}{l} \omega_1 = \omega + \tilde{\omega}_1 \\ \omega_2 = \omega + \tilde{\omega}_2 \end{array}\right\} \Rightarrow \omega_1 + \omega_2 - \omega = \tilde{\omega}_1 + \tilde{\omega}_2 + \omega.$$

Note that $-\omega_1\omega_2 + (\omega_1 + \omega_2)\omega = -(\omega + \tilde{\omega}_1)(\omega + \tilde{\omega}_2) + (\tilde{\omega}_1 + \tilde{\omega}_2 + \omega)\omega = -\tilde{\omega}_1\tilde{\omega}_2$. Equation (10.32) can now be written, after dropping the tildes,

$$I = \int_0^1 g(\zeta) \int_{-\infty}^{\infty} \int_{-\infty}^{\infty} \hat{U}(\omega_1 + \omega, Z) \hat{U}(\omega_2 + \omega, Z)$$

$$\times \hat{U}^*(\omega_1 + \omega_2 + \omega, Z) \left(\frac{1}{2\pi}\right)^2 e^{i\omega_1\omega_2 C(\zeta)} d\omega_1 \, d\omega_2 \, d\zeta. \quad (10.33)$$

We now define the DMNLS kernel $r$:

$$r(x) \equiv \frac{1}{(2\pi)^2} \int_0^1 g(\zeta) e^{ixC(\zeta)} \, d\zeta.$$

Then (10.33) becomes

$$I = \int_{-\infty}^{\infty} \int_{-\infty}^{\infty} r(\omega_1\omega_2) \hat{U}(\omega_1 + \omega, Z)$$

$$\times \hat{U}(\omega_2 + \omega, Z) \hat{U}^*(\omega_1 + \omega_2 + \omega, Z) \, d\omega_1 \, d\omega_2,$$

which is the third term in (10.30), i.e., (10.27).

For the lossless case, i.e., $g(\zeta) = 1$ and a two-step map, $r(x)$, the analysis can be considerably simplified. We now give details of this calculation. First, for a two-step map with $C(0) = 0$

$$C(\zeta) = \int_0^\zeta \Delta(\zeta') \, d\zeta' = \begin{cases} \Delta_1 \zeta, & 0 < \zeta < \theta/2 \\ \Delta_2 \zeta + C_1, & \theta/2 < \zeta < 1 - \theta/2 \\ \Delta_1 \zeta + C_2, & 1 - \theta/2 < \zeta < 1 \end{cases}$$

$$C_1 = \Delta_1 \theta/2 - \Delta_2 \theta/2 = (\Delta_1 - \Delta_2)\theta/2$$
$$C_2 = -\Delta_1,$$

where $C_{1,2}$ are obtained by requiring continuity at $\zeta = \theta/2$, $1-\theta/2$, respectively. Recall that since $\langle \Delta \rangle = 0$, we have $\Delta_1\theta + \Delta_2(1-\theta) = 0$. Thus,

$$I = \int_0^1 e^{ixC(\zeta)}\, d\zeta$$

$$= \int_0^{\theta/2} e^{i\Delta_1\zeta x}\, d\zeta + \int_{\theta/2}^{1-\theta/2} e^{i(\Delta_2\zeta+C_1)x}\, d\zeta + \int_{1-\theta/2}^1 e^{i(\Delta_1\zeta+C_2)x}\, d\zeta$$

$$I = \frac{e^{i\Delta_1\theta x/2} - 1}{i\Delta_1 x}$$

$$+ e^{iC_1x}\left[\frac{e^{i\Delta_2(1-\theta/2)x} - e^{i\Delta_2\theta x/2}}{i\Delta_2 x}\right] + e^{iC_2x}\left[\frac{e^{i\Delta_1 x} - e^{i\Delta_1(1-\theta/2)x}}{i\Delta_1 x}\right].$$

Using $C_1 = (\Delta_1 - \Delta_2)\theta/2$ and $C_2 + \Delta_1 = 0$,

$$I = \frac{e^{i\Delta_1\theta x/2} - e^{-i\Delta_1\theta/2}}{i\Delta_1 x} + \frac{e^{i(\Delta_2(1-\theta/2)x+(\Delta_1-\Delta_2)\theta/2)} - e^{i(\Delta_2\theta x/2 + (\Delta_1-\Delta_2)\theta/2)}}{i\Delta_2 x}$$

$$I = \frac{e^{i\Delta_1(\theta/2)x} - e^{-i\Delta_1(\theta/2)x}}{i\Delta_1 x} + \frac{e^{i\Delta_2(1-\theta)x+\Delta_1\theta/2} - e^{-i\Delta_1(\theta/2)x}}{i\Delta_2 x}.$$

Then using $\Delta_2(1-\theta) = -\Delta_1\theta$, we get

$$I = \frac{e^{i\Delta_1(\theta/2)x} - e^{-i\Delta_1(\theta/2)x}}{i\Delta_1 x} - \frac{e^{i\Delta_1(\theta/2)x} - e^{-i\Delta_1(\theta/2)x}}{i\Delta_2 x}.$$

Setting $s = \dfrac{\Delta_1\theta - (1-\theta)\Delta_2}{4}$, we have $s = \dfrac{\Delta_1\theta}{2}$ and hence

$$\int_0^1 e^{iC(\zeta)x}\, d\zeta = \frac{1}{ix}\left[I = \frac{e^{isx} - e^{-isx}}{\Delta_1} - \frac{e^{isx} - e^{-isx}}{\Delta_2}\right]$$

$$= \frac{2\sin sx}{x}\left(\frac{1}{\Delta_1} - \frac{1}{\Delta_2}\right)$$

$$= \frac{2\sin sx}{x}\left(\frac{1}{\Delta_1} + \frac{1-\theta}{\theta\Delta_1}\right)$$

$$= \frac{2\sin sx}{\Delta_1\theta x} = \frac{\sin sx}{sx};$$

thus when $g = 1$

$$\int_0^1 e^{iC(\zeta)x}\, d\zeta = \frac{\sin sx}{sx}. \tag{10.34}$$

Next we investigate the inverse Fourier transform of (10.30). Using

$$U = \frac{1}{2\pi}\int_{-\infty}^{\infty} \hat{U}e^{i\omega t}\, d\omega$$

and taking the inverse Fourier transform of (10.30) gives

$$iU_Z + \frac{\langle d \rangle}{2} U_{tt} + \mathcal{F}^{-1} \left\langle g\mathcal{F} \left\{ |u_0|^2 u_0 \right\} e^{i\omega^2 C(\zeta)/2} \right\rangle = 0.$$

Now,

$$\mathcal{F}^{-1} \left\langle g\mathcal{F} \left\{ |u_0|^2 u_0 \right\} e^{i\omega^2 C(\zeta)/2} \right\rangle$$

$$= \int_{-\infty}^{\infty} e^{i\omega t} \left[ \int_{-\infty}^{\infty} \int_{-\infty}^{\infty} r(\omega_1 \omega_2) \hat{U}(\omega_1 + \omega, Z) \hat{U}(\omega_2 + \omega, Z) \right.$$

$$\left. \times \hat{U}^*(\omega_1 + \omega_2 + \omega, Z) \, d\omega_1 d\omega_2 \right] d\omega$$

$$= \int_{-\infty}^{\infty} e^{i\omega t} \int_{-\infty}^{\infty} \int_{-\infty}^{\infty} r(\omega_1 \omega_2) \left[ \int_{-\infty}^{\infty} U(t_1) e^{-i(\omega_1 + \omega)t_1} \, dt_1 \right]$$

$$\times \left[ \int_{-\infty}^{\infty} U(t_2) e^{-i(\omega_2 + \omega)t_2} \, dt_2 \right]$$

$$\times \left[ \int_{-\infty}^{\infty} U^*(t_3) e^{i(\omega_1 + \omega_2 + \omega)t_3} \, dt_3 \right] d\omega_1 \, d\omega_2 d\omega$$

$$= \int_{-\infty}^{\infty} \int_{-\infty}^{\infty} d\omega_1 d\omega_2 r(\omega_1 \omega_2) \int_{-\infty}^{\infty} dt_1 U(t_1) \, dt_2 \int_{-\infty}^{\infty} U(t_2) \, dt_3$$

$$\times \int_{-\infty}^{\infty} U^*(t_3) e^{i\omega_1(t_3 - t_1)} e^{i\omega_2(t_3 - t_2)} \underbrace{\frac{1}{2\pi} \int_{-\infty}^{\infty} e^{i\omega(t - t_1 - t_2 + t_3)} \, d\omega}_{\text{use } \delta(t - t_1 - t_2 + t_3)}$$

$$= \int_{-\infty}^{\infty} \int_{-\infty}^{\infty} d\omega_1 d\omega_2 r(\omega_1 \omega_2)$$

$$\times \int_{-\infty}^{\infty} \int_{-\infty}^{\infty} dt_1 \, dt_2 U(t_1) e^{i\omega_1(t_2 - t)} U(t_2) e^{i\omega_2(t_1 - t)} U^*(t_1 + t_2 - t).$$

Thus,

$$\mathcal{F}^{-1} \left\langle g\mathcal{F} \left\{ |u_0|^2 u_0 \right\} e^{i\omega^2 C(\zeta)/2} \right\rangle$$

$$= \int_{-\infty}^{\infty} \int_{-\infty}^{\infty} R(t_1, t_2) U(t_1) U(t_2) U^*(t_1 + t_2 - t) \, dt_1 \, dt_2,$$

where

$$R(t_1, t_2) \equiv \int_{-\infty}^{\infty} \int_{-\infty}^{\infty} r(\omega_1 \omega_2) e^{i\omega_1(t_2 - t)} e^{i\omega_2(t_1 - t)} \, d\omega_1 \, d\omega_2.$$

With $t_1 = \tilde{t}_1 + t$ and $t_2 = \tilde{t}_2 + t$ (and dropping the $\sim$) we have,

$$\mathcal{F}^{-1}\left\langle g\mathcal{F}\left\{|u_0|^2\, u_0\right\} e^{i\omega^2 C(\zeta)/2}\right\rangle$$
$$= \int_{-\infty}^{\infty}\int_{-\infty}^{\infty} R(t_1, t_2) U(t+t_1) U(t+t_2) U^*(t+t_1+t_2)\, dt_1\, dt_2,$$

and

$$R(t_1, t_2) = \int_{-\infty}^{\infty}\int_{-\infty}^{\infty} r(\omega_1 \omega_2) e^{i\omega_1 t_1 + i\omega_2 t_2}\, d\omega_1\, d\omega_2, \qquad (10.35)$$

the last equality being obtained by interchanging the roles of $\omega_1$ and $\omega_2$, i.e., $\omega_1 \to \omega_2$ and $\omega_2 \to \omega_1$.

The expression for $R(t_1, t_2)$ can also be further simplified when $g(\zeta) = 1$. From (10.34) and (10.35),

$$R(t_1, t_2) = \frac{1}{(2\pi)^2}\int_{-\infty}^{\infty}\int_{-\infty}^{\infty} \frac{\sin s\omega_1 \omega_2}{s\omega_1 \omega_2} e^{i\omega_1 t_1 + i\omega_2 t_2}\, d\omega_1\, d\omega_2.$$

Letting, $\omega_1 = u_1/t_1$, $\omega_2 = u_2/t_2$

$$R(t_1, t_2) = \frac{1}{(2\pi)^2}\int_{-\infty}^{\infty}\int_{-\infty}^{\infty} \underbrace{\operatorname{sgn} t_1 \operatorname{sgn} t_2}_{\operatorname{sgn}(t_1 t_2)}\, \frac{\sin\left(\frac{s u_1 u_2}{t_1 t_2}\right)}{s u_1 u_2} e^{i u_1 + i u_2}\, du_1\, du_2$$

$$= \frac{1}{(2\pi)^2 s}\int_{-\infty}^{\infty}\int_{-\infty}^{\infty} \frac{\sin\left(\frac{u_1 u_2}{\gamma}\right)}{u_1 u_2} e^{i u_1 + i u_2}\, du_1\, du_2,$$

where $\gamma = \left|\dfrac{t_1 t_2}{s}\right|$. Then

$$R(t_1, t_2) = \frac{1}{(2\pi)^2 s}\int_{-\infty}^{\infty}\int_{-\infty}^{\infty} \frac{\left(e^{i u_1 u_2/\gamma} - e^{-i u_1 u_2/\gamma}\right)}{2 i u_1 u_2} e^{i(u_1 + u_2)}\, du_1\, du_2,$$

$$= \frac{1}{2 i s (2\pi)^2}\int_{-\infty}^{\infty}\int_{-\infty}^{\infty} \frac{e^{i u_1}}{u_1} \frac{\left(e^{i u_2(u_1/\gamma + 1)} - e^{-i u_2(u_1/\gamma - 1)}\right)}{u_2}\, du_1\, du_2.$$

Let us look at the second integral:

$$I = \int_{-\infty}^{\infty} \frac{\left(e^{i u_2(u_1/\gamma + 1)} - e^{-i u_2(u_1/\gamma - 1)}\right)}{u_2}\, du_2.$$

Denoting the two terms as $I_1$ and $I_2$, respectively, and using contour integration yields

$$I_1 = \int_{-\infty}^{\infty} \frac{e^{i u_2(u_1/\gamma + 1)}}{u_2}\, du_2 = +i\pi \operatorname{sgn}\left(\frac{u_1}{\gamma} + 1\right)$$

$$I_2 = \int_{-\infty}^{\infty} \frac{e^{i u_2(1 - u_1/\gamma)}}{u_2}\, du_2 = +i\pi \operatorname{sgn}\left(1 - \frac{u_1}{\gamma}\right),$$

where $\int_{-\infty}^{\infty}$ denotes the Cauchy principal value integral. Thus we have

$$R(t_1, t_2) = \frac{i\pi}{2is\,(2\pi)^2} \int_{-\infty}^{\infty} \frac{e^{iu_1}}{u_1} \left[ \text{sgn}\left(\frac{u_1}{\gamma} + 1\right) - \text{sgn}\left(1 - \frac{u_1}{\gamma}\right) \right] du_1$$

$$= \frac{1}{(4\pi s)\cdot 2} \left[ \left( \int_{-\gamma}^{\infty} \frac{e^{iu_1}}{u_1}\, du_1 - \int_{-\infty}^{-\gamma} \frac{e^{iu_1}}{u_1}\, du_1 \right) \right.$$

$$\left. - \left( \int_{-\infty}^{\gamma} \frac{e^{iu_1}}{u_1}\, du_1 - \int_{\gamma}^{\infty} \frac{e^{iu_1}}{u_1}\, du_1 \right) \right]$$

$$= \frac{1}{(4\pi s)\cdot 2} \left[ -\int_{-\infty}^{-\gamma} \frac{e^{-iu_1}}{u_1}\, du_1 - \int_{-\infty}^{-\gamma} \frac{e^{iu_1}}{u_1}\, du_1 \right.$$

$$\left. + \int_{\gamma}^{\infty} \frac{e^{iu_1}}{u_1}\, du_1 + \int_{\gamma}^{\infty} \frac{e^{-iu_1}}{u_1}\, du_1 \right]$$

$$= \frac{1}{4\pi s} \left[ \int_{\gamma}^{\infty} \frac{\cos u_1}{u_1}\, du_1 - \int_{-\infty}^{-\gamma} \frac{\cos u_1}{u_1}\, du_1 \right]$$

$$= \frac{1}{2\pi s} \int_{\gamma}^{\infty} \frac{\cos u_1}{u_1}\, du_1$$

$$= \frac{1}{2\pi s} \int_{1}^{\infty} \frac{\cos \gamma y}{y}\, dy \equiv \frac{1}{2\pi s} C_i\left(\frac{|t_1 t_2|}{s}\right).$$

where we used $\gamma = \left| \dfrac{t_1 t_2}{s} \right|$.

### 10.4.3 Dispersion-managed solitons

DM solitons are stationary solutions of the DMNLS equation, (10.30). We now look for solutions of the form

$$\hat{U}(\omega, Z) = e^{i\lambda^2 Z/2}\hat{F}(\omega),$$

where the only $z$-dependence is a linearly evolving phase. With this substitution, (10.30) becomes

$$\left(\frac{\lambda^2 + \langle d\rangle\,\omega^2}{2}\right)\hat{F}(\omega) - \int_{-\infty}^{\infty}\int_{-\infty}^{\infty} r(\omega_1\omega_2)\hat{F}(\omega + \omega_1)$$

$$\times \hat{F}(\omega + \omega_2)\hat{F}^*(\omega + \omega_1 + \omega_2)\,d\omega_1\,d\omega_2 = 0$$

or

$$\hat{F}(\omega) = -\frac{2}{\lambda^2 + \langle d\rangle\,\omega^2} \int_{-\infty}^{\infty}\int_{-\infty}^{\infty} r(\omega_1\omega_2)\hat{F}(\omega + \omega_1)$$

$$\times \hat{F}(\omega + \omega_2)\hat{F}^*(\omega + \omega_1 + \omega_2)\,d\omega_1\,d\omega_2. \quad (10.36)$$

Thus we wish to find fixed points of the integral equation (10.36). When $r(x)$ is not constant, obtaining closed-form solutions of this nonlinear integral equation is difficult. However we can readily obtain numerical solutions for $\langle d \rangle > 0$. Ablowitz and Biondini (1998) obtained numerical solutions using a fixed point iteration technique similar to that originally introduced by Petviashvili (1976). Below we first discuss a modification of this method called spectral renormalization, which has proven to be useful in a variety of nonlinear problems where obtaining localized solutions are of interest.

As we discussed earlier, another formulation of the non-local term is

$$\int_{-\infty}^{\infty} \int_{-\infty}^{\infty} r(\omega_1 \omega_2) \hat{F}(\omega + \omega_1) \hat{F}(\omega + \omega_2) \hat{F}^*(\omega + \omega_1 + \omega_2) \, d\omega_1 \, d\omega_2 =$$

$$= \left\langle g(z) \mathcal{F} \left\{ |u_0|^2 u_0 \right\} e^{i\omega^2 C(\zeta)/2} - i\lambda^2 Z/2 \right\rangle$$

which is useful in the numerical schemes mentioned below. It should also be noted from the multiple-scales perturbation expansion, that $u_0$ is obtained by the inverse Fourier transform of

$$\hat{u}_0(\omega, \zeta, Z) = F(\omega) e^{i\lambda^2 Z/2} e^{-i\omega^2 C(\zeta)/2}.$$

The naive iteration procedure is given by

$$\hat{F}^{(m+1)}(\omega) = \frac{2}{\lambda^2 + \langle d \rangle \omega^2} \left\langle g(z) \mathcal{F} \left\{ |u_0|^2 u_0 \right\}^{(m)} e^{i\omega^2 C(\zeta)/2 - i\lambda^2 Z/2} \right\rangle \equiv \text{Im}[\hat{F}],$$

$$(10.37)$$

where $m = 0, 1, 2, \ldots$ and $\hat{u}_0^{(m)} = \hat{F}^m(\omega) e^{-i\omega^2 C(\zeta)/2}$. At $m = 0$, $\hat{F}^{(0)}(\omega)$ is taken to be the inverse Fourier transform of a Gaussian or "sech"-type function. However, if we iterate (10.37) directly, the iteration generally diverges so we renormalize as follows.

In (10.37) we substitute $\hat{F} = \mu \hat{G}$. We then find the following iteration equations

$$\hat{G}_{n+1} = \mu_n^2 I[\hat{G}_n]$$

with

$$\mu_n^2 = \frac{(\hat{G}_n, \hat{G}_n)}{(\hat{G}_n, I[\hat{G}_n])}$$

where $(G_n, I_n) \equiv \int \hat{G}_n^*, I_n) \, d\omega$ and $G_0(t)$ can be a localized function, i.e., Gaussian or sech profile. In either case, the iterations converge rapidly. More details can be found in Ablowitz and Musslimani (2005) and Ablowitz and Horikis (2009a).

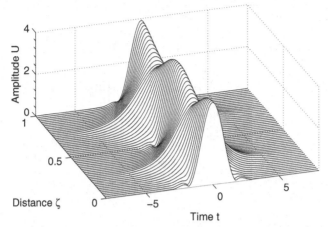

Figure 10.6 Fast evolution, $u(z, t)$, of the stationary solitons of DMNLS, reconstructed from the perturbation expansion, with $\langle d \rangle = 1$, $s = 1$ and $\lambda = 4$.

In Figure 10.6 a typical DM soliton is represented in physical space ($u \sim u^{(0)}(z, t)$). Figure 10.7 compares typical DM solitons in the Fourier and temporal domains in logarithmic scales. The dashed line is a classical soliton with the same FWHM. Figure 10.8 depicts a class of DM solitons for varying values of the map strength, $s$. In the figures note that $u(z, t) \sim u^{(0)}$. Hence from $\hat{u}^{(0)}(\omega) = \hat{U}(\omega, Z)e^{-iC(\zeta)\omega^2/2}$, with $\hat{U}(\omega, Z) = F(\omega)e^{i\lambda^2 Z/2}$, then $u^{(0)}(t) = \frac{1}{2\pi} \int_{-\infty}^{\infty} \hat{u}^{(0)}(\omega)e^{i\omega t} d\omega$. The latter yields the behavior in physical space.

Originally these methods where introduced by Petviashvili (1976) to find solutions of the KP equations. In the context of DMNLS theory (see Ablowitz et al., 2000b; Ablowitz and Musslimani, 2003) the following implementation of Petviashvili's method is employed

$$\hat{F}(\omega) = \left( \frac{S_L[\hat{F}]}{S_R[\hat{F}]} \right)^p K[\hat{F}],$$

where

$$S_L = \int_{-\infty}^{\infty} |\hat{F}|^2 \, d\omega,$$

$$S_R = \int_{-\infty}^{\infty} \hat{F}^* \cdot K[\hat{F}] \, d\omega$$

and

$$K[\hat{F}] \equiv \int_{-\infty}^{\infty} \int_{-\infty}^{\infty} r(\omega_1 \omega_2)\hat{F}(\omega + \omega_1)\hat{F}(\omega + \omega_2)\hat{F}^*(\omega + \omega_1 + \omega_2) \, d\omega_1 d\omega_2.$$

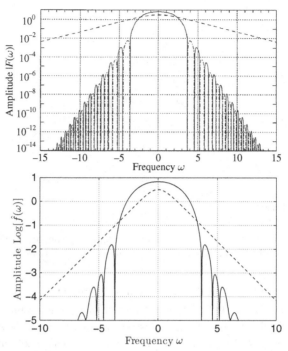

Figure 10.7 Shape of the stationary pulses in the Fourier and temporal domains for $s = 1$ and $\lambda = 4$.

The power $p$ is usually chosen such that the degree of homogeneity of the right-hand side is zero, i.e.,

$$\deg\{\text{homog.}\}(\text{RHS}) = \frac{\hat{F}^{2p}}{\hat{F}^{4p}}\hat{F}^3 = \hat{F}^{3-2p} \Rightarrow p = \frac{3}{2}.$$

With this choice for $p$, we employ the iteration scheme

$$\hat{F}^{(m+1)}(\omega) = \left(\frac{S_L[\hat{F}^{(m)}]}{S_R[\hat{F}^{(m)}]}\right)^{3/2} K[\hat{F}^{(m)}]. \tag{10.38}$$

A difficulty with this "homogeneity" method is how to handle more complex nonlinearities that do not have a fixed nonlinearity, such as mixed nonlinear terms with polynomial nonlinearity, or more complex nonlinearities, such as saturable nonlinearity $u/(1 + |u|^2)$ or transcendental nonlinearities.

We further note:

- DM solitons are well approximated by a Gaussian, $\hat{F}(\omega) \simeq ae^{-b\omega^2/2}$, for a wide range of $s$ and $\lambda$.

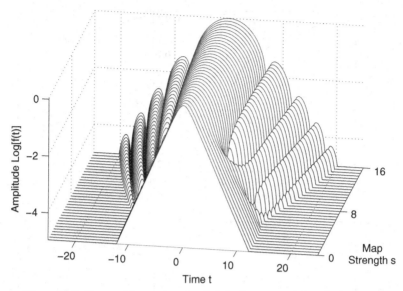

Figure 10.8 Shape of the stationary pulses in the temporal domain for $\lambda = 1$ and various values of $s$.

- The parameters $a$ and $b$ are dependent upon $s$ and $\lambda$, $a = a(s, \lambda), b = b(s, \lambda)$.
- For $s \gg 1$, $|\hat{F}|$ grows $|\hat{F}|(\omega) \sim s^{1/2}$.

See also Lushnikov (2001).

Figure 10.9 shows "chirp" versus amplitude phase plane plots for four different DM soliton systems. This figure shows how the multi-scale approximation improves as $z_a = \varepsilon \to 0$. Here the chirp is calculated as

$$c(z) = \frac{\int_{-\infty}^{\infty} t \, \text{Im} \{u^* u_t\} \, dt}{\int_{-\infty}^{\infty} t^2 |u|^2 \, dt}, \qquad (10.39)$$

where $\text{Im}(x)$ represents the imaginary part of $x$. If we approximate the DMNLS solution as a Gaussian exponential, then

$$\hat{U}_{\text{DMNLS}} = ae^{-b\omega^2/2} e^{-iC(\zeta)\omega^2/2},$$

so that in the time domain, $u(t) = \dfrac{a}{\sqrt{2\pi(b + iC)}} e^{-t^2/2(b+iC)}$, then the chirp $c(z)$ is found to be given by $c(z) = C(\zeta) = C(z/z_a)$; that is also given by (10.39) after integration.

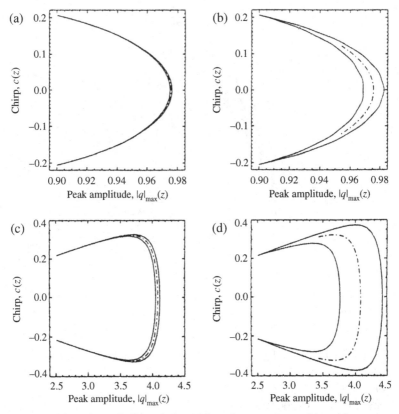

Figure 10.9 The periodic evolution of the pulse parameters: chirp $c(z)$ versus the peak amplitude. The dot–dashed line represents the leading-order approximation, $u_0(\zeta, t)$, as reconstructed from the DMNLS pulses and (10.26); the solid lines represent numerical solitions of the PNLS equation (10.19) with $g = 1$ and the same pulse energy. The four cases correspond to: (a) $s = 1$, $\lambda = 1$, $z_a = 0.02$; (b) $s = 1$, $\lambda = 1$, $z_a = 0.2$; (c) $s = 4$, $\lambda = 4$, $z_a = 0.02$; (d) $s = 4$, $\lambda = 4$, $z_a = 0.2$. The pulse energies are $\|f\|^2 = 2.27$ for cases (a) and (b), and $\|f\|^2 = 38.9$ for cases (c) and (d). See also Ablowitz et al. (2000b).

In Figure 10.10 we plot the relative $L^2$ norm of the difference between the stationary soliton solution of the DMNLS and the numerical solution of the original perturbed (PNLS) equation (10.19); equations with the same energy, sampled at the mid-point of the fiber segment. The difference is normalized to the $L^2$-norm of the numerical solution of the PNLS equation; that is

$$\frac{\|f_{\text{DMNLS}} - f_{\text{PNLS}}\|_2^2}{\|f_{\text{PNLS}}\|} = \frac{\int_{-\infty}^{\infty} dt \, |f_{\text{DMNLS}} - f_{\text{PNLS}}|^2}{\int_{-\infty}^{\infty} dt \, |f_{\text{PNLS}}|^2}.$$

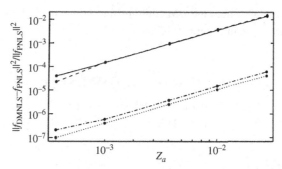

Figure 10.10 Comparison between stationary solutions of the DMNLS equation $f(t)$ and the solutions $u(z, t)$ of the PNLS equation (10.19) with the same energy taken at the mid-point of the anomalous fiber segment for $g = 1$, $\langle d \rangle = 1$ and five different values of $z_a$: 0.01, 0.02, 0.05, 0.1 and 0.2. Solid line: $s = 4$, $\lambda = 4$, $\|f\|^2 = 38.9$; dashed line: $s = 1$, $\lambda = 4$, $\|f\|^2 = 20.7$; dot–dashed line: $s = 4$, $\lambda = 1$, $\|f\|^2 = 3.1$; dotted line: $s = 1$, $\lambda = 1$, $\|f\|^2 = 2.27$. See also (Ablowitz et al., 2000b).

We also note there is an existence proof of DMNLS soliton solutions as ground states of a Hamiltonian. Furthermore, it can be proven that the DMNLS equation (10.30) is asymptotic, i.e., its solution remains close to the solution of the original NLS equation (10.21), see Zharnitsky et al. (2000).

### 10.4.4 Numerical "averaging" method

To find soliton solutions from the original PNLS equation (10.19) we may use a numerical averaging method (Nijhof et al., 1997, 2002). In this method one starts with the non-dimensional PNLS equation, (10.19):

$$iu_z + \frac{d(z)}{2}u_{tt} + g|u|^2 u = 0,$$

and an initial guess for the DM soliton, e.g., a Gaussian, $u_g(z = 0, t) = u_g(t) = u^{(0)}$ with $E_0 = \int_{-\infty}^{\infty} |u_g|^2 \, dt$. Next, numerically integrate from $z = 0$ to $z = z_a$ to get $u(z = z_a, t)$; call this $\tilde{u}(t) = |u(t)| e^{i\theta(t)}$. Then one takes the following "average":

$$\tilde{\tilde{u}}(t) = \frac{u_g(t) + \tilde{u}(t)e^{-i\theta(0)}}{2}.$$

Then $u^{(1)}$ is obtained by renormalizing the energy: $u^{(1)} = \tilde{\tilde{u}}(t)\sqrt{\frac{E_0}{\tilde{\tilde{E}}}}$, where $\tilde{\tilde{E}} = \int_{-\infty}^{\infty} \left|\tilde{\tilde{u}}^2\right| dt$. This procedure is repeated until $u^{(m)}(t)$ converges. This method can also be used to obtain the DM soliton solutions of the PNLS equation and the solid line curves in Figure 10.9.

### 10.4.5 The higher-order DMNLS equation

The analysis of Section 10.4 can be taken to the next order in $z_a$ (Ablowitz et al., 2002b). In the Fourier domain we write the more general DMNLS equation in the form

$$i\hat{U}_z - \frac{\langle d \rangle}{2}\omega^2 \hat{U}$$

$$+ \overbrace{\int_{-\infty}^{\infty} \int_{-\infty}^{\infty} d\omega_1\, d\omega_2\, \hat{U}(\omega+\omega_1)\hat{U}(\omega+\omega_2)\hat{U}^*(\omega+\omega_1+\omega_2)r(\omega_1\omega_2)}^{n_0(\omega,z)}$$

$$= z_a \hat{n}_1(\omega, z) + \cdots .$$

The method is similar to those described earlier for water waves and nonlinear optics. We find $\hat{n}_1$ such that the secular terms at higher order are removed. Recalling $u = u_0 + \epsilon u_1 + \epsilon^2 u_2 + \cdots$ where $\epsilon = z_a$, we substitute this into the perturbed NLSE equation (10.21), and as before employ multiple scales. We find

$$O\left(\epsilon^{-1}\right): \quad \mathcal{L}u_0 = i\frac{\partial u_0}{\partial \zeta} + \frac{\Delta(z)}{2}\frac{\partial^2 u_0}{\partial t^2} = 0$$

$$O(1): \quad \mathcal{L}u_1 = -\left(i\frac{\partial u_0}{\partial Z} + \frac{\langle d \rangle}{2}\frac{\partial u_0}{\partial t^2} + g(\zeta)\,|u_0|^2\, u_0\right)$$

$$O(\epsilon): \quad \mathcal{L}u_2 = -\left(i\frac{\partial u_1}{\partial Z} + \frac{\langle d \rangle}{2}\frac{\partial u_1}{\partial t^2} + g(\zeta)\left(2\,|u_0|^2\, u_1 + u_0^2 u_1^*\right)\right).$$

The first two terms are the same as what we found earlier for the DMNLS equation. At $O\left(\epsilon^{-1}\right)$,

$$u_0 = \hat{U}(\omega, Z)e^{-i\omega^2 C(\zeta)/2}, \quad C(\zeta) = \int_0^\zeta \Delta(\zeta')\, d\zeta',$$

and at $O(1)$

$$i\frac{\partial \hat{u}_1}{\partial \zeta} - \frac{\Delta(\zeta)}{2}\omega^2 \hat{u}_1 = -\left(i\frac{\partial \hat{u}_0}{\partial Z} - \frac{\langle d \rangle}{2}\omega^2 \hat{u}_0 + g(\zeta)\mathcal{F}\left\{|u_0|^2\, u_0\right\}\right)$$

$$= -\left(i\frac{\partial \hat{U}}{\partial Z} - \omega^2 \frac{\langle d \rangle}{2}\hat{U}\right)e^{-i\omega^2 C(\zeta)/2} - g\mathcal{F}\left\{|u_0|^2\, u_0\right\}.$$

Thus,

$$\frac{\partial}{\partial \zeta}\left(i\hat{u}_1 e^{iC(\zeta)\omega^2/2}\right) = -\left(i\frac{\partial \hat{U}}{\partial Z} - \omega^2 \frac{\langle d \rangle}{2}\hat{U}\right)$$
$$- g(\zeta)e^{iC(\zeta)\omega^2/2}\mathcal{F}\left\{|u_0|^2 u_0\right\} \equiv \hat{F}_1.$$

To remove secular terms we enforce periodicity of the right-hand side, i.e., we require $\langle \hat{F}_1 \rangle = 0$, where $\langle \hat{F}_1 \rangle = \int_0^1 \hat{F}_1\, d\zeta$. This implies

$$i\frac{\partial \hat{U}}{\partial Z} - \omega^2 \frac{\langle d \rangle}{2}\hat{U} + \hat{n}_0(\omega, Z) = 0,$$

with

$$\hat{n}_0(\omega, Z) = \left\langle g(\zeta)e^{i\omega^2 C(\zeta)/2}\mathcal{F}\left\{|u_0|^2 u_0\right\}\right\rangle.$$

This is the DMNLS equation (10.25) that can be transformed to (10.30).
We now take

$$i\frac{\partial \hat{U}}{\partial Z} - \omega^2 \frac{\langle d \rangle}{2}\hat{U} + \hat{n}_0(\omega, Z) = \epsilon \hat{n}_1(\omega, Z) + \cdots . \tag{10.40}$$

Once we obtain $\hat{n}_1(\omega, Z)$ then (10.40) will be the higher-order DMNLS equation. However we need to solve for $u_1$ since it will enter into the calculation for $n_1$. Using an integrating factor in the $O(1)$ equation, adding and subtracting a mean term and using the DMNLS equation, we find

$$\frac{\partial}{\partial \zeta}\left(i\hat{u}_1 e^{iC(\zeta)\omega^2/2}\right) = -\left[\left(i\frac{\partial U}{\partial Z} - \omega^2 \frac{\langle d \rangle}{2}U\right) + \left\langle g(\zeta)e^{i\omega^2 C(\zeta)/2}\mathcal{F}\left\{|u_0|^2 u_0\right\}\right\rangle\right]$$
$$- g(\zeta)e^{i\omega^2 C(\zeta)/2}\mathcal{F}\left\{|u_0|^2 u_0\right\} + \left\langle g(\zeta)e^{i\omega^2 C(\zeta)/2}\mathcal{F}\left\{|u_0|^2 u_0\right\}\right\rangle$$

$$i\hat{u}_1 e^{iC(\zeta)\omega^2/2} = -\int_0^\zeta g(\zeta)e^{i\omega^2 C(\zeta)/2}\mathcal{F}\left\{|u_0|^2 u_0\right\} d\zeta$$
$$+ \zeta\left\langle g(\zeta)e^{i\omega^2 C(\zeta)/2}\mathcal{F}\left\{|u_0|^2 u_0\right\}\right\rangle + \hat{U}_{1H}(\omega, Z).$$

Therefore,

$$\hat{u}_1 e^{iC(\zeta)\omega^2/2} = i\int_0^\zeta g(\zeta)e^{i\omega^2 C(\zeta)/2}\mathcal{F}\left\{|u_0|^2 u_0\right\} d\zeta$$
$$- i\zeta\left\langle g(\zeta)e^{i\omega^2 C(\zeta)/2}\mathcal{F}\left\{|u_0|^2 u_0\right\}\right\rangle - i\hat{U}_{1H}(\omega, Z),$$

where $\hat{U}_{1H}(\omega, Z)$ is an integration "constant" to be determined. We choose $U_1(\omega, Z)$ such that $\left\langle u_1(\omega, Z)e^{iC(\zeta)\omega^2/2}\right\rangle = \int_0^1 u_1 e^{iC(\zeta)\omega^2/2}\, d\zeta = 0$. This is done to ensure that $u_1$ does not produce secular terms at next order. Therefore

$$\hat{U}_{1H} = \left\langle \int_0^\zeta g(\zeta')e^{i\omega^2 C(\zeta')/2}\mathcal{F}\left\{|u_0|^2\, u_0\right\} d\zeta'\right\rangle - \frac{1}{2}\left\langle g(\zeta)e^{i\omega^2 C(\zeta)/2}\mathcal{F}\left\{|u_0|^2\, u_0\right\}\right\rangle.$$

This fixes $\hat{u}_1$ and thus $u_1$. Then $u_2$ is given by

$$\frac{\partial}{\partial\zeta}\left(i\hat{u}_2 e^{iC(\zeta)\omega^2/2}\right) = -\left\{\hat{n}_1(\omega, Z) + e^{iC(\zeta)\omega^2/2}\left[i\frac{\partial\hat{u}_1}{\partial Z} - \right.\right.$$
$$\left.\left. - \frac{\langle d\rangle}{2}\omega^2\hat{u}_1 + g(\zeta)\hat{P}_1(\zeta, Z, \omega)\right]\right\} \equiv \hat{F}_2,$$

where $\hat{P}_1 = \mathcal{F}\left\{2|u_0|^2\, u_1 + u_0^2 u_1^*\right\}$. To eliminate secularities we require $\left\langle\hat{F}_2\right\rangle = 0$. This condition determines $\hat{n}_1$ and we obtain the higher-order DMNLS equation. Thus, since $\left\langle\hat{u}_1 e^{iC(\zeta)\omega^2/2}\right\rangle = 0$, we have

$$\hat{n}_1 = -\left\langle g(\zeta)e^{i\omega^2 C(\zeta)/2}\hat{P}_1(\omega, Z)\right\rangle. \qquad (10.41)$$

With $\hat{n}_0$ and $\hat{n}_1$ we now have the higher-order DMNLS equation (10.40). The form (10.41) is useful numerically. But as with the standard DMNLS theory we also note that we can write $\hat{n}_1$ as a multiple integral in the Fourier domain. After manipulation we find,

$$\hat{n}_1 = \frac{-i}{(2\pi)^4}\int_{-\infty}^\infty\int_{-\infty}^\infty\int_{-\infty}^\infty\int_{-\infty}^\infty d\omega_1\, d\omega_2\, d\Omega_1\, d\Omega_2\; r_1(\omega_1\omega_2, \Omega_1\Omega_2)$$
$$\left\{2\hat{U}(\omega + \omega_1)\times\hat{U}^*(\omega + \omega_1 + \omega_2)\hat{U}(\omega + \omega_2 + \Omega_1)\hat{U}(\omega + \omega_2 + \Omega_2)\right.$$
$$\times\hat{U}^*(\omega + \omega_2 + \Omega_1 + \Omega_2) - \hat{U}(\omega + \omega_1)\hat{U}(\omega + \omega_2)\hat{U}^*(\omega + \omega_1 + \omega_2 + \Omega_1)$$
$$\left.\times\hat{U}(\omega + \omega_1 + \omega_2 - \Omega_2)\hat{U}(\omega + \omega_1 + \omega_2 + \Omega_1 - \Omega_2)\right\},$$
$$(10.42)$$

where the kernel is given by

$$r_1(x, y) = \left\langle g(\zeta)K(\zeta, x)\int_0^\zeta g(\zeta')K(\zeta', y)\, d\zeta'\right\rangle$$
$$- \langle g(\zeta)K(\zeta, x)\rangle\left\langle\int_0^\zeta g(\zeta')K(\zeta', y)\, d\zeta'\right\rangle$$
$$- \left\langle(\zeta - \tfrac{1}{2})g(\zeta)K(\zeta, x)\right\rangle\langle g(\zeta)K(\zeta, y)\rangle,$$

with $K(\zeta, z) = e^{iC(\zeta)z}$. When $g(\zeta) = 1$ (i.e., a lossless system), it is found that

$$r_1(x, y) = \frac{i(2\theta - 1)}{2s^3(xy)^2(x + y)} \Big[ sxy(y \cos sx \sin sy - x \cos sy \sin sx)$$
$$+ (x^2 - y^2) \sin sx \sin sy \Big].$$

Note that when $g = 1$, $r_1$ depends on $s$ and $\theta$ separately, whereas when $g = 1$ the kernel $r$ in the DMNLS equation only depends on $s$ (recall: $r(x) = \frac{sx}{(2\pi)^2 sx}$). The higher-order DMNLS (HO-DMNLS) equation (10.40) with the terms, $\hat{n}_0(\omega, Z)$, $\hat{n}_0(\omega, Z)$, derived in this section has a class of interesting solutions. With the HO-DMNLS, a wide variety of solutions including multi-hump soliton solutions can be obtained. Interestingly, single-humped, two-humped, three-humped, four-humped, etc., solutions are found. They also come with phases, for example, bi-solitons with positive and negative humps, as schematically illustrated in Figure 10.11. It has been conjectured that there exists an infinite family of such DMNLS solitons (Ablowitz et al., 2002b); see also Maruta et al. (2002).

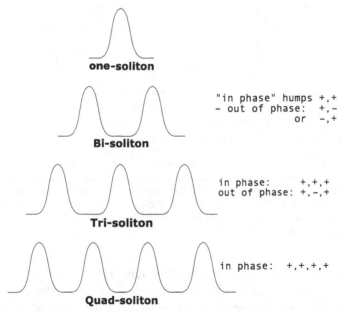

Figure 10.11 The HO-DMNLS equation has different families of soliton solutions, including the single soliton, also found in the DMNLS equation, and new multi-soliton solutions. Amplitude is plotted; + denotes a positive hump, whereas − denotes a negatives hump.

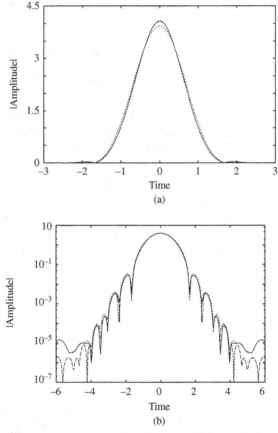

Figure 10.12 The solid curve is the approximate solution $u(t)$ = $u_0(t) + z_a u_1(t)$, which is calculated from (10.40)–(10.42) with typical parameters and $g = 1$. The dashed curve represents a DM soliton obtained from the averaging method (see Section 10.4.4) on the PNLS equation (10.19) with $g = 1$; it is indistinguishable from the solid curve. The dotted curve is $U_0(t)$, which is the inverse Fourier transform of $\hat{U}_0(\omega)$, the leading-order solution of the HO-DMNLS. The waveforms are shown in (a) linear and (b) log scale. In (a) the dashed curve is indistinguishable from the solid curve. See also Ablowitz et al. (2002b).

Shown in Figures 10.12 and 10.13 are typical comparisons of the HO-DMNLS solutions and the DMNLS solutions (here: $z_a = 0.06$, $\langle d \rangle = 1.65$, $s = 0.54$, $\theta = 0.8$). These figures show the extra detail gained by including the higher-order terms. Figure 10.14 shows a bi-soliton solution of the HO-DMNLS, which cannot be found from the first-order DMNLS equation because the parameter values require both $s$ and $\theta$ to be given separately. Note

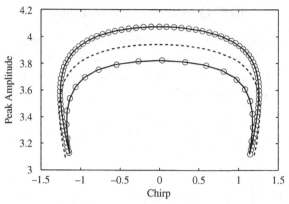

Figure 10.13 Evolution of chirp and peak amplitude of a DM soliton for typical parameters within a DM period. The solid curve and dashed curve denote the approximate solution $u(t) = u_0(t) + z_a u_1(t)$ and leading-order solution $u_0(t)$, respectively. The circles represent the evolution on the PNLS equation (10.19).

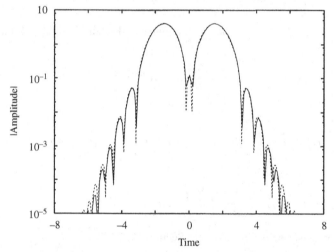

Figure 10.14 Bi-soliton solution with $\tau_s = 3t_{\text{FWHM}}$, which is obtained by the averaging method (see Section 10.4.4) on the HO-DMNLS equation for $s = 0.56$, $\theta = 0.8$, $\langle d \rangle = 1.65$, and $M = 0.340$ (solid curve); recall $M = s/\langle d \rangle$ is the reduced map strength. The dotted curve shows the same solution that is given by the averaging method on the PNLS equation (here $g = 1$) (10.19).

that $M = s/\langle d \rangle$ is called the reduced map strength obtained by rescaling the HO-DMNLS equation (with $g = 1$) and finding that it can be rewritten by taking $\langle d \rangle = 1$ and replacing $s$ by $M$ in the HO-DMNLS equation (Ablowitz et al., 2002b).

## 10.5  Quasilinear transmission

A different type of pulse, termed "quasilinear", exists in strongly dispersion-managed transmission systems ($s \gg 1$) (Ablowitz et al., 2001b; Ablowitz and Hirooka, 2002). The analysis of this transmission mode begins with the DMNLS equation, discussed in Section 10.4.1,

$$i\hat{U}_Z - \frac{\langle d \rangle}{2}\omega^2 \hat{U} + \hat{N}\left[\hat{U}(\omega, Z)\right] = 0,$$

where

$$\hat{N}[\hat{U}(\omega, Z)] = \int_{-\infty}^{\infty} \int_{-\infty}^{\infty} d\omega_1\, d\omega_2\, \hat{U}(\omega + \omega_1)$$
$$\times \hat{U}(\omega + \omega_2)\hat{U}^*(\omega + \omega_1 + \omega_2)r(\omega_1\omega_2),$$

and

$$r(x) = \frac{1}{(2\pi)^2}\int_0^1 g(\zeta)\exp\left(-ixC(\zeta)\right)d\zeta.$$

Alternatively, as shown earlier, in the physical domain the DMNLS equation is given by

$$iU_z + \frac{\langle d \rangle}{2}U_{tt} + N[U] = 0,$$

where

$$N[U] = \int_{-\infty}^{\infty} \int_{-\infty}^{\infty} R(t_1, t_2)U(t + t_1)U(t + t_2)U^*(t + t_1 + t_2)\,dt_1\,dt_2,$$

and

$$R(t_1, t_2) = \frac{1}{(2\pi)^2}\int_{-\infty}^{\infty} \int_{-\infty}^{\infty} r(\omega_1\omega_2)e^{i\omega_1 t_1 + \omega_2 t_2}\,d\omega_1\,d\omega_2.$$

In the lossless case we have seen in Section 10.4.2 that

$$R(t_1, t_2) = \frac{1}{2\pi s}\mathrm{Ci}\left(\left|\frac{t_1 t_2}{s}\right|\right)$$
$$= \frac{1}{2\pi s}\int_{\left|\frac{t_1 t_2}{s}\right|}^{\infty} \frac{\cos u}{u}\,du,$$

where $Ci(x) \equiv \int_x^{\infty} \dfrac{\cos u}{u}\,du$ is the cosine integral.

In the quasilinear regime $s \gg 1$, so that in the asymptotic limit $s \to \infty$ we have $x = \left|\dfrac{t_1 t_2}{s}\right| \to 0$. We then use (see Section 10.5.1 below for further details),

$$\int_x^\infty \frac{\cos u}{u} du = -\gamma - \log x + O(x)$$

as $x \to 0$; $\gamma$ is the so-called Euler constant. Thus,

$$R(t_1, t_2) \sim \frac{1}{2\pi s}\left(-\gamma - \log\left|\frac{t_1 t_2}{s}\right|\right) + O\left(s^{-1}\right),$$

and therefore

$$N(U) \sim \frac{1}{2\pi s}\int_{-\infty}^\infty \int_{-\infty}^\infty \left(-\gamma - \log\left|\frac{t_1 t_2}{s}\right|\right) U(t + t_1)$$
$$\times U(t + t_2)U^*(t + t_1 + t_2)\, dt_1\, dt_2.$$

Hence, taking the Fourier transform

$$\hat{N}\left[\hat{U}(\omega, Z)\right] \sim \frac{1}{2\pi s}\left\{(\log s - \gamma)|\hat{U}|^2\hat{U}\right.$$
$$\left. - \mathcal{F}\left\{\log|t_1 t_2|\, U(t + t_1)U(t + t_2)U^*(t + t_1 + t_2)\right\}\right\}.$$

With the Fourier transform of the last integral (see Section 10.5.1), we find

$$\mathcal{F}\left\{\log|t_1 t_2|\, U(t + t_1)U(t + t_2)U^*(t + t_1 + t_2)\right\}$$
$$= \frac{1}{\pi}\hat{U}(\omega)\int_{-\infty}^\infty dt\, \log|t|\, e^{i\omega t}\left[\int_{-\infty}^\infty |\hat{U}(\omega')|^2 e^{-i\omega' t}\, d\omega'\right].$$

Therefore

$$i\hat{U}_z - \frac{\langle d\rangle}{2}\omega^2\hat{U} + \Phi\left(|\hat{U}|^2\right)\hat{U} = 0, \qquad (10.43)$$

where

$$\Phi\left(|\hat{U}|^2\right) = \frac{1}{2\pi s}\left[(\log s - \gamma)|\hat{U}(\omega)|^2 - \int_{-\infty}^\infty f(\omega - \omega')|\hat{U}(\omega')|^2\, d\omega'\right]$$

and $f(\omega) = \frac{1}{\pi}\int \log|t|\, e^{i\omega t} dt$. From (10.43) we find that $|\hat{U}(\omega, z)|^2 = |\hat{U}(\omega, 0)|^2 = |\hat{u}_0(\omega)|^2$. Therefore,

$$\hat{U}_z = -i\left[\frac{\langle d\rangle}{2}\omega^2 - \Phi\left(|\hat{U}_0|^2\right)\right]\hat{U},$$

and after integrating

$$\hat{U}(\omega, z) = \hat{U}(\omega, 0)\exp\left[-i\frac{\langle d\rangle}{2}\omega^2 z + i\Phi\left(|\hat{U}_0|^2\right)z\right]. \qquad (10.44)$$

We refer to (10.44) as a quasilinear mode. When $g \neq 1$ the analysis is somewhat more complex and is derived in detail by Ablowitz and Hirooka (2002).

### 10.5.1 Analysis of the cosine integral and associated quasilinear Fourier integral

First we discuss an asymptotic limit of the cosine integral. Namely,

$$\text{Ci}(x) = \int_x^\infty \frac{\cos t}{t}\, dt, \quad \text{as} \quad x \to 0+.$$

Note that

$$\int_x^\infty \frac{\cos t}{t}\, dt = \text{Re}\left\{ \int_x^\infty \frac{e^{it}}{t}\, dt \right\}.$$

where $\text{Re}(x)$ represents the real part of $x$. Using the contour indicated in Figure 10.15 and appealing to Cauchy's theorem ($x > 0$),

$$\int_x^\infty \frac{e^{it}}{t}\, dt + \int_{C_R} \frac{e^{iz}}{z}\, dz + \int_{i\infty}^{-ix} \frac{e^{iz}}{z}\, dz + \int_{C_\varepsilon} \frac{e^{iz}}{z}\, dz = 0.$$

Since $\text{Im}\{z\} > 0$ on $C_R$, we have

$$\lim_{R \to \infty} \int_{C_R} \frac{e^{iz}}{z}\, dz = 0.$$

On the quarter circle at the origin, we get

$$\lim_{\varepsilon \to 0} \int_{C_\varepsilon} \frac{e^{iz}}{z}\, dz = \lim_{\varepsilon \to 0} \int_{\pi/2}^0 \frac{e^{i\epsilon e^{i\theta}}}{\epsilon e^{i\theta}} i\epsilon e^{i\theta}\, d\theta = -i\frac{\pi}{2}.$$

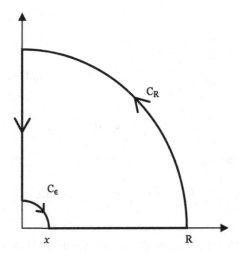

Figure 10.15 Integration contour.

Therefore we have

$$\int_x^\infty \frac{e^{it}}{t}\,dt = \int_{ix}^{i\infty} \frac{e^{iz}}{z}\,dz + i\frac{\pi}{2}$$

$$= \int_x^\infty \frac{e^{-y}}{y}\,dy + i\frac{\pi}{2};$$

thus,

$$I \equiv \int_x^\infty \frac{\cos t}{t}\,dt = \Re\left\{\int_x^\infty \frac{e^{it}}{t}\,dt\right\} = \int_x^\infty \frac{e^{-y}}{y}\,dy.$$

Now,

$$I = \int_x^\infty \left(\frac{e^{-y}}{y} - \frac{1}{y(y+1)}\right)dy + \int_x^\infty \frac{1}{y(y+1)}\,dy$$

$$= \int_0^\infty \frac{\left(e^{-y} - \frac{1}{y+1}\right)}{y}\,dy + \int_0^x \frac{\left(e^{-y} - \frac{1}{y+1}\right)}{y}\,dy + \int_x^\infty \frac{1}{y(y+1)}\,dy.$$

Note that

$$\frac{e^{-y} - \frac{1}{y+1}}{y} \sim e^{-y} - (1 - y + y^2 - \cdots)$$

$$\sim \left[1 - y + \frac{y^2}{2} - \cdots - (1 - y + y^2 - \cdots)\right]/y$$

$$\sim -\frac{y}{2} + \cdots,$$

showing the integral $\int_0^\infty \frac{e^{-y} - \frac{1}{y+1}}{y}\,dy$ is convergent as $y \to 0$ (the integral clearly converges at the infinite upper limit) and it is well known that

$$\int_0^\infty \frac{\left(e^{-y} - \frac{1}{y+1}\right)}{y}\,dy \equiv -\gamma,$$

where $\gamma$ is the Euler constant $\gamma \approx 0.57722$. Also

$$\int_0^x \frac{\left(e^{-y} - \frac{1}{y+1}\right)}{y}\,dy \underset{x\to 0}{\sim} -\frac{x^2}{2} + \cdots$$

Similarly,

$$\int_x^\infty \frac{1}{y(y+1)}\,dy = \int_x^\infty \left(\frac{1}{y} - \frac{1}{y+1}\right)dy$$

$$= \log(1+x) - \log(x) \underset{x\to 0}{\sim} x - \log x.$$

Putting this all together implies, as $x \to 0+$,

$$I \sim -\gamma - \log x + x + O(x^2).$$

We now consider the Fourier integral

$$F = \mathcal{F}\left\{ \int_{-\infty}^{\infty} \int_{-\infty}^{\infty} \log|t_1 t_2| \, U(t+t_1) U(t+t_2) U^*(t+t_1+t_2) \, dt_1 \, dt_2 \right\}$$

$$= \frac{1}{(2\pi)^3} \int_{-\infty}^{\infty} e^{-i\omega t} \, dt \iint dt_1 \, dt_2 \, \log|t_1 t_2| \left[ \int_{-\infty}^{\infty} \hat{U}(\omega_1) e^{i\omega_1(t+t_1)} \, d\omega_1 \right]$$

$$\times \left[ \int_{-\infty}^{\infty} \hat{U}(\omega_2) e^{i\omega_2(t+t_2)} \, d\omega_2 \right] \left[ \int_{-\infty}^{\infty} \hat{U}^*(\omega_3) e^{-i\omega_3(t+t_1+t_2)} \, d\omega_3 \right].$$

Noting that the function is symmetric in $t_1$, $t_2$ and using $\int dt_2 \, e^{i\omega_2 t_2} e^{-i\omega_3 t_2} = 2\pi\delta(\omega_2 - \omega_3)$, we have

$$F = \frac{2}{(2\pi)^2} \int_{-\infty}^{\infty} dt \int_{-\infty}^{\infty} dt_1 \, \log|t_1|$$

$$\int_{-\infty}^{\infty} \int_{-\infty}^{\infty} \int_{-\infty}^{\infty} d\omega_1 \, d\omega_2 \, d\omega_3 \, e^{i(\omega_1-\omega_3)t_1} \, \hat{U}(\omega_1) \hat{U}(\omega_2) \hat{U}^*(\omega_3)$$

$$\times \delta(\omega_2 - \omega_3) e^{i(\omega_1+\omega_2-\omega_3-\omega)t}$$

$$= \frac{1}{2\pi^2} \int_{-\infty}^{\infty} dt \int_{-\infty}^{\infty} dt_1 \, \log|t_1| \int_{-\infty}^{\infty} d\omega_1 d\omega_3 \, \hat{U}(\omega_1) \hat{U}(\omega_3) \hat{U}^*(\omega_3)$$

$$\times e^{i(\omega_1-\omega)t} e^{i(\omega_1-\omega_3)t_1}.$$

Using $\int_{-\infty}^{\infty} dt \, e^{i(\omega_1-\omega)t} = 2\pi\delta(\omega_1 - \omega)$ implies

$$F = \frac{1}{\pi} \int_{-\infty}^{\infty} dt_1 \, \log|t_1| \int_{-\infty}^{\infty} d\omega_3 \, \hat{U}(\omega) \left| \hat{U}(\omega_3) \right|^2 e^{i(\omega-\omega_3)t_1}.$$

Or finally,

$$F = \hat{U}(\omega) \int_{-\infty}^{\infty} \left| \hat{U}(\omega') \right|^2 f(\omega - \omega') \, d\omega',$$

where $f(x)$ is the generalized function,

$$f(x) = \int_{-\infty}^{\infty} dt \, \frac{\log|t|}{\pi} e^{ixt}.$$

For calculation, it is preferable to use $F$ in the form

$$F = \frac{1}{\pi} \hat{U}(\omega) \int_{-\infty}^{\infty} dt \, \log|t| \, e^{i\omega t} \left[ \int_{-\infty}^{\infty} \left| \hat{U}(\omega') \right|^2 e^{-i\omega' t} \, d\omega' \right].$$

When a Gaussian shape is taken, $|\hat{U}(\omega)| = \alpha e^{-\beta\omega^2}$, then the inner integral decays rapidly as $|t| \to \infty$.

## 10.6 WDM and soliton collisions

As mentioned earlier, in a WDM system solitons are transmitted at many different wavelengths, or frequencies; these different wavelengths are often called "channels". Due to the dispersion in the fiber, solitons at different wavelengths travel at different velocities. This leads to inevitable collisions between solitons in different wavelength channels.

Considering the case of only two WDM channels, centered at the normalized radial frequencies $\Omega_1$ and $\Omega_2$ we take two pulses $u_1$ and $u_2$ in channel 1 and 2 respectively. The average frequency of the pulses can be calculated as

$$\langle \omega \rangle = \mathrm{Im}\left\{ \int_{-\infty}^{\infty} u_j^* \frac{\partial u_j}{\partial t}\, dt \right\} / W_j = \frac{1}{2\pi W_j} \int_{-\infty}^{\infty} \omega |\hat{u}_j(\omega)|^2\, d\omega, \qquad (10.45)$$

where $\hat{u}_j(\omega)$ is the Fourier transform of $u(t)$, $W_j = \int_{-\infty}^{\infty} |u_j(t)|^2\, dt$, the energy of pulse $u_j$, and $j \in \{1, 2\}$ is the channel number. The right-hand side of (10.45) is obtained by replacing $u(t)$ by its Fourier transform.

In Section 10.2.1, we saw that the NLS equation with XPM contributions [see (10.12) and (10.13)] for channel 1 $(u_1)$ is

$$i\frac{\partial u_1}{\partial z} + \frac{1}{2}\frac{\partial^2 u_1}{\partial t^2} + g(z)|u_1|^2 u_1 = -2g(z)|u_2|^2 u_1 \qquad (10.46)$$

and for channel 2 $(u_2)$ is

$$i\frac{\partial u_2}{\partial z} + \frac{1}{2}\frac{\partial^2 u_2}{\partial t^2} + g(z)|u_2|^2 u_2 = -2g(z)|u_1|^2 u_2.$$

We assume the signals $u_1$ and $u_2$ have negligible overlap in the frequency domain and that the FWM contribution is small. This assumption enables us to separate the components at different frequencies in the original perturbed NLS (PNLS) equation; i.e., one considers each equation as being valid in separate regions of Fourier space.

Assuming $W_j$ is constant, the change in average frequency can be found from (10.45):

$$\begin{aligned} W_j \frac{d}{dz}\langle \omega_j \rangle &= \frac{\partial}{\partial z}\, \mathrm{Im}\left\{ \int_{-\infty}^{\infty} \frac{\partial u_j}{\partial t} u_j^*\, dt \right\}, \\ &= \mathrm{Im}\left\{ \int_{-\infty}^{\infty} \left( \frac{\partial^2 u_j}{\partial t \partial z} u_j^* + \frac{\partial u_j}{\partial t}\frac{\partial u_j^*}{\partial z} \right) dt \right\}. \end{aligned} \qquad (10.47)$$

The NLS equation for channel $u_1$ with a perturbation $F_1$ is

$$i u_{1z} + \frac{d}{2} u_{1tt} + g|u_1|^2 u_1 = F_1.$$

Operating on the above equation with $-iu_1^* \partial_t$ and adding to the result $iu_{1t}$ times the complex conjugate of the above equation, we find

$$\frac{\partial^2 u_j}{\partial t \partial z} u_j^* + \frac{\partial u_j}{\partial t}\frac{\partial u_j^*}{\partial z} = \frac{id}{2}\left(u_1^* u_{1ttt} - u_{1t}u_{1tt}^*\right)$$

$$+ ig\left(u_1^* \partial_t(|u_1|^2 u_1) - |u_1|^2 u_1^* u_{1t}\right) + i\left(u_{1t}F_1^* - u_1^* F_{1t}\right).$$

Substituting this result into (10.47) and integrating the first two terms by parts (assuming $u_1$ and its derivatives vanish at $t = \pm\infty$), we find

$$W_1\frac{d\langle\omega_1\rangle}{dz} = \mathrm{Im}\left\{\int_{-\infty}^{\infty} i\left(F_1 u_{1t}^* + F_1^* u_{1t}\right) dt\right\} = \int_{-\infty}^{\infty}\left(F_1 u_{1t}^* + F_1^* u_{1t}\right) dt.$$

If we take $F_1 = -2g(z)|u_2|^2 u_1$ as indicated by (10.46), then

$$W_1\frac{d\langle\omega_1\rangle}{dz} = 2g(z)\int_{-\infty}^{\infty} |u_1|^2 \partial_t |u_2|^2 \, dt,$$

or more generally

$$\frac{d\langle\omega_j\rangle}{dz} = \frac{2g(z)}{W_j}\int_{-\infty}^{\infty} |u_j|^2 \partial_t |u_{3-j}|^2 \, dt, \quad j = 1, 2. \tag{10.48}$$

The frequency shift is found by integrating (10.48). Similarly, the timing shift may be found from the definition of the average position in time, $\langle t_j\rangle$, $j = 1, 2$:

$$W_j\langle t_j\rangle = \int_{-\infty}^{\infty} t\,|u_j|^2 \, dt.$$

Thus, again assuming $W_j$ is constant,

$$W_1\frac{d}{dz}\langle t_1\rangle = \frac{\partial}{\partial z}\left(\int_{-\infty}^{\infty} t\,|u_1|^2 \, dt\right)$$

$$= \int_{-\infty}^{\infty} t\left(u_1^*\frac{\partial u_1}{\partial z} + \frac{\partial u_1^*}{\partial z}u_1\right) dt$$

$$= \int_{-\infty}^{\infty} t\left\{u_1^*\left[\frac{id}{2}u_{1tt} + ig(z)u_1^2 u_1^* - iF_1\right]\right\} dt$$

$$+ \int_{-\infty}^{\infty} t\left\{u_1\left[-\frac{id}{2}u_{1tt}^* - ig(z)\left(u_1^*\right)^2 u_1 + iF_1^*\right]\right\} dt$$

$$= -i\frac{d}{2}\int_{-\infty}^{\infty}\left(u_1^* u_{1tt} - u_1 u_{1tt}^*\right) dt - i\int_{-\infty}^{\infty} t\left(u_1^* F_1 - u_1 F_1^*\right) dt.$$

Recalling

$$W_j \langle \omega_j \rangle = \text{Im} \left\{ \int_{-\infty}^{\infty} \frac{\partial u_j}{\partial t} u_j^* \, dt \right\} = \frac{1}{2i} \int_{-\infty}^{\infty} \left( \frac{\partial u_j}{\partial t} u_j^* - \frac{\partial u_j^*}{\partial t} u_j \right) dt,$$

we can write

$$W_1 \frac{d}{dz} \langle t_1 \rangle = d(z) W_1 \langle \omega_1 \rangle - i \int_{-\infty}^{\infty} t \left( u_1^* F_1 - u_1 F_1^* \right) dt.$$

Using $F_1 = -2g(z)|u_2|^2 u_1$, the integral $\int_{-\infty}^{\infty} t \left( u_1^* F_1 - u_1 F_1^* \right) dt = 0$ and we obtain

$$\frac{d}{dz} \langle t_1 \rangle = d(z) \langle \omega_1 \rangle. \tag{10.49}$$

The timing shift may be obtained by integrating (10.49) with respect to $z$.

In the context of RZ (return to zero) soliton or quasilinear communications systems it is useful to define the residual frequency shift as the relative frequency shift of the pulse at the end of the communications system. Using a system starting at $z = 0$ and of length $L$, with this definition the residual frequency shift of a pulse in channel $j$ is given by

$$\Delta\Omega_{\text{res}}^{(j)} = \langle \omega_j \rangle (L) - \langle \omega \rangle (0) = \int_0^L \frac{d\langle \omega_j \rangle}{dz} dz.$$

We are also interested in the relative timing shift at the end of the collision process. The total timing shift, (10.49), yields

$$\frac{d}{dz} \langle t_j \rangle (z) = d(z) \left[ \langle \omega_j \rangle (z) - \langle \omega_j \rangle (0) \right] + d(z) \langle \omega_j \rangle (0)$$

$$= d(z) \int_0^z \frac{d\langle \omega_j \rangle}{dz} dz + d(z) \langle \omega_j \rangle (0).$$

Defining $\bar{d}(z) = \int_0^z d(z') \, dz'$, and integrating the above equation from $z = 0$ to $z = L$,

$$\langle t_j \rangle (L) - \langle t_j \rangle (0) = \int_0^L d(z) \int_0^z \frac{d\langle \omega \rangle_j}{dz'} dz' \, dz + \bar{d}(L) \langle \omega_j \rangle (0).$$

Note that $\langle t_j \rangle (0) + \bar{d}(L) \langle \omega_j \rangle (0)$ is invariant. We can also define the relative timing shift as

$$\Delta t_j(L) \equiv \langle t_j \rangle (L) - \langle t_j \rangle (0) - \bar{d}(L) \langle \omega_j \rangle (0) = \int_0^L d(z) \int_0^z \frac{d\langle \omega \rangle_j}{dz'} dz' \, dz.$$

This can now be simplified by changing the order of integration:

$$\Delta t_j(L) = \int_0^L d(z) \int_0^z \frac{d\langle\omega\rangle_j}{dz'} dz'\, dz \qquad (10.50)$$

$$= \int_0^L \frac{d\langle\omega\rangle_j}{dz'} \int_{z'}^L D(z)\, dz\, dz'$$

$$= \int_0^L \frac{d\langle\omega\rangle_j}{dz'} \left[\bar{d}(L) - \bar{d}(z')\right] dz'$$

$$= \bar{d}(L) \int_0^L \frac{d\langle\omega\rangle_j}{dz'} dz' - \int_0^L \frac{d\langle\omega\rangle_j}{dz'} \bar{d}(z')\, dz'$$

$$= \bar{d}(L)\Delta\Omega_{\text{res}}^{(j)} - \int_0^L \frac{d\langle\omega\rangle_j}{dz'} \bar{d}(z')\, dz'.$$

Thus, the relative timing shift can be written as the difference of two terms:

$$\Delta t_j(L) = \bar{d}(L)\Delta\Omega_{\text{res}}^{(j)} - \Delta t_{\text{res}}^{(j)},$$

where the so-called residual timing shift is

$$\Delta t_{\text{res}}^{(j)} \equiv \int_0^L \frac{d\langle\omega\rangle_j}{dz} \bar{d}(z)\, dz.$$

## 10.7 Classical soliton frequency and timing shifts

In this section we obtain the frequency and time shift for classical solitons; see also Mollenauer et al. (1991). Using the classical soliton shape

$$u_j = \eta_j \operatorname{sech}[\eta_j(t - \Omega_j z)]\, e^{i\Omega_j t + i(\eta_j^2 - \Omega^2)/2 + i\phi_j},$$

we substitute this into (10.48) to find

$$\frac{d\langle\omega_j\rangle}{dz} = \frac{g(z)}{\eta_j} \int_{-\infty}^\infty \frac{d}{dt} \left\{\eta_{3-j}^2 \operatorname{sech}^2\left[\eta_{3-j}\left(t - \Omega_{3-j}(z - z_0)\right)\right]\right\}$$
$$\times \eta_j^2 \operatorname{sech}^2\left[\eta_j\left(t - \Omega_j(z - z_0)\right)\right] dt.$$

Note that the classical soliton frequency is $\langle\omega_j\rangle = \Omega_j$. For $|\Omega_j| \gg 1$ we take $\Omega_j$ approximately being a constant within the integral. Then we can transform the derivative in $t$ to one in $z$:

$$\frac{d\Omega_j}{dz} = \frac{g(z)\eta_{3-j}^2 \eta_j}{\Omega_{3-j}} \int_{-\infty}^\infty \frac{d}{dz} \left\{\operatorname{sech}^2\left[\eta_{3-j}\right.\right.$$
$$\left.\left. \times \left(t - \Omega_{3-j}(z - z_0)\right)\right]\right\} \operatorname{sech}^2\left[\eta_j\left(t - \Omega_j(z - z_0)\right)\right] dt.$$

To simplify this integral we assume that the solitons are both of equal energy (amplitude) with $\eta_1 = \eta_2 = \eta$ and we take a frame of reference where $\Omega_1 = -\Omega_2 = \Omega$. For an arbitrary frame of reference we may define $\Omega$, without loss of generality, as half the frequency separation:

$$\Omega \equiv \frac{1}{2}(\Omega_1 - \Omega_2).$$

With this substitution the average frequency change can be simplified to

$$\frac{d\Omega}{dz} = \frac{g(z)\eta^3}{2\Omega} \frac{d}{dz} \int_{-\infty}^{\infty} \text{sech}^2 \left[ \eta \left( t - \Omega(z - z_0) \right) \right]$$
$$\times \text{sech}^2 \left[ \eta \left( t + \Omega(z - z_0) \right) \right] dt.$$

Performing the integration in time analytically, we obtain an expression for the frequency shift, $\Delta\Omega = \Omega(z) - \Omega(-\infty)$ (we take the "initial" frequency to be given at large negative values of $z$ that is taken to be $-\infty$):

$$\Delta\Omega(z) = \frac{2\eta^2}{\Omega}$$
$$\times \int_{-\infty}^{z} g(z') \frac{d}{dz'} \frac{2\eta\Omega(z' - z_0) \cosh 2\eta\Omega(z' - z_0) - \sinh 2\eta\Omega(z' - z_0)}{\sinh^3 2\eta\Omega(z' - z_0)} dz'.$$

We can similarly calculate the timing shift. First we assume a constant dispersion of $d(z) = 1$. Then we can calculate the residual timing shift by using (10.50) (and integration by parts)

$$\Delta t(z) = \int_{-\infty}^{z} \Delta\Omega(z') \, dz'.$$

From these equations we can calculate the frequency and timing shifts for a classical soliton collision. A typical collision is shown in Figure 10.16, where the propagating path is shown explicitly; the soliton is injected into the system at $z = 0$ at a time position of $t_0$ for $u_1$ and $-t_0$ for $u_2$; they meet and collide with a collision center of $z_0$ and propagate onward until the end of the system at $z = L$. The initial location of the solitons, $t_0$, is related to the center of collision by

$$t_0 = \Omega D z_0.$$

The frequency shift is shown in Figure 10.17a over the distance of the collision relative to its center, $z_0$. The frequency shift is maximal at the center of the collision and returns to zero after the collision for the lossless case ($g = 1$). For the "lossy" case, $g \neq 1$ (i.e., with damping and amplification included

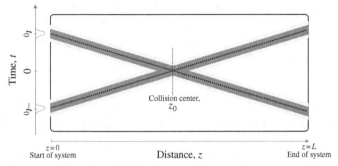

Figure 10.16 Two classical solitons colliding at $z_0$; the solitons have an equal but opposite velocity. The inset to the left shows the initial pulse envelope with solitons centered at $t_0$ and $-t_0$.

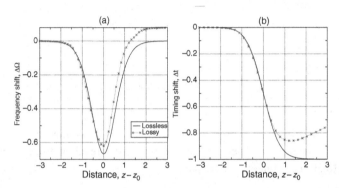

Figure 10.17 Typical frequency, (a), and timing, (b), shifts versus the distance over the collision $z - z_0$, with the collision center being at $z_0$ shown for both the lossless and lossy cases. Here $\Omega = 1$, ($g = 1$) for the lossless case and ($g \neq 1$) for the lossy case.

with the term $g(z)$) there is a considerable frequency shift after the collision process is complete. This is the residual frequency shift that in the total timing shift grows with $z$, see (10.50). This is especially damaging to classical soliton transmission.

The timing shifts for the same collisions are shown in Figure 10.17. Here it can be seen that after a collision the timing shift does not return to zero. When there is a residual frequency shift the timing shift continues to change after the collision. The total timing shift at the end of the communications channel is a combination of a fixed residual timing shift and a growing shift from the residual frequency shift; see (10.50).

## 10.8 Characteristics of DM soliton collisions

Collisions also occur in DM soliton systems. Solitons in different wavelength channels have a different *average* velocity and can therefore overlap and collide. Dispersion-management is characterized by rapidly varying large opposing dispersions that give the DM soliton (or any pulse in a DM system) large local velocities. This results in the "zig-zag" path characteristic of DM systems, as illustrated in Figure 10.18. These diversions from the average path become more pronounced as the DM map strength is increased and make DM soliton collisions quite different from collisions in a classical soliton system.

When two DM solitons collide they go through a collision process consisting of many individual "mini-collisions", where the solitons will meet and pass through each other completely in one half of the map only to change velocity and pass through each other on the next half, performing two mini-collisions per map period.

The collision center is defined as the center of the average paths of the solitons. For classical solitons changing $z_0$ changed the position of the collision in the system but did not change the outcome of the collision. In a DM system changing $z_0$ not only changes the position of the collision process, it also changes the position of the mini-collisions within the DM map. This in turn changes the nature of the collision process, and consequently the final frequency and timing shifts are dependent on $z_0$.

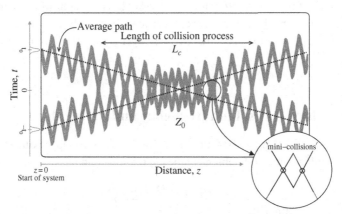

Figure 10.18 The movements of two colliding DM solitons, with an enlarged view of the soliton centers over a single DM map period, showing the intersections that are the mini-collisions.

As with classical soliton collisions the initial time displacement at the start of the channel can be calculated from the collision center, defined as

$$t_0 = \Omega_j \langle d \rangle z_0,$$

with $j \in \{1, 2\}$ indicating one of the two frequency channels.

The center of a DM soliton satisfies $t_c^{(n)} - \Omega_j \bar{d}(z) = 0$ for $t_c^{(n)}$ and $z$, with $n \in \{1, 2\}$ indicating one of the two frequency channels. An approximate analytical formula for a DM soliton is given in the next section; see (10.51). Using this we can find the locations of the mini-collisions, those values of $z$ at which the centers of both solitons are coincident, namely $t_c^{(1)} = t_c^{(2)}$:

$$\Omega_1 \bar{d}(z) = \Omega_2 \bar{d}(z)$$

so that

$$\bar{d}(z) = 0.$$

Therefore mini-collision locations are found from the solution of $\bar{d}(z) = \langle d \rangle (z - z0) + C(z) = 0$, and are not influenced by the average frequencies of either soliton. The range of the chirp, $C(z)$, is limited; it periodically oscillates between $\min\{C(z)\}$ and $\max\{C(z)\}$. As a consequence $\bar{d}(z) = 0$ only has solutions over a limited domain, which corresponds to the extent of the collision process. When $\langle d \rangle (z - z_0)$ is larger or smaller than the range of $C(z)$, there are no mini-collisions, and we can therefore estimate the extent of the collision as

$$- \max\{C(z)\} < \langle d \rangle (z - z_0) < - \min\{C(z)\}$$
$$z_0 - s / \langle d \rangle < z < z_0 + s / \langle d \rangle,$$

where $\max\{C(z)\} = - \min\{C(z)\} = s = \theta \Delta_1 / 2$. The length of the collision $L_C = 2s / \langle d \rangle$ is the difference of the upper and lower collision limits, and we notice that the larger the map strength the longer is the DM soliton's collision process.

## 10.9 DM soliton frequency and timing shifts

We have seen in Section 10.4 that $\hat{u} = \hat{U} e^{-iC\omega^2/2}$. If we approximate $\hat{U}$ by $\hat{U} = \alpha e^{-\beta \omega^2 / 2}$, then a DM soliton can be approximated in physical space by

$$u_1(z, t) = \frac{\alpha}{\sqrt{2\pi(\beta + iC(z))}} \exp \left\{ -\frac{[t - \Omega_1 \bar{d}(z)]^2}{2[\beta + iC(z)]} \right\}$$
$$\times \exp \left\{ \frac{i}{2} \left[ \lambda^2 z + 2\Omega_1 t - \Omega_1^2 \bar{d}(z) \right] \right\}, \quad (10.51)$$

where $\bar{d}(z) - \langle d \rangle z_0 = \int_{z_0}^{z} d(z')\,dz' = (z - z_0)\langle d \rangle + C(z)$. Substituting the above DM soliton representation for $u_1$ and $u_2$ into (10.48) we get

$$\frac{d\langle \omega_j \rangle}{dz} = -\frac{g(z)\alpha_j \alpha_{3-j} \beta_{3-j}}{2\pi \left[\beta_j^2 + C^2(z)\right]^{1/2}\left[\beta_{3-j}^2 + C^2(z)\right]^{5/2}}$$

$$\times \int_{-\infty}^{\infty} [t - \Omega_{3-j}\bar{d}(z)] \exp\left\{-\frac{\beta_j[t - \Omega_j\bar{d}(z)]^2}{2\left[\beta_j^2 + C^2(z)\right]} - \frac{\beta_{3-j}[t - \Omega_{3-j}\bar{d}(z)]^2}{2\left[\beta_{3-j}^2 + C^2(z)\right]}\right\} dt$$

for $j = 1, 2$. We further assume that the solitons in both channels have the same shape; that is, $\alpha_1 = \alpha_2$ and $\beta_1 = \beta_2$. With this assumption and noting $\langle \omega_j \rangle = \Omega_j$, we obtain an analytic formula for DM soliton collision-induced frequency shifts,

$$\frac{d\Omega_1}{dz} = (A/2)(\Omega_1 - \Omega_2)\frac{\bar{d}(z)g(z)}{B^3(z)}\exp\left(-(\beta/2)\left[\frac{\bar{d}(z)}{B(z)}(\Omega_1 - \Omega_2)\right]^2\right), \quad (10.52)$$

where $d\Omega_1/dz = -d\Omega_2/dz$ and in order to simplify this and equations that follow, we use the following definitions:

$$A = \frac{4W\beta^{3/2}}{\sqrt{2\pi}},$$

$$B^2(z) = \beta^2 + C^2(z/z_a).$$

We remark that in the case of quasilinear pulses the only difference turns out to be the definition of $B(z)$; for quasilinear pulses, we use $B^2(z) = \beta^2 + (\bar{d})^2(z)$. Because the two-channel system is symmetric, we can, without loss of generality, define $\Omega = (\Omega_1 - \Omega_2)/2$. The residual frequency shift, $\Delta\Omega(z) = \Omega(z) - \Omega(-\infty)$, is obtained by integration of (10.52). Assuming $\Omega(z)$ is a slowly varying function (i.e., $|d\Omega/dz| \ll 1$) which must be the case for the validity of the perturbation method (and is verified *a posteriori*), we can treat $\Omega$ on the right-hand side of (10.52) as a constant, obtaining

$$\Delta\Omega(z) = A\Omega\int_0^z \frac{g(z)\bar{d}(z)}{B^3(z)}\exp\left[-2\beta\left(\frac{\bar{d}(z)}{B(z)}\Omega\right)^2\right]dz. \quad (10.53)$$

We can also obtain the timing shift from (10.50):

$$\Delta t(L) = \bar{d}(L)\Delta\Omega(L) - \int_0^L \frac{d\langle \omega \rangle}{dz'}\bar{d}(z')\,dz'. \quad (10.54)$$

Ablowitz et al. (2002a, 2003b), Docherty (2003), and Ahrens (2006) the details of various types of collisions describe. There it is shown that direct numerical computation of the NLS equation agrees with numerical evaluation of (10.53) and (10.54), which also agree with asymptotic evaluation of (10.53) and (10.54) by employing the asymptotic analysis of integrals using a modification of the Laplace method.

# 11

## Mode-locked lasers

In the previous chapter we have seen that both constant dispersion and dispersion-managed solitons are important pulses in the study of long-distance transmission/communications. It is noteworthy that experiments reveal that solitons or localized pulses are also present in mode-locked (ML) lasers. Femtosecond solid-state lasers, such as those based on the Ti:sapphire (Ti:S) gain medium, and fiber ring lasers have received considerable attention in the field of ultra-fast science. In the past decade, following the discovery of mode-locking, the improved performance of these lasers has led to their widespread use, cf. Cundiff et al. (2008). In most cases interest in ultra-short pulse mode-locking has been in the net anomalous dispersive regime. But mode-locking has also been demonstrated in fiber lasers operating in the normal regime. Mode-locking operation has been achieved with relatively large pulse energies (Ilday et al., 2004b,a; Chong et al., 2008b).

In our investigations we have employed a distributive model, termed the power-energy saturation (PES) equation (cf. Ablowitz et al., 2008; Ablowitz and Horikis, 2008, 2009b; Ablowitz et al., 2009c). This model goes beyond the well-known master laser equation, cf. Haus (1975, 2000), in that it contains saturable power (intensity) terms; i.e., terms that saturate due to large field amplitudes. This equation has localized pulses that propagate and mode-lock in both an anomalous and a normally dispersive laser for both in the constant as well as dispersion-managed system. This is consistent with recent experimental observations (Ilday et al., 2004b; Chong et al., 2008a). We also note that it has been shown that dispersion-managed models with saturable power (intensity) energy saturation terms are in good agreement with experimental results in Ti:sapphire lasers (Sanders et al., 2009b). Pulse formation in an ultra-short pulse laser is typically dominated by the interplay between dispersion and nonlinearity. Suitable gain media and an effective saturable absorber are essential for initiating the mode-locking operation from intra-cavity background

313

fields and subsequent stabilization of the pulse. In the anomalous regime the mode-locked soliton solutions are found to approximately satisfy nonlinear Schrödinger (NLS)-type equations. The main difference between the NLS and PES equations, for both the constant and dispersion-managed cases, is that when gain and loss are introduced as in the PES equation, only one value of the propagation constant is allowed. Thus unlike the classical NLS equation there is not a full parameter family of solutions. In the normal dispersive regime the pulses are found to be significantly chirped and much broader, i.e., they are slowly varying, than those in the anomalous regime. These NLS equations with gain and loss terms are interesting and natural to study in the context of nonlinear waves.

In order to obtain localized or solitary waves, or as called in the physical literature, solitons, we often employ numerical methods. The numerical method, which we employed earlier in the study of communications (Ablowitz and Biondini, 1998; Ablowitz et al., 2000b), for numerically obtaining soliton solutions, is based upon taking the Fourier transform of the nonlinear equation, introducing a convergence factor, and then iterating the resulting equation until convergence to a fixed point in function space. The method was first introduced in 1976 (Petviashvili, 1976). The convergence factor depends on the homogeneity of the nonlinear terms. The technique works well for problems with a single polynomial nonlinear term (Pelinovsky and Stepanyants, 2004). However, many interesting systems are more complex. Recently, another way of finding localized waves was introduced (Ablowitz and Musslimani, 2005). The main ideas are to go to Fourier space (this part is the same as Petviashvili, 1976), then renormalize variables and obtain an algebraic system coupled to the nonlinear integral equation. We have found the method of coupling to be effective and straightforward to implement. The localized mode is determined from a convergent fixed point iteration scheme. The numerical technique is called spectral renormalization (SPRZ); it finds localized waves to a variety of nonlinear problems that arise in nonlinear optics and fluid dynamics. See also Chapter 10.

## 11.1 Mode-locked lasers

A laser is essentially an optical oscillator and as such it requires amplification, feedback and loss in its operation. The amplification is provided by stimulated emission in the gain medium. Feedback is provided by the laser "cavity", which is often a set of mirrors that allow the light to reflect back on itself. One of the mirrors transmits a small fraction of the incident light to provide output.

The loss that counteracts gain is due to a variety of factors such as a field cutoff or polarization effects, etc.

A mode-locked pulse refers to the description of how ultra-short pulses are generated by a laser. The formation of a mode means that the electromagnetic field is essentially unchanged after one round trip in the laser. This implies that lasing occurs for those frequencies for which the cavity length is an integer number of wavelengths. If multiple modes lase at the same time, then a short pulse can be formed if the modes are also phase-locked; we say the laser is mode-locked, cf. Cundiff (2005). In terms of the mathematics of nonlinear waves, mode-locking usually relates to specific pulse solutions admitted by the underlying nonlinear equations.

A distinguishing feature of such a laser is the presence of an element in the cavity that causes it to mode-lock. This can be an active or a passive element. Passive mode-locking has produced shorter pulses and has a saturable absorber, i.e., one where the absorption saturates. In this case, higher intensity is less attenuated than a lower intensity.

An advantage of Ti:s lasers is their large-gain bandwidth, which is necessary for supporting ultra-short pulses (it is useful to think of this as the Fourier relationship). The gain band is typically quoted as extending from 700 to 1000 nm, although lasing can be achieved well beyond 1000 nm. The Ti:s crystal provides the essential mode-locking mechanism in these lasers. This is due to a strong cubic nonlinear index of refraction (the Kerr effect), which is manifested as an increase of the index of refraction as the optical intensity increases. Together with a correctly positioned effective aperture, the nonlinear Kerr-lens can act as a saturable absorber, i.e., high intensities are focused and hence transmit fully through the aperture while low intensities experience losses. Since short pulses produce higher peak powers, they experience a lower loss that also favors a mode-locked operation. This mode-locking mechanism has the advantage of being essentially instantaneous. However, it has the disadvantage of often being not self-starting; typically it requires a critical misalignment from optimum continuous wave operation. Ti:s lasers can produce pulses on the order of a few femtoseconds. An application of interest is producing a very long string of equally spaced mode-locked pulses: this has the potential to be used as an "optical clock" where each tick of the clock corresponds to a pulse, cf. Cundiff (2005). A schematic diagram of a typical Ti:s laser system is given in Figure 11.1 with typical units: pulsewidth: $\tau = 10\,\text{fs} = 10^{-14}\,\text{s}$ and repetition time: $\text{Tr} = 10\,\text{ns} = 10^{10}\,\text{s}$ or repetition frequency $\text{fr} = \frac{1}{\text{Tr}} = 100\,\text{MHz}$. In this setup the prism pair is used to compensate for the net normal dispersion of the Ti:s crystal. In the crystal we have strong nonlinear effects, but in the prism the nonlinearity is very weak. So we assume

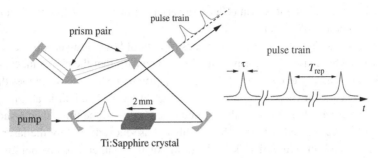

Figure 11.1 Left: Ti:sapphire laser; right: the pulse train.

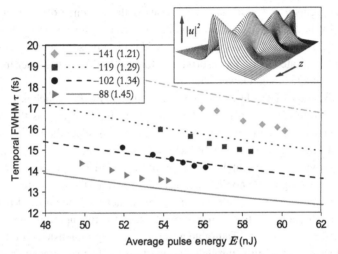

Figure 11.2 Curves are given from DMNLS theory. Symbols: experiments; inset: Dispersion-managed soliton.

a linear state in the prisms. This laser system is dispersion-managed and the nonlinearity is also varying along the laser cavity.

Quraishi et al. (2005) produced ultrashort pulses for a number of average net anomalous dispersions. The results were compared with theory based upon the dispersion-managed NLS equation (i.e., the DMNLS equation) discussed in the previous section. The results were favorable and are indicated in Figure 11.2. The different symbols correspond to different average dispersions times the cavity length: $\langle k''(z) \rangle \, l_m$. The graph and figures relate the pulsewidth $\tau$ to the energy of a pulse.

However the important feature of mode-locking is left out of the DMNLS theory. The DMNLS equation predicts a continuous curve for each average dispersion value. But the experiments indicate that there is not a continuous curve, rather for each value of energy there is one mode-locked configuration. This is discussed next as part of the mathematical model.

## 11.2 Power-energy-saturation equation

The distributed model that we use to describe the propagation of pulses in a laser cavity is given in dimensionless form by

$$iu_z + \frac{d(z)}{2}u_{tt} + n(z)|u|^2u = \frac{ig}{1 + E/E_0}u + \frac{i\tau}{1 + E/E_0}u_{tt} - \frac{il}{1 + P/P_0}u, \quad (11.1)$$

where $d(z)$, $n(z)$ are the dispersion and coefficient of nonlinearity, each of which vary rapidly in $z$, $E(z) = \int_{-\infty}^{\infty} |u|^2\, dt$ is the pulse energy, $E_0$ is related to the saturation energy, $P(z,t) = |u|^2$ is the instantaneous pulse power, $P_0$ is related to the saturation power, and the parameters $g$, $\tau$, $l$, are all positive, real constants. The first term on the right-hand side represents saturable gain, the second is spectral filtering and the third saturable loss that reflects the fast saturable absorber: $\frac{l}{1+P/P_0}$.

Notice that gain and filtering are saturated with energy while loss is saturated with power. The gain and filtering mechanisms are related to the energy of the pulse while the loss is related to the power (intensity) of the pulse. Saturation terms can be expected to prevent the pulse from reaching a singular state; i.e., "infinite" energy or a blow-up in amplitude. Indeed, if blow-up were to occur that would suggest that both the amplitude and the energy of the pulse would become large, which in turn, make the perturbing effects small thus reducing the equation to the unperturbed NLS, which admits a stable, finite solution.

We refer to (11.1) as the power-energy-saturation equation or simply the PES equation. The right-hand side of (11.2) is what we will refer to as the perturbing contribution; it is denoted hereafter by $R[u]$.

The PES model reflects many developments over the years. In order to model the effects of nonlinearity, dispersion, bandwidth limited gain, energy saturation and intensity discrimination in a laser cavity the so-called master equation was introduced, cf. Haus et al. (1992); Haus (2000). The master equation is a generalization of the classical NLS equation, modified to contain gain, filtering and loss terms; the normalized master equation, with dispersion management included is given below.

$$iu_z + \frac{d(z)}{2}u_{tt} + n(z)|u|^2u = \frac{ig}{1 + E/E_0}u + \frac{i\tau}{1 + E/E_0}u_{tt} - i(l - \beta|u|^2)u.$$

The master equation is obtained from the PES model by Taylor expanding the power saturation term and putting $\beta = l/P_0$. As with the PES model, gain and filtering are saturated by energy (i.e., the time integral of the pulse power), but in the master equation the loss is converted into a linear and a cubic nonlinear term. For certain values of the parameters this equation exhibits a range of phenomena including: mode-locking evolution; pulses that disperse into radiation; some that evolve to a non-localized quasiperiodic state; and some whose amplitude grows rapidly (Kapitula et al., 2002). In the latter case, if the nonlinear gain is too high, the linear attenuation terms are unable to prevent the pulse from blowing up; i.e., the master mode-locking model breaks down (Kutz, 2006). However, unlike what is observed in experiments, there is only a small window of parameter space that allows for the generation of stable mode-locked pulses. In particular, the model is highly sensitive to the nonlinear loss/gain parameter.

We have shown that the PES model yields mode-locking for wide ranges of the parameters (cf. Ablowitz et al., 2008; Ablowitz and Horikis, 2008, 2009b; Ablowitz et al., 2009c). As mentioned above, this model is a distributed equation. There are also interesting lumped models that have been studied (cf. Ilday et al., 2004b,a; Chong et al., 2008b); e.g., in some laser models, loss is introduced in the form of fast saturable power absorbers that are placed periodically. It has been found (Ablowitz and Horikis, 2009a), however, that all features are essentially the same in both lumped and distributive models thus indicating that distributive models are very good descriptions of modes in mode-locked lasers. Lumped models reflect sharp changes in the parameters/coefficients due to corresponding elements in the system; mathematically these models are often dealt with by Dirac delta function transitions (see also Chapter 10). Mathematically it is also more convenient to work with distributed models. We note that Haus (1975) derived models of fast saturable absorbers in two-level media that are similar to the ones we are studying here. However in order to obtain analytical results, Haus Taylor-expanded and therefore obtained the cubic nonlinear model of a fast saturable absorber.

We also mention that to overcome the sensitivity inherent in the master equation, other types of terms, such as quintic terms, can be added to the master equation in order to stabilize the solutions. This increases the parameter range for mode-locking somewhat (instabilities may still occur); but it also adds another parameter to the model. Cubic and quintic nonlinear models are based on Ginzburg–Landau-type (GL) equations (cf. Akhmediev and Ankiewicz, 1997). In fact, if the pulse energy is taken to be constant the master equation reduces to a GL-type system. In general, GL-type equations exhibit a wide spectrum of interesting phenomena and pulses that exhibit complex and chaotic dynamics and ones whose amplitude grows rapidly.

### 11.2.1 The anomalous dispersion regime

We first discuss the constant dispersive case in which case the laser cavity is described, in dimensionless form, by (called here the constant dispersion PES equation)

$$iu_z + \frac{d_0}{2}u_{tt} + |u|^2 u = \frac{ig}{1 + E/E_0}u + \frac{i\tau}{1 + E/E_0}u_{tt} - \frac{il}{1 + P/P_0}u \qquad (11.2)$$

with $d_0$ representing the constant dispersion, $n(z) = 1$, and remaining parameters defined above.

In what follows we keep all terms constant and only change the gain parameter $g$. More precisely, we take typical values: $E_0 = P_0 = 1$, $\tau = l = 0.1$, assume anomalous dispersion: $d_0 = 1$ ($d_0 > 0$), and in the evolution studies we take a unit Gaussian initial condition. For stable soliton solutions to exist the gain parameter $g$ needs only to be sufficiently large to counter the two loss terms. It is also noted that we employ a fourth-order Runge–Kutta method to evolve (11.2) in $z$.

The evolution of the pulse peak for different values of the gain parameter $g$ is shown in Figure 11.3. When $g = 0.1$ the pulse vanishes quickly due to excessive loss with no noticeable oscillatory behavior; the pulse simply decays, yielding damped evolution. When $g = 0.2, 0.3$, due to the loss in the system, the pulse initially undergoes a sharp decrease relative to its amplitude. However, it rapidly recovers and evolves into a stable solution. Much like the damped evolution, the amplitude is initially decreased but the resulting evolution is stable. Interestingly, e.g., when $g = 0.7, 1$, and the perturbations can no longer be considered small, a stable evolution is nevertheless again obtained, although somewhat different from the case above. Now with considerable gain in the system, the pulse amplitude increases rapidly and then a steady state is reached typically after some oscillations. The only major difference between the resulting modes is the resulting amplitude and the width of the pulse: i.e., for larger $g$ the pulse is larger and narrower (Ablowitz and Horikis, 2008).

The above suggests that in the PES model, the mode-locking effect is generally present for $g \geq g^*$, a critical gain value. Without enough gain i.e., $g < g^*$, pulses dissipate to the trivial zero state. Furthermore, for this class of initial data and parameters studied, there are no complex radiation states or states

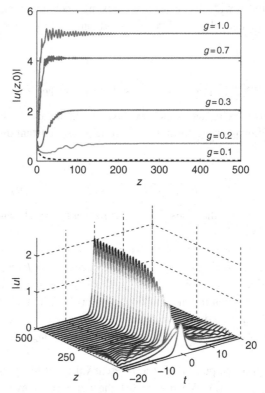

Figure 11.3 Evolution of the pulse peak, $|u(z, 0)|$, of an arbitrary initial profile under PES with different values of gain. The damped pulse-peak evolution is shown with a dashed line. In the bottom figure, a typical evolution is given for $g = 0.3$.

whose amplitudes grow without bound. In terms of solutions, (11.2) admits soliton states for all values of $g \geq g^* > l$ (recall, here $l = 0.1$). This can also be understood and confirmed by analytical methods (soliton perturbation theory) (Ablowitz et al., 2009a). Mode-locking of the solution is found to tend to soliton states. By a soliton state we mean a solution of the form $u(t, z) = f(t) \exp(-i\mu z)$, where $\mu$ is usually referred to as the propagation constant and $f(t)$ is the soliton shape.

As mentioned above, as the gain parameter increases so does the amplitude, and the pulse becomes narrower. In fact, it is observed that the energy changes according to $E \sim \sqrt{\mu}$. Indeed, from the soliton theory of the classical NLS equation this is exactly the way a classical soliton's energy changes, the key difference between the PES and pure NLS equation being that in

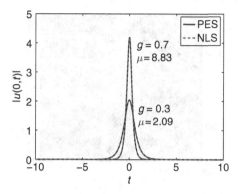

Figure 11.4 Solitons of the perturbed and unperturbed equations.

the pure NLS equation a semi-infinite set of $\mu$ exists, whereas now for the PES equation $\mu$ is unique for the given set of parameters. Once mode-locking occurs we find the solutions of the two equations, PES and NLS, are comparable. In Figure 11.4 we plot the two solutions for different values of $g$. In each case the same value of $\mu$ is used. The amplitudes match so closely that they are indistinguishable in the figure, meaning the perturbing effect is strictly the mode-locking mechanism, i.e., its effect is to mode-lock to a soliton of the pure NLS with the appropriate propagation constant. The solitons of the unperturbed NLS system are well known in closed analytical form, i.e., they are expressed in terms of the hyperbolic secant function, $u = \sqrt{2\mu}\,\text{sech}\left(\sqrt{2\mu}\,t\right)\exp(-i\mu z)$, and therefore describe solitons of the PES to a good approximation.

### Soliton strings

As mentioned above, solitons are obtained when the gain is above a certain critical value, $g > l$, otherwise pulses dissipate and eventually vanish (Ablowitz and Horikis, 2008). As gain becomes stronger, additional soliton states are possible and two, three, four or more coupled pulses are found to be supported; we call these states soliton strings. As above, we set $\tau = l = 0$, $E_0 = P_0 = d_0 = n = 1$ and vary the gain parameter $g$. The value of $\Delta\xi/\alpha$, where $\Delta\xi$ and $\alpha$ are the pulse separation and pulsewidth (the full width of half-maximum (FWHM) for pulsewidth is used) respectively, is a useful parameter. We measure $\Delta\xi$, between peak values of two neighboring pulses, and $\Delta\phi$ is the phase difference between the peak amplitudes. With sufficient gain ($g = 0.5$) equation (11.2) is evolved starting from unit Gaussians with initial peak separation $\Delta\xi = 10$ and $\Delta\xi/\alpha = 8.5$. The evolution and final state are depicted in Figure 11.5. For the

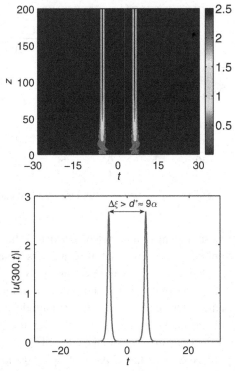

Figure 11.5 Top: Mode-locking evolution for an in phase two-soliton state of the anomalously constant dispersive PES equation; Bottom: the resulting soliton profile at $z = 300$. Here $g = 0.5$.

final state we find $\Delta \xi / \alpha \approx 10$. The resulting individual pulses are similar to the single soliton mode-locking case (Ablowitz and Horikis, 2008), i.e., individual pulses are approximately solutions of the unperturbed NLS equation, namely hyperbolic secants. We also note that the individual pulse energy is smaller then that observed for the single-soliton mode-locking case for the same choice of $g$, while the total energy of the two-soliton state is larger. This is due to the non-locality of energy saturation in the gain and filtering terms (Ablowitz et al., 2009c).

To investigate the minimum distance, $d^*$, between the solitons in order for no interactions to occur (in a prescribed distance) we evolve the equation starting with two solitons. If the initial two pulses are sufficiently far apart then the propagation evolves to a two-soliton state and the resulting pulses have a constant phase difference. If the distance between them is smaller than a critical value then the two pulses interact in a way characterized by the difference in phase between the peaks amplitudes: $\Delta\phi$. When initial conditions are symmetric (in phase) two pulses are found to merge into a single soliton of (11.2). When the initial conditions are antisymmetric (out of phase by $\pi$) then they repel each other until their separation is larger than this critical distance while retaining the same difference in phase, resulting in an effective two-pulse high-order soliton state. The above situation is found with numerous individual pulses ($N = 2, 3, 4, \ldots$). This situation does not occur in the pure NLS equation where two out-of-phase pulses strongly repel each other. The interaction phenomena can be is described by perturbation theory (Ablowitz et al., 2009a). In the constant dispersion case this critical distance is found to be $\Delta\xi = d^* \approx 9\alpha$ (see Figure 11.5) where effectively no interaction occurs for $z < z_*$, corresponding to soliton initial conditions. Interestingly, this is consistent with the experimental observations of Tang et al. (2001). To further illustrate, we plot the evolution of in and out of phase cases in Figure 11.6. At $z = 500$ for the repelling solitons we find $\Delta\xi/\alpha = 8.9$.

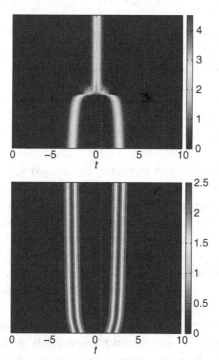

Figure 11.6 Two-pulse interaction when $\Delta\xi < d^*$. Initial pulses ($z = 0$) in phase: $\Delta\phi = 0$, $\Delta\xi/\alpha \approx 7$ (top) merge while those out of phase by $\Delta\phi = \pi$ with $\Delta\xi/\alpha \approx 6$ (bottom) repel. Here $g = 0.5$.

## 11.2.2 Dispersion-managed PES equation

Next we will discuss the situation when we employ the dispersion-managed PES equation (11.1). As in the communications application in the previous section, $d(z)$ is large and both $d(z)$ and $n(z)$ are rapidly varying. We take

$$d(z) = d_0 + \frac{\Delta(z/z_a)}{z_a}, \quad n(z) = n(z/z_a), \quad 0 < z_a \ll 1$$

where $\Delta = \Delta_1 < 0$, $n = 1$ inside the crystal; elsewhere $\Delta = \Delta_2 > 0$, $n = 0$ (linear). The analysis follows that in communications (see Chapter 10): we introduce multiple scales $\zeta = z/z_a$ (short), $Z = z$ (long), and expand the solution in powers of $z_a$: $u = u^{(0)} + z_a u^{(1)} + \cdots$, $0 < z_a \ll 1$. At $O(1/z_a)$ we have a linear equation:

$$i u_\zeta^{(0)} + \frac{\Delta(\zeta)}{2} u_{tt}^{(0)} = 0,$$

which we solve via Fourier transforms in $t$ and find, in the Fourier domain, that

$$\hat{u}^{(0)} = \hat{U}_0(Z, \omega) \exp\left[-i\frac{\omega^2}{2}C(\zeta)\right], \quad C(\zeta) = \int_0^\zeta \Delta(\zeta')\,d\zeta',$$

where $\hat{U}_0(Z, \omega)$ is free at this stage. Proceeding to the next order we find a forced linear equation. In order to remove unbounded growth, i.e., remove secularities, we find an equation for $\hat{U}_0$ that is given below:

$$i\frac{\partial \hat{U}_0}{\partial Z} - \frac{d_0}{2}\omega^2 \hat{U}_0 +$$

$$\int_0^1 \exp\left[i\frac{\omega^2}{2}C(\zeta)\right]\left(\mathcal{F}\left\{n(\zeta)|u^{(0)}|^2 u^{(0)} - R[u^{(0)}]\right\}\right) d\zeta = 0, \qquad (11.3)$$

where $\mathcal{F}$ represents the Fourier transform; i.e., $\mathcal{F}w(\omega) = \hat{w} = \int w(t)e^{-i\omega t}\,dt$. This equation is referred to here as the DM-PES equation or simply the averaged equation. The strength of the dispersion-management is usually measured by the map length, which we take in the normal and anomalous domains to be equal ($\theta = 1/2$ in the notation of Chapter 10) and so $s = |\Delta_1|/4$; we take the distance in the crystal to be the same as the distance in the prisms. The equation is a nonlinear integro-differential equation; as in the communications application it can be solved numerically. Similar to the constant dispersive case, for $d_0 > 0$ we find mode-locking. Fixing the parameters as $d_0 = E_0 = P_0 = 1$, $z_a = \tau = l = 0.1$ and varying the gain, we find that with sufficient gain strength mode-locking occurs. The mode-locking also depends on the strength of the dispersion-management: given a map strength, $s$, and where $g_* = g_*(s)$, we find: for $g < g_*$, pulses damp; for $g > g_*$, mode-locking. In Figure 11.7 is a typical example of the evolution of a unit Gaussian with map strength $s = 1$.

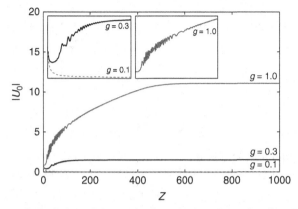

Figure 11.7 Evolution of the peak of a unit Gaussian using the DM-PES equation with $s = 1$ and gain parameters $g = 0.1$ (damped evolution), 0.3 and 1.0. In the inset, the early evolution of the peaks is shown.

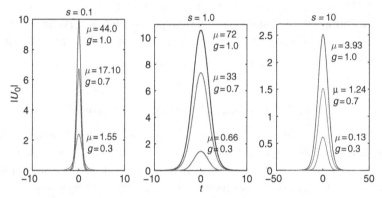

Figure 11.8 DM-PES solitons for various values of gain, $g$, and propagation constant, $\mu$, with map strength $s = 1$.

The final mode-locking state corresponds to a soliton that can be obtained from spectral renormalization methods (discussed in Chapter 10). A soliton satisfies a fixed point integral equation obtained after substituting $\hat{U}_0(Z, \omega) \to \hat{U}_0(\omega)e^{i\mu Z}$ into the DM-PES equation (11.3). In Figure 11.8 are some typical solitons corresponding to different values of gain $g$ and propagation constant $\mu$. As in the constant dispersive case, DM-PES solitons correspond to specific values of $\mu$, which is the soliton analog of mode-locking. This is also in contrast to the unperturbed case where solutions exist for all $\mu > 0$. We see that as map strength increases the solitons become smaller in amplitude but much more broad. As $\mu$ increases the modes become sharper and taller.

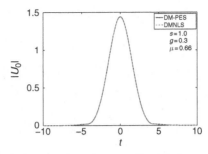

Figure 11.9 Typical comparison between the solutions of the DM-PES equation (11.3) and the pure DMNLS equation for $s = 1$ and $g = 0.3$.

Each mode-locked state is affected by strong saturation effects. Hence the soliton is essentially a solution of the unperturbed DM-PES equation (10.25) (where $g(\zeta)$ is replaced by $n(\zeta)$ in the notation of this section); i.e., this is the DMNLS equation that we obtain when we omit the gain and damping terms. This is indicated in Figure 11.9 for typical parameters. The main effect of the gain-damping is to isolate the parameters for which mode-locking occurs.

### 11.2.3 Remarks on perturbation theory

We have studied the behavior of both single solitons and soliton strings in ML lasers in detail via perturbation theory (Ablowitz et al., 2009a). Here we briefly mention some of the key results in the case of constant dispersion, taking $d_0 = 1$. Since the pulses have been found to be well approximated by solutions of classical NLS equation, we assume a soliton solution of the NLS form written as

$$u = ue^{i\phi}, \quad u = \eta \, \text{sech}(\eta\theta), \quad \xi = \int_0^z V \, dz + t_0 \qquad (11.4a)$$

$$\theta = t - \xi, \quad \sigma = \int_0^z \left[\mu + \frac{V^2}{2}\right] dz + \sigma_0, \quad \phi = V\theta + \sigma, \qquad (11.4b)$$

with parameters for the height ($\eta$, $\mu = \eta^2/2$), velocity ($V$), temporal shift ($t_0$) and phase shift ($\sigma_0$) assumed to vary adiabatically, i.e., slowly, in $z$.

We consider the effects of gain, filtering and loss as perturbations of the NLS equation, taking $d_0 = n = 1$ in the PES equation,

$$iu_z + \frac{1}{2}u_{tt} + |u|^2 u = R[u] \qquad (11.5)$$

where the right-hand side $F$ contains the gain-damping terms and is assumed small. Recall

$$R = \frac{gu}{1 + E/E_0} + \frac{\tau u_{tt}}{1 + E/E_0} - \frac{lu}{1 + P/P_0}$$

and $E = \int_{-\infty}^{\infty} |u|^2 \, dt$, $P = |u|^2$. While the individual contributions of terms in $F$ may not always be small, the combined effects are small; as discussed above they provide the mode-locking mechanism (Ablowitz and Horikis, 2008).

We find the equations for the evolution of the parameters $\eta$, $V$, $t_0$, $\sigma_0$ by requiring the leading-order solution, (11.4), to satisfy a set of integrated quantities related to the (unperturbed) conservation laws suitably modified by the terms in $R[u]$. These can be derived directly from (11.5) as

$$\frac{d}{dz} \int |u|^2 = 2 \, \mathrm{Im} \int R[u]u^* \tag{11.6a}$$

$$\frac{d}{dz} \, \mathrm{Im} \int uu_t^* = 2 \, \mathrm{Re} \int R[u]u_t^* \tag{11.6b}$$

$$\frac{d}{dz} \int t|u|^2 = - \, \mathrm{Im} \int uu_t^* + 2 \, \mathrm{Im} \int tR[u]u^* \tag{11.6c}$$

$$\mathrm{Im} \int u_z u_\mu^* = -\frac{1}{2} \, \mathrm{Re} \int u_t u_{t\mu}^* + \mathrm{Re} \int |u|^2 \, uu_\mu^* - \mathrm{Re} \int R[u]u_\mu^*, \tag{11.6d}$$

where all integrals are taken over $-\infty < t < \infty$.

When (11.4) are substituted into (11.6) the modified integral equations reduce to a system of differential equations that govern the evolution and mode-locking of soliton pulses:

$$\frac{d\eta}{dz} = g \left[ \frac{2\eta}{2\eta + 1} \right] - \tau \frac{1}{2\eta + 1} \left[ \frac{2}{3}\eta^3 + 2V^2 \eta \right] + l \left[ 2\eta \frac{1}{a - b} \log \left( \frac{a}{b} \right) \right] \tag{11.7a}$$

$$\frac{dV}{dz} = -\tau V \left[ \frac{4}{3} \frac{\eta^2}{2\eta + 1} \right] \tag{11.7b}$$

$$\frac{dt_0}{dz} = 0 \tag{11.7c}$$

$$\frac{d\sigma_0}{dz} = 0; \tag{11.7d}$$

here $a$ and $b$ are the roots of the polynomial $x^2 + 2(1 + 2\eta^2)x + 1 = 0$ (we can choose $a$ and $b$ to be either root). Note, the first two equations are independent of $t_0$ and $\sigma_0$, both of which are constant. We only consider parameters in the domain $\eta \geq 0$ (since $\eta < 0$ can be represented as a positive $\eta$ with a $\pi$ phase shift).

Consider the case $V = 0$; this is the only equilibrium for $V$ and it is stable for $\eta > 0$. In this case we find an equation for $\eta$ independent of any other parameters

$$\frac{d\eta}{dz} = g\left[\frac{2\eta}{2\eta+1}\right] - \tau\frac{1}{2\eta+1}\left[\frac{2}{3}\eta^3\right] + l\left[2\eta\frac{1}{a-b}\log\left(\frac{a}{b}\right)\right],$$

which accurately predicts the mode-locking behavior found numerically. Clearly, $\eta$ has at least one equilibrium at 0, which can be proved to be stable for $g < l$ and unstable for $g > l$. For $g > l$ there is a second equilibrium that is stable and corresponds to the mode-locked solution, as indicated in Figure 11.10.

Weakly interacting solitons can also be studied analytically by considering the interaction terms as a small perturbation (Karpman and Solov'ev, 1981). We refer the interested reader to Ablowitz et al. (2009a) for further details.

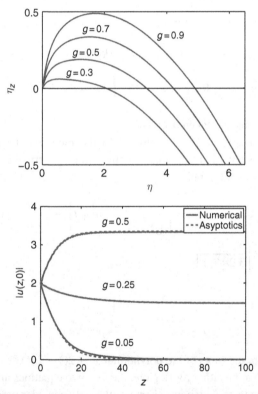

Figure 11.10 Phase portrait of the amplitude equation for several values of $g > l$ illustrating the single stable equilibrium (top). Comparison of asymptotic and numerical results for mode-locking pulses (bottom).

### 11.2.4 The normal regime

The dynamics of pulses evolving under the PES equation in the normal regime ($d_0 < 0$) are studied next. As above, all terms are kept constant and only the gain parameter $g$ is changed. Typical values are taken: $d_0 = -1$ (i.e., the normal regime), $n = 1$, $E_0 = P_0 = 1$, $\tau = l = 0.1$. We study the evolution of an initial unit Gaussian pulse peak for different values of the gain parameter $g$; see Figure 11.11. When $g = 0.1$ the pulse decays quickly due to excessive loss with no noticeable oscillatory or chaotic behavior; the pulse exhibits damped evolution. When $g = 0.5$, due to the loss in the system the pulse undergoes an initial decrease and then very slowly dissipates. When $g = 1.5$ a localized evolution is obtained. With sufficient gain in the system, the pulse amplitude initially

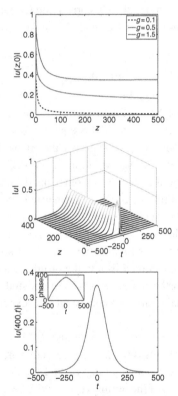

Figure 11.11 Evolution of the pulse peak of an initial profile under the PES equation, in the normal regime ($d_0 = -1$) with different values of gain. Top: the damped pulse-peak evolution is shown with a dashed line. Middle: the complete evolution is given for $g = 1.5$. Bottom: the resulting soliton and the corresponding phase in the inset.

decreases but then a steady state is reached. The sharp and narrow initial Gaussian rapidly evolves into a wide but finite-energy pulse. Three regimes are observed: (a) when the loss is much greater than the gain the pulse decays to zero; (b) when the loss is again a prominent effect but sufficient gain exists in the system to sustain a very slowly decaying evolution resulting in a "quasi-soliton" state; and (c) the soliton regime above a certain value of gain. When a localized state results in the normal regime, there is a strong phase effect, as indicated in Figure 11.11. This is unlike the anomalous regime where the pulse's temporal phase is constant and the pulses have relatively much narrower width.

As indicated above, the values for the gain, loss and filtering parameters provide operating regions of the laser system that depend critically on the gain parameter. The parameters $d_0$, $E_0$, $P_0$, $g$, $\tau$, $l$ were chosen to have typical values. Changing these values to more closely correspond to experimental data shows that the above observations are quite general (see Ablowitz and Horikis, 2008, 2009b).

## Bi-solitons

As mentioned above, individual pulses of (11.2) in the normal regime exhibit strong chirp and cannot be identified as the solutions of the "unperturbed" NLS equation. Indeed, the classical NLS equation does not exhibit decaying "bright" soliton solutions in the normal regime. If we begin with an initial Gaussian, $u(0, t) = \exp(-t^2)$, the evolution of the PES equation mode-locks into a fundamental soliton state. On the other hand, we can obtain a higher-order antisymmetric soliton, i.e., an antisymmetric bi-soliton, one that has its peak amplitudes differing in phase by $\pi$. Such a state can be obtained if we start with an initial state of the form $u(0, t) = t \exp(-t^2)$ (i.e., a Gauss–Hermite polynomial). The evolution results in an antisymmetric bi-soliton and is shown in Figure 11.12 along with a comparison to the profile of the single soliton. This is a true bound state. Furthermore, the results of our study finding antisymmetric solitons in the normal regime are consistent with experimental observations (Chong et al., 2008a).

It is interesting that the normal regime also exhibits higher soliton states when two general initial pulses (e.g., unit Gaussians) are taken sufficiently far apart. The resulting pulses, shown in Figure 11.13, individually have a similar shape to the single soliton of the normal regime with lower individual energies. These pulses if initially far enough apart can have independent chirps that may be out of phase by an arbitrary constant. We again find that the minimum distance required for the two solitons to remain apart is $d^* \approx 9\alpha$ which is a good estimate for the required pulse separation just as in the constant anomalous dispersive case! These pulses are "effective bound states" in that after a long

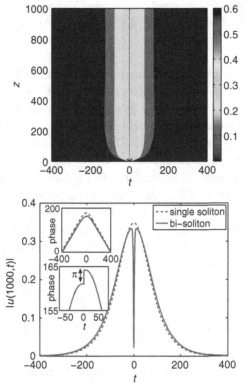

Figure 11.12 Evolution of the antisymmetric soliton (top) and the antisymmetric bi-soliton, in the normally dispersive regime, superimposed with the corresponding single soliton (bottom) at $z = 1000$. Here $g = 1.5$.

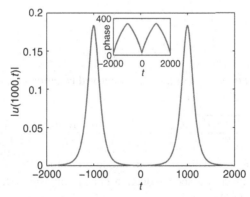

Figure 11.13 Symmetric two-soliton state of the normal regime with in-phase pulses. The phase structure is depicted in the inset. Here $g = 1.5$.

distance they separate very slowly. Additional higher-order 3-, 4-, ... soliton
states can also be found. We refer the reader to Ablowitz et al. (2009c) for
further details.

### Asymptotic approximations of normal solitons

Using the spectral renormalization method, we have found localized pulses in
the normal dispersive regime for a broad range of parameters, provided that
sufficient gain is present in the system. These pulses are slowly varying in $t$,
with a large phase. This suggests that by assuming a slow time-scale in the
equation and using perturbation theory (i.e., a WKB-type expansion) the soliton system can be reduced to simpler ordinary differential equations for the
amplitude and the phase of the pulse. The first step before performing perturbation theory is to write the solution of (11.2) in the form $u(z, t) = R \exp(i\mu z + i\theta)$,
where $R = R(t)$ and $\theta = \theta(t)$ are the pulse amplitude and phase, respectively.
Substituting into the equation and equating real and imaginary parts and calling
$\epsilon = 1/E_0, \delta = 1/P_0$, we get

$$-\mu R - \frac{d_0}{2}\left(R_{tt} - R\theta_t^2\right) + R^3 = -\frac{\tau}{1 + \epsilon E}(2R_t\theta_t + R\theta_{tt}) \qquad (11.8a)$$

$$-\frac{d_0}{2}(2R_t\theta_t + R\theta_{tt}) = \frac{g}{1 + \epsilon E}R + \frac{\tau}{1 + \epsilon E} - \frac{l}{1 + \delta R^2}R. \qquad (11.8b)$$

We then take the characteristic time length of the pulse to be such that we can
define a scaling in the independent variable of the form $v = T/t$ or $T = vt$
so that $R = R(vt)$, $\theta_t = O(1)$ (i.e., $\theta$ is large) and $\theta_{tt} = O(v)$ where $v \ll 1$.
Then (11.8a) becomes

$$-\mu R + \frac{d_0}{2}R\theta_t^2 + R^3 = \frac{d_0}{2}v^2 R_{TT} - \frac{\tau}{1 + \epsilon E}(2vR_T\theta_t + R\theta_{tt}).$$

The leading-order equation is

$$\theta_t^2 = \frac{2}{d_0}(\mu - R^2). \qquad (11.9)$$

Using the same argument and this newly derived equation (11.9) we then find
that (11.8b) to leading order reads

$$R_t = -\operatorname{sgn}(t)\frac{\sqrt{2(\mu - R^2)/d_0}}{3R^2 - 2\mu}\left(\frac{g}{1 + \epsilon E} - \frac{4\tau(\mu - R^2)/d_0}{1 + \epsilon E} - \frac{l}{1 + \delta R^2}\right). \qquad (11.10)$$

This is now a non-local first-order differential equation for $R = R(t)$, since
$E = \int_{-\infty}^{\infty} R^2\, dt$. From the above equation (11.9), imposing $\theta_t(t = 0) = 0$, it
follows that $\mu \approx R^2(0)$.

To remove the non-locality another condition is needed and it is based on the
singular points of (11.10). Recall that $R(t)$ is a decaying function in $t \in [0, +\infty)$

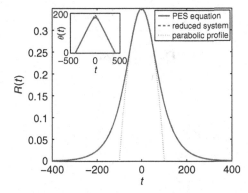

Figure 11.14 Solutions of the complete PES equation and the reduced system, in the normally dispersive regime. The relative parabolic profile is also shown. Here $g = 1.5$.

and $\mu \approx R^2(0) \geq R(t)^2$. Thus there exists a point in $t$ such that the denominator in the equation becomes zero. To remove the singularity we require that the numerator of the equation is also zero at the same point; that leads to

$$1 + \epsilon E = \frac{1}{l}\left(g - \frac{2\tau}{3d_0}\mu\right)\left(1 + \frac{2\delta}{3}\mu\right).$$

Thus (11.10) is now a first-order equation that can be solved by standard numerical methods and analyzed by phase plane methods. The resulting solutions from the PES equation and the reduced equations are compared in Figure 11.14. Notice that here $\mu = 0.1216$ while $R^2(0) = 0.1218$.

# References

Ablowitz, M.J., Bakirtas, I., and Ilan, B. 2005. Wave collapse in nonlocal nonlinear Schrödinger systems. *Physica D*, **207**, 230–253.

Ablowitz, M.J., and Biondini, G. 1998. Multiscale pulse dynamics in communication systems with strong dispersion-management. *Opt. Lett.*, **23**, 1668–1670.

Ablowitz, M.J., Biondini, G., Biswas, A., Docherty, A., Hirooka, T., and Chakravarty, S. 2002a. Collision-induced timing shifts in dispersion-managed soliton systems. *Opt. Lett.*, **27**, 318–320.

Ablowitz, M.J., Biondini, G., and Blair, S. 1997. Multi-dimensional pulse propagation in non-resonant $\chi^{(2)}$ materials. *Phys. Lett. A*, **236**, 520–524.

Ablowitz, M.J., Biondini, G., and Blair, S. 2001a. Nonlinear Schrödinger equations with mean terms in non-resonant multi-dimensional quadratic materials. *Phys. Rev. E*, **63**, 605–620.

Ablowitz, M.J., Biondini, G., Chakravarty, S., and Horne, R.L. 2003a. Four wave mixing in dispersion-managed return-to-zero systems. *J. Opt. Soc. Am. B*, **20**, 831–845.

Ablowitz, M.J., Biondini, G., Chakravarty, S., Jenkins, R.B., and Sauer, J.R. 1996. Four-wave mixing in wavelength-division-multiplexed soliton systems: damping and amplification. *Opt. Lett.*, **21**, 1646–1648.

Ablowitz, M.J., Biondini, G., and Olson, E. 2000b. On the evolution and interaction of dispersion-managed solitons. In: *Massive WDM and TDM Soliton Transmission Systems*, edited by Akira Hasegawa. Kyoto, Japan: Kluwer Academic, pp. 362–367.

Ablowitz, M.J., and Clarkson, P.A. 1991. *Solitons, Nonlinear Evolution Equations and Inverse Scattering*. Cambridge: Cambridge University Press.

Ablowitz, M.J., Docherty, A., and Hirooka, T. 2003b. Incomplete collisions in strongly dispersion-managed return-to-zero communication systems. *Opt. Lett.*, **28**, 1191–1193.

Ablowitz, M.J., and Fokas, A.S. 2003. *Complex Variables: Introduction and Applications*. Second edition. Cambridge: Cambridge University Press.

Ablowitz, M.J., Fokas, A.S., and Musslimani, Z. 2006. On a new nonlocal formulation of water waves. *J. Fluid Mech.*, **562**, 313–344.

Ablowitz, M.J., and Haberman, R. 1975. Resonantly coupled nonlinear evolution equations. *Phys. Rev. Lett.*, **38**, 1185–1188.

Ablowitz, M.J., Hammack, J., Henderson, D., and Schober, C.M. 2000a. Modulated periodic waves in deep water. *Phys. Rev. Lett.*, **84**, 887–890.

Ablowitz, M.J., Hammack, J., Henderson, D., and Schober, C.M. 2001. Long time dynamics of the modulational instability of deep water waves. *Physica D*, **152**, 416–433.

Ablowitz, M.J., and Haut, T.S. 2009a. Asymptotic expansions for solitary gravity–capillary waves in two and three dimensions. *Proc. R. Soc. Lond. A*, **465**, 2725–2749.

Ablowitz, M.J., and Haut, T.S. 2009b. Coupled nonlinear Schrödinger equations from interfacial fluids with a free surface. *Theor. Math. Phys.*, **159**, 689–697.

Ablowitz, M.J., and Haut, T.S. 2010. Asymptotic expansions for solitary gravity-capillary waves in two dimensions. *J. Phys. Math Theor.*, **43**, 434005.

Ablowitz, M.J., Hirooka, T., and Biondini, G. 2001b. Quasi-linear optical pulses in strongly dispersion-managed transmission systems. *Opt. Lett.*, **26**, 459–461.

Ablowitz, M.J., and Hirooka, T. 2002. Managing nonlinearity in strongly dispersion-managed optical pulse transmission. *J. Opt. Soc. Am. B*, **19**, 425–439.

Ablowitz, M.J., Hirooka, T., and Inoue, T. 2002b. Higher order asymptotic analysis of dispersion-managed transmission systems: solitons and their characteristics. *J. Opt. Soc. Am. B*, **19**, 2876–2885.

Ablowitz, M.J., and Horikis, T.P. 2008. Pulse dynamics and solitons in mode-locked lasers. *Phys. Rev. A*, **78**, 011802.

Ablowitz, M.J., and Horikis, T.P. 2009a. Solitons and spectral renormalization methods in nonlinear optics. *Eur. Phys. J. Special Topics*, **173**, 147–166.

Ablowitz, M.J., and Horikis, T.P. 2009b. Solitons in normally dispersive mode-locked lasers. *Phys. Rev. A*, **79**, 063845.

Ablowitz, M.J., Horikis, T.P., and Ilan, B. 2008. Solitons in dispersion-managed mode-locked lasers. *Phys. Rev. A*, **77**, 033814.

Ablowitz, M.J., Horikis, T.P., and Nixon, S.D. 2009c. Soliton strings and interactions in mode-locked lasers. *Opt. Comm.*, **282**, 4127–4135.

Ablowitz, M.J., Horikis, T.P., Nixon, S.D., and Zhu, Y. 2009a. Asymptotic analysis of pulse dynamics in mode-locked lasers. *Stud. Appl. Math.*, **122**, 411–425.

Ablowitz, M.J., Ilan, B., and Cundiff, S.T. 2004a. Carrier-envelope phase slip of ultra-short dispersion-managed solitons. *Opt. Lett.*, **29**, 1818–1820.

Ablowitz, M.J., Kaup, D.J., Newell, A.C., and Segur, H. 1973a. Method for solving sine–Gordon equation. *Phys. Rev. Lett.*, **30**, 1262–1264.

Ablowitz, M.J., Kaup, D.J., Newell, A.C., and Segur, H. 1973b. Nonlinear-evolution equations of physical significance. *Phys. Rev. Lett.*, **31**, 125–127.

Ablowitz, M.J., Kaup, D.J., Newell, A.C., and Segur, H. 1974. Inverse scattering transform—Fourier analysis for nonlinear problems. *Stud. Appl. Math.*, **53**, 249–315.

Ablowitz, M.J., Kruskal, M.D., and Segur, H. 1979. A note on Miura's transformation. *J. Math. Phys.*, **20**, 999–1003.

Ablowitz, M.J., Manakov, S.V., and Schultz, C.L. 1990. On the boundary conditions of the Davey–Stewartson equation. *Phys. Lett. A*, **148**, 50–52.

Ablowitz, M.J., and Musslimani, Z. 2003. Discrete spatial solitions in a diffraction-managed nonlinear waveguide array: a unified approach. *Physica D*, **184**, 276–303.

Ablowitz, M.J., and Musslimani, Z. 2005. Spectral renormalization method for computing self-localized solutions to nonlinear systems. *Opt. Lett.*, **30**, 2140–2142.

Ablowitz, M.J., Nixon, S.D., and Zhu, Y. 2009b. Conical diffraction in honeycomb lattices. *Phys. Rev. A*, **79**, 053830.

Ablowitz, M.J., Prinari, B., and Trubatch, A.D. 2004b. *Discrete and Continuous Nonlinear Schrödinger Systems*. Cambridge: Cambridge University Press.

Ablowitz, M.J., and Segur, H. 1979. On the evolution of packets of water waves. *J. Fluid Mech.*, **92**, 691–715.

Ablowitz, M.J., and Segur, H. 1981a. *Solitons and the Inverse Scattering Transform*. Philadelphia: SIAM.

Ablowitz, M.J., and Villarroel, J. 1991. On the Kadomtsev–Petviashili equation and associated constraints. *Stud. Appl. Math.*, **85**, 195–213.

Ablowitz, M.J., and Wang, X.P. 1997. Initial time layers in Kadomtsev–Petviashili type equations. *Stud. Appl. Math.*, **98**, 121–137.

Abramowitz, M., and Stegun, I.R. 1972. *Handbook of Mathematical Functions*. Tenth edition. New York: Dover.

Agrawal, G.P. 2001. *Nonlinear Fiber Optics*. New York: Academic Press.

Agrawal, G.P. 2002. *Fiber-Optic Communication Systems*. New York: Wiley-Interscience.

Ahrens, C. 2006. *The asymptotic analysis of communications and wave collapse problems in nonlinear optics*. Ph.D. thesis, University of Colorado.

Airy, G.B. 1845. Tides and waves. *Encyc. Metrop.*, **192**, 241–396.

Akhmediev, N.N., and Ankiewicz, A. 1997. *Solitons, Nonlinear Pulses and Beams*. London: Chapman & Hall.

Akhmediev, N.N., Soto-Crespo, J.M., and Town, G. 2001. Pulsating solitons, chaotic solitons, period doubling, and pulse coexistence in mode-locked lasers: Complex Ginzburg–Landau equation approach. *Phys. Rev. E*, **63**, 056602.

Alfimov, G.L., Kevrekidis, P.G., Konotop, V.V., and Salerno, M. 2002. Wannier functions analysis of the nonlinear Schrödinger equation with a periodic potential. *Phys. Rev. E*, **66**, 046608.

Angulo, J., Bona, J.L., Linares, F., and Scialom, F. 2002. Scaling, stability and singularities for nonlinear, dispersive wave equations: the critical case. *Nonlinearity*, **15**, 759–786.

Batchelor, G.K. 1967. *An Introduction to Fluid Dynamics*. Cambridge: Cambridge University Press.

Beals, R., and Coifman, R.R. 1984. Scattering and inverse scattering for 1st order systems. *Comm. Pure Appl. Math.*, **37**, 39–90.

Beals, R., and Coifman, R.R. 1985. Inverse scattering and evolution-equations. *Comm. Pure Appl. Math.*, **38**, 29–42.

Beals, R., Deift, P., and Tomei, C. 1988. *Direct and Inverse Scattering on the Line*. Providence, RI: AMS.

Bender, C.M., and Orszag, S.A. 1999. *Advanced Mathematical Methods for Scientists and Engineers*. Berlin: Springer.

Benjamin, T.B., and Feir, J.F. 1967. The disintegration of wave trains on deep water. Part 1. Theory. *J. Fluid Mech.*, **27**, 417–430.

Benjamin, T.B., Bona, J.L., and Mahony, J.J. 1972. Model equations for long waves in nonlinear dispersive systems. *Phil. Trans. Roy. Soc. A*, **227**, 47–78.

Benney, D.J. 1966a. Long non-linear waves in fluid flows. *J. Math. and Phys.*, **45**, 52–63.

Benney, D.J. 1966b. Long waves on liquid films. *J. Math. and Phys.*, **45**, 150–155.

Benney, D.J., and Luke, J.C. 1964. Interactions of permanent waves of finite amplitude. *J. Math. Phys.*, **43**, 309–313.

Benney, D.J., and Newell, A.C. 1967. The propagation of nonlinear envelopes. *J. Math. and Phys.*, **46**, 133–139.

Benney, D.J., and Roskes, G.J. 1969. Wave instabilities. *Stud. Appl. Math.*, **48**, 377–385.

Bleistein, N. 1984. *Mathematical Methods for Wave Phenomena*. New York: Academic Press.

Bleistein, N., and Handelsman, R.A. 1986. *Asymptotic Expansions of Integrals*. New York: Dover.

Bogoliubov, N.N., and Mitropolsky, Y.A. 1961. *Asymptotic Methods in the Theory of Non-linear Oscillations*. Second edition. Russian monographs and texts on advanced mathematics and physics, vol. 10. London: Taylor & Francis.

Boussinesq, J.M. 1871. Théorie de l'intumescence appelée onde solitaire ou de translation se propageant dans un canal rectangulaire. *Comptes Rendues, Acad. Sci. Paris*, **72**, 755–759.

Boussinesq, J.M. 1872. Théorie des ondes et des remous qui se propagent le long d'un canal rectangulaire horizontal, en communiquant au liquide contenu dans ce canal des vitesses sensiblemant parielles de la surface au fond. *J. Math. Pures Appl. Ser. (2)*, **17**, 55–108.

Boussinesq, J.M. 1877. *Essai sur la théorie des eaux courantes. Memoires présenté par divers savantes à l'Acad. des Sci.* Inst. NAT. France XXIII, 1–680.

Boyd, R.W. 2003. *Nonlinear Optics*. New York: Academic Press.

Byrd, P.F., and Friedman, M.D. 1971. *Handbook of Elliptic Integrals for Engineers and Physicists*. Berlin: Springer.

Calogero, F., and Degasperis, A. 1982. *Spectral Transform and Solitons*. Amsterdam: Elsevier.

Calogero, F., and DeLillo, S. 1989. The Burgers equation on the semi-infinite and finite intervals. *Nonlinearity*, **2**, 37–43.

Camassa, R., and Holm, D.D. 1993a. An integrable shallow water equation with peaked solitons. *Phys. Rev. Lett.*, **71**, 1661–1664.

Camassa, R., and Holm, D.D. 1993b. An integrable shallow water equation with peaked solitons. *Phys. Rev. Lett.*, **11**, 1661–1664.

Caudrey, P.J. 1982. The inverse problem for a general $N \times N$ spectral equation. *Physica D*, **6**, 51–66.

Chapman, S., and Cowling, T. 1970. *Mathematical Theory of Nonuniform Gases*. Cambridge: Cambridge University Press.

Chen, M., Tsankov, M.A., Nash, J.M., and Patton, C.E. 1994. Backward volume wave microwave envelope solitons in yttrium iron garnet films. *Phys. Rev. B*, **49**, 12773–12790.

Chester, C., Friedman, B., and Ursell, F. 1957. An extension of the method of steepest descents. *Proc. Camb. Phil. Soc.*, **53**, 599–611.

Chong, A., Renninger, W.H., and Wise, F.W. 2008a. Observation of antisymmetric dispersion-managed solitons in a mode-locked laser. *Opt. Lett.*, **33**, 1717–1719.

Chong, A., Renninger, W.H., and Wise, F.W. 2008b. Properties of normal-dispersion femtosecond fiber lasers. *J. Opt. Soc. Am. B*, **25**, 140–148.

Christodoulides, D.N., and Joseph, R.I. 1998. Discrete self-focusing in nonlinear arrays of coupled waveguides. *Opt. Lett.*, **13**(9), 794–796.

Clarkson, P.A., Fokas, A.S., and Ablowitz, M.J. 1989. Hodograph transformations of linearizable partial differential equations. *SIAM J. Appl. Math.*, **49**, 1188–1209.

Cole, J.D. 1951. On a quasilinear parabolic equation occurring in aerodynamics. *Quart. Appl. Math.*, **9**, 225–236.

Cole, J.D. 1968. *Perturbation Methods in Applied Mathematics*. London: Ginn and Co.

Copson, E.T. 1965. *Asymptotic Expansions*. Cambridge: Cambridge University Press.

Courant, R., Friedrichs, K., and Lewy, H. 1928. On the partial differential equations of mathematical physics. *Math. Ann.*, **100**, 32–74. English Translation, *IBM Journal*, 11:215–234, 1967.

Courant, R., and Hilbert, D. 1989. *Methods of Mathematical Physics*. New York: John Wiley.

Craig, W., and Groves, M.D. 1994. Hamiltonian long-wave approximations to the water-wave problem. *Wave Motion*, **19**, 367–389.

Craig, W., and Sulem, C. 1993. Numerical simulation of gravity-waves. *J. Comp. Phys.*, **108**, 73–83.

Crasovan, L.C., Torres, J.P., Mihalache, D., and Torner, L. 2003. Arresting wave collapse by self-rectification. *Phys. Rev. Lett.*, **9**, 063904.

Cundiff, S.T. 2005. Soliton dynamics in mode-locked lasers. *Lect. Notes Phys.*, **661**, 183–206.

Cundiff, S.T., Ye, J., and Hall, J. 2008. Rulers of light. *Scientific American*, April, 74–81.

Davey, A., and Stewartson, K. 1974. On three dimensional packets of surface waves. *Proc. Roy. Soc. London A*, **338**, 101–110.

Deift, P., and Trubowitz, E. 1979. Inverse scattering on the line. *Comm. Pure Appl. Math.*, **32**, 121–251.

Deift, P., Venakides, S., and Zhou, X. 1994. The collisionless shock region for the long-time behavior of solutions of the KdV equation. *Comm. Pure Appl. Math.*, **47**, 199–206.

Deift, P., and Zhou, X. 1992. A steepest descent method for oscillatory Riemann–Hilbert problems. *Bull. Amer. Math. Soc.*, **26**, 119–123.

Dickey, L.A. 2003. *Soliton Equations and Hamiltonian Systems*. Singapore: World Scientific.

Djordjevic, V.D., and Redekopp, L.G. 1977. On two-dimensional packets of capillary–gravity waves. *J. Fluid Mech.*, **79**, 703–714.

Docherty, A. 2003. *Collision induced timing shifts in wavelength-division-multiplexed optical fiber communications systems*. Ph.D. thesis, University of New South Wales.

Dodd, R.K., Eilbeck, J.C., Gibbon, J.D., and Morris, H.C. 1984. *Solitons and Nonlinear Wave Equations*. New York: Academic Press.

Douxois, T. 2008. Fermi–Pasta–Ulam and a mysterious lady, *Physics Today*, Jan. 2008, 55–57.

Dysthe, K.B. 1979. Note on a modification to the nonlinear Schrödinger equation for application to deep water waves. *Proc. Roy. Soc. London A*, **369**, 105–114.

Elgin, J.N. 1985. Inverse scattering theory with stochastic initial potentials. *Phys. Lett. A*, **110**, 441–443.

Erdelyi, A. 1956. *Asymptotic Expansions*. New York: Dover.

Faddeev, L.D. 1963. The inverse problem in the quantum theory of scattering. *J. Math. Phys.*, **4**, 72–104.

Faddeev, L.D., and Takhtajan, L.A. 1987. *Hamiltonian Methods in the Theory of Solitons*. Berlin: Springer.

Falcon, E., Laroche, C., and Fauve, S. 2002. Observation of depression solitary surface waves on a thin fluid layer. *Phys. Rev. Lett.*, **89**, 204501.

Fermi, E., Pasta, S., and Ulam, S. 1955. *Studies of nonlinear problems, I.* Tech. rept. Los Alamos Report LA1940. [Reproduced in "Nonlinear Wave Motion," proceedings, Potsdam, New York, 1972, ed. A.C. Newell, *Lect. Appl. Math.*, **15**, 143–156, A.M.S., Providence, RI, (1974).].

Fibich, G., and Papanicolaou, G. 1999. Self-focusing in the perturbed and unperturbed nonlinear Schrödinger equation in critical dimension. *SIAM J. App. Math.*, **60**, 183–240.

Fokas, A.S. 2008. *A Unified Approach to Boundary Value Problems*. Philadelphia: SIAM.

Fokas, A.S., and Anderson, R.L. 1982. On the use of isospectral eigenvalue problems for obtaining hereditary symmetries for Hamiltonian-systems. *J. Math. Phys.*, **23**, 1066–1073.

Fokas, A.S., and Fuchssteiner, B. 1981. Symplectic structures, their Bäcklund transformations and hereditary symmetries. *Physica D*, **4**, 47–66.

Fokas, A.S., and Santini, P.M. 1989. Coherent structures in multidimensions. *Phys. Rev. Lett.*, **63**, 1329–1333.

Fokas, A.S., and Santini, P.M. 1990. Dromions and a boundary value problem for the Davey–Stewartson I equation. *Physica D*, **44**, 99–130.

Forsyth, A.R. 1906. *Theory of Differential Equations. Part IV—Partial Differential Equations*. Cambridge: Cambridge University Press.

Gabitov, I.R., and Turitsyn, S.K. 1996. Averaged pulse dynamics in a cascaded transmission system with passive dispersion compensation. *Opt. Lett.*, **21**, 327–329.

Garabedian, P. 1984. *Partial Differential Equations*. New York: Chelsea.

Gardner, C.S., Greene, J., Kruskal, M., and Miura, R.M. 1967. Method for solving the Korteweg–de Vries equation. *Phys. Rev. Lett.*, **19**, 1095–1097.

Gardner, C.S., and Su, C.S. 1969. The Korteweg–de Vries equation and generalizations. III. *J. Math. Phys.*, **10**, 536–539.

Gardner, C.S., Greene, J.M., Kruskal, M.D., and Miura, R.M. 1974. Korteweg–de Vries and generalizations. VI. Methods for exact solution. *Commun. Pure Appl. Math.*, **27**, 97–133.

Gel'fand, I.M., and Dickii, L.A. 1977. Resolvants and Hamiltonian systems. *Func. Anal. Appl.*, **11**, 93–104.

Goldstein, H. 1980. *Classical Mechanics*. Reading, MA: Addison Wesley.

Gordon, J.P., and Haus, H.A. 1986. Random walk of coherently amplified solitons in optical fiber transmission. *Opt. Lett.*, **11**, 665–667.

Hasegawa, A., and Kodama, Y. 1991a. Guiding-center soliton. *Phys. Rev. Lett.*, **66**, 161–164.

Hasegawa, A., and Kodama, Y. 1991b. Guiding-center soliton in fibers with periodically varying dispersion. *Opt. Lett.*, **16**, 1385–1387.

Hasegawa, A., and Kodama, Y. 1995. *Solitons in Optical Communications.* Oxford: Oxford University Press.

Hasegawa, A., and Matsumoto, M. 2002. *Optical Solitons in Fibers.* Berlin: Springer.

Hasegawa, A., and Tappert, F. 1973a. Transmission of stationary nonlinear optical pulses in dispersive dielectric fibres. I. Anomalous dispersion. *Appl. Phys. Lett.*, **23**, 142–144.

Hasegawa, A., and Tappert, F. 1973b. Transmission of stationary nonlinear optical pulses in dispersive dielectric fibres. II. Normal dispersion. *Appl. Phys. Lett.*, **23**, 171–172.

Haus, H.A. 1975. Theory of mode locking with a fast saturable absorber. *J. Appl. Phys.*, **46**, 3049–3058.

Haus, H.A. 2000. Mode-locking of lasers. *IEEE J. Sel. Topics Q. Elec.*, **6**, 1173–1185.

Haus, H.A., Fujimoto, J.G., and Ippen, E.P. 1992. Analytic theory of additive pulse and Kerr lens mode locking. *IEEE J. Quant. Elec.*, **28**, 2086–2096.

Haut, T.S., and Ablowitz, M.J. 2009. A reformulation and applications of interfacial fluids with a free surface. *J. Fluid Mech.*, **631**, 375–396.

Hopf, E. 1950. The partial differential equation $u_t + uu_x = \mu u_{xx}$. *Comm. Pure Appl. Math.*, **3**, 201–230.

Ilday, F.Ö., Buckley, J.R., Clark, W.G., and Wise, F.W. 2004b. Self-similar evolution of parabolic pulses in a laser. *Phys. Rev. Lett.*, **92**, 213901.

Ilday, F.Ö., Wise, F.W., and Kaertner, F.X. 2004a. Possibility of self-similar pulse evolution in a Ti:sapphire laser. *Opt. Express*, **12**, 2731–2738.

Infeld, E., and Rowlands, G. 2000. *Nonlinear Waves, Solitons and Chaos.* Cambridge: Cambridge University Press.

Ishimori, Y. 1981. On the modified Korteweg–deVries soliton and the loop soliton. *J. Phys. Soc. Jpn.*, **50**, 2471–2472.

Ishimori, Y. 1982. A relationship between the Ablowitz–Kaup–Newell–Segur and Wadati–Konno–Ichikawa schemes of the inverse scattering method. *J. Phys. Soc. Jpn.*, **51**, 3036–3041.

Jackson, J.D. 1998. *Classical Electrodynamics.* New York: John Wiley.

Jeffreys, H., and Jeffreys, B.S. 1956. *Methods of Mathematical Physics.* Cambridge: Cambridge University Press.

Kadomtsev, B.B., and Petviashvili, V.I. 1970. On the stability of solitary waves in weakly dispersive media. *Sov. Phys. Doklady*, **15**, 539–541.

Kalinikos, B.A., Kovshikov, N.G., and Patton, C.E. 1997. Decay free microwave envelope soliton pulse trains in yttrium iron garnet thin films. *Phys. Rev. Lett.*, **78**, 2827–2830.

Kalinikos, B.A., Scott, M.M., and Patton, C.E. 2000. Self generation of fundamental dark solitons in magnetic films. *Phys. Rev. Lett.*, **84**, 4697–4700.

Kapitula, T., Kutz, J.N., and Sandstede, B. 2002. Stability of pulses in the master mode-locking equation. *J. Opt. Soc. Am. B*, **19**, 740–746.

Karpman, V.I., and Solov'ev, V.V. 1981. A perturbation theory for soliton systems. *Physica D*, **3**, 142–164.

Kay, I., and Moses, H.E. 1956. Reflectionless transmission through dielectrics and scattering potentials. *J. Appl. Phys.*, **27**, 1503–1508.

Kevorkian, J., and Cole, J.D. 1981. *Perturbation Methods in Applied Mathematics.* Berlin: Springer.

Konno, K., and Jeffrey, A. 1983. Some remarkable properties of two loop soliton solutions. *J. Phys. Soc. Jpn.*, **52**, 1–3.

Konopelchenko, B.G. 1993. *Solitons in Multidimensions*. Singapore: World Scientific.

Korteweg, D., and de Vries, G. 1895. On the change of a form of long waves advancing in a rectangular canal and a new type of long stationary waves. *Phil. Mag.*, 5th Series, 422–443.

Kruskal, M.D. 1963. Asymptotology. In: *Proceedings of Conference on Mathematical Models on Physical Sciences*, edited by S. Drobot. Upper Saddle River, NJ: Prentice-Hall.

Kruskal, M.D. 1965. Asymptotology in numerical computation: Progress and plans on the Fermi–Pasta–Ulam problem. In: *IBM Scientific Computing Symposium on Large-Scale Problems in Physics*, pp. 43–62

Krylov, N.M., and Bogoliubov, N.N. 1949. *Introduction to Non-Linear Mechanics*. Annals of Mathematics Studies, vol. 11. Princeton: Princeton University Press.

Kutz, J.N. 2006. Mode-locked soliton lasers. *SIAM Rev.*, **48**, 629–678.

Kuzmak, G.E. 1959. Asymptotic solutions of nonlinear second order differential equations with variable coefficients. *PMM*, **23**(3), 515–526.

Lamb, H. 1945. *Hydrodynamics*. New York: Dover.

Landau, L.D., and Lifshitz, L.M. 1981. *Quantum Mechanics: Non-relativistic Theory*. London: Butterworth–Heinemann.

Landau, L.D., Lifshitz, E.M., and Pitaevskii, L.P. 1984. *Electrodynamics of Continuous Media*. London: Butterworth–Heinemann.

Lax, P.D. 1968. Integrals of nonlinear equations of evolution and solitary waves. *Commun. Pure Appl. Math.*, **21**, 467–490.

Lax, P.D. 1987. *Hyperbolic Systems of Conservation Laws and the Mathematical Theory of Shock Waves*. Philadelphia: SIAM.

Lax, P.D., and Levermore, C.D. 1983a. The small dispersion limit of the Korteweg–de Vries equation. III. *Commun. Pure Appl. Math.*, **36**, 809–829.

Lax, P.D., and Levermore, C.D. 1983b. The small dispersion limit of the Korteweg–de Vries equation. I. *Commun. Pure Appl. Math.*, **36**, 253–290.

Lax, P.D., and Levermore, C.D. 1983c. The small dispersion limit of the Korteweg–de Vries equation. II. *Commun. Pure Appl. Math.*, **36**, 571–593.

LeVeque, R.J. 2002. *Finite Volume Methods for Hyperbolic Problems*. Cambridge: Cambridge University Press.

Lighthill, M.J. 1958. *Introduction to Fourier Analysis and Generalized Functions*. Cambridge: Cambridge University Press.

Lighthill, M.J. 1978. *Waves in Fluids*. Cambridge: Cambridge University Press.

Lin, C., Kogelnik, H., and Cohen, L.G. 1980. Optical-pulse equalization of low-dispersion transmission in single-mode fibers in the $1.3–1.7\mu m$ spectral region. *Opt. Lett.*, **5**, 476–478.

Luke, J.C. 1966. A perturbation method for nonlinear dispersive wave problems. *Proc. R. Soc. Lond. A*, **292**, 403–412.

Lushnikov, P.M. 2001. dispersion-managed solitons in the strong dispersion map limit. *Opt. Lett.*, **26**, 1535–1537.

Mamyshev, P.V., and Mollenauer, L.F. 1996. Pseudo-phase-matched four-wave mixing in soliton wavelength-division multiplexing transmission. *Opt. Lett.*, **21**, 396–398.

Marchenko, V.A. 1986. *Sturm–Liouville Operators and Applications*. Basel: Birkhauser.

Maruta, A., Inoue, T., Nonaka, Y., and Yoshika, Y. 2002. Bi-solitons propagating in dispersion-managed transmission systems. *IEEE J. Sel. Top. Quant. Electron.*, **8**, 640–650.

Matveev, V.B., and Salle, M.A. 1991. *Darboux Transformations and Solitons*. Berlin: Springer.

Melin, A. 1985. Operator methods for inverse scattering on the real line. *Commun. Part. Diff. Eqns.*, **10**, 677–766.

Merle, F. 2001. Existence of blow-up solutions in the energy space for the critical generalized KdV equation. *J. Amer. Math. Soc.*, **14**, 555–578.

Mikhailov, A.V. 1979. Integrability of a two-dimensional generalization of the Toda chain. *Sov. Phys. JETP Lett.*, **30**, 414–418.

Mikhailov, A.V. 1981. The reduction problem and the inverse scattering method. *Physica D*, **3**, 73–117.

Miles, J.W. 1981. The Korteweg–de Vries equation: An historical essay. *J. Fluid Mech.*, **106** (focus issue), 131–147.

Miura, R.M. 1968. Korteweg–de Vries equation and generalizations. I. A remarkable explicit nonlinear transformation. *J. Math. Phys.*, **9**, 1202–1204.

Miura, R.M., Gardner, C.S., and Kruskal, M.D. 1968. Korteweg–de Vries equations and generalizations. II. Existence of conservation laws and constants of motion. *J. Math. Phys.*, **9**, 1204–1209.

Mollenauer, L.F., Evangelides, S.G., and Gordon, J.P. 1991. Wavelength division multiplexing with solitons in ultra-long distance transmission using lumped amplifiers. *J. Lightwave Technol.*, **9**, 362–367.

Molleneauer, L.F., and Gordon, J.P. 2006. *Solitons in Optical Fibers: Fundamentals and Applications to Telecommunications*. New York: Academic Press.

Mollenauer, L.F., Stolen, R.H., and Gordon, J.P. 1980. Experimental observation of picosecond pulse narrowing and solitons in optical fibers. *Phys. Rev. Lett.*, **45**, 1095–1098.

Newell, A.C. 1978. The general structure of integrable evolution equations. *Proc. Roy. Soc. Lond. A.*, **365**, 283–311.

Newell, A.C. 1985. *Solitons in Mathematics and Physics*. Philadelphia: SIAM.

Nijhof, J.H.B., Doran, N.J., Forysiak, W., and Knox, F.M. 1997. Stable soliton-like propagation in dispersion-managed systems with net anomalous, zero and normal dispersion. *Electron. Lett.*, **33**, 1726–1727.

Nijhof, J.H.B., Forysiak, W., and Doran, N.J. 2002. The averaging method for finding exactly periodic dispersion-managed solitons. *IEEE J. Sel. Topics Q. Elec.*, **6**, 330–336.

Novikov, S.P., Manakov, S.V., Pitaevskii, L.P., and Zakharov, V.E. 1984. *Theory of Solitons: The Inverse Scattering Method*. New York: Plenum.

Ostrovsky, L.A., and Potapov, A.S. 1986. *Modulated Waves: Theory and Applications*. Baltimore: The John Hopkins University Press.

Papanicolaou, G., Sulem, C., Sulem, P.L., and Wang, X.-P. 1994. The focusing singularity of the Davey-Stewartson equations for gravity-capillary surface waves. *Physica D*, **72**, 61–86.

Patton, C.E., Kabos, P., Xia, H., Kolodin, P.A., Zhang, H.Y., Staudinger, R., Kalinikos, B.A., and Kovshikov, N.G. 1999. Microwave magnetic envelope solitons in thin ferrite films. *J. Mag. Soc. Japan*, **23**, 605–610.

Pelinovsky, D.E., and Stepanyants, Y.A. 2004. Convergence of Petviashvili's iteration method for numerical approximation of stationary solutions of nonlinear wave equations. *SIAM J. Numer. Anal.*, **42**, 1110–1127.

Pelinovsky, D.E., and Sulem, C. 2000. Spectral decomposition for the Dirac system associated to the DSII equation. *Inverse Problems*, **16**, 59–74.

Petviashvili, V.I. 1976. Equation of an extraordinary soliton. *Sov. J. Plasma Phys.*, **2**, 257–258.

Phillips, O.M. 1977. *The Dynamics of the Upper Ocean*. Cambridge: Cambridge University Press.

Prinari, B., Ablowitz, M.J., and Biondini, G. 2006. Inverse scattering for the vector nonlinear Schrödinger equation with non-vanishing boundary conditions. *J. Math. Phys.*, **47**, 1–33.

Quraishi, Q., Cundiff, S.T., Ilan, B., and Ablowitz, M.J. 2005. Dynamics of nonlinear and dispersion-managed solitons. *Phys. Rev. Lett.*, **94**, 243904.

Rabinovich, M.I., and Trubetskov, D.I. 1989. *Oscillations and Waves in Linear and Nonlinear Systems*. Dordsecht: Kluwer Academic.

Rayleigh, Lord. 1876. On waves. *Phil. Mag.*, **1**, 257–279.

Remoissenet, M. 1999. *Waves Called Solitons*. Berlin: Springer.

Rogers, C., and Schief, W.K. 2002. *Bäcklund and Darboux Transformations*. Cambridge: Cambridge University Press.

Russell, J.S. 1844. Report on Waves. In: *Report of the 14th meeting of the British Association for the Advancement of Science*. London: John Murray, pp. 311–390.

Sanders, J.A., Verhulst, F., and Murdock, J. 2009a. *Averaging Methods in Nonlinear Dynamical Systems*. Berlin: Springer.

Sanders, M.Y., Birge, J., Benedick, A., Crespo, H.M., and Kärtner, F.X. 2009b. Dynamics of dispersion-managed octave-spanning titanium: sapphire lasers. *J. Opt. Soc. Am. B*, **26**, 743–749.

Satsuma, J., and Ablowitz, M.J. 1979. Two-dimensional lumps in nonlinear dispersive systems. *J. Math. Phys.*, **20**, 1496–1503.

Shimizu, T., and Wadati, M. 1980. A new integrable nonlinear evolution equation. *Prog. Theor. Phys.*, **63**, 808–820.

Stokes, G.G. 1847. On the theory of oscillatory waves. *Camb. Trans.*, **8**, 441–473.

Sulem, C., and Sulem, P. 1999. *The Nonlinear Schrödinger Equation: Self-Focusing and Wave Collapse*. Berlin: Springer.

Tang, D.Y., Man, W.S., Tam, H.Y., and Drummond, P.D. 2001. Observation of bound states of solitons in a passively mode-locked fiber laser. *Phys. Rev. A*, **64**, 033814.

Taniuti, T., and Wei, C.C. 1968. Reductive perturbation method in nonlinear wave propagation I. *J. Phys. Soc. Japan*, **24**, 941–946.

Taylor, G.I. 1950. The formation of a blast wave by a very intense explosion. I. Theoretical Discussion. *Proc. Roy. Soc. A*, **201**, 159–174.

Tsankov, M.A., Chen, M., and Patton, C.E. 1994. Forward volume wave microwave envelope solitons in yttrium iron garnet films-propagation, decay, and collision. *J. Appl. Phys.*, **76**, 4274–4289.

Venakides, S. 1985. The zero dispersion limit of the Korteweg–de Vries equation for initial potentials with non-trivial reflection coefficient. *Commun. Pure Appl. Math.*, **38**, 125–155.

Villarroel, J., and Ablowitz, M.J. 2002. The Cauchy problem for the Kadomtsev–Petviashili II equation with nondecaying data along a line. *Stud. Appl. Math*, **109**, 151–162.

Villarroel, J., and Ablowitz, M.J. 2003. On the discrete spectrum of systems in the plane and the Davey-Stewartson II equation. *SIAM J. Math. Anal.*, **34**, 1253–1278.

Vlasov, S., Petrishchev, V., and Talanov, V. 1970. Averaged description of wave beams in linear and nonlinear media. *Radiophys. Quantum Electronics*, **14**, 1062.

Wadati, M. 1974. The modified Korteweg–de Vries equation. *J. Phys. Soc. Jpn.*, **34**, 1289–1296.

Wadati, M., Konno, K., and Ichikawa, Y.-H. 1979a. A generalization of inverse scattering method. *J. Phys. Soc. Jpn.*, **46**, 1965–1966.

Wadati, M., Konno, K., and Ichikawa, Y.H. 1979b. New integrable nonlinear evolution equations. *J. Phys. Soc. Jpn.*, **47**, 1698–1700.

Wadati, M., and Sogo, K. 1983. Gauge transformations in soliton theory. *J. Phys. Soc. Jpn.*, **52**, 394–398.

Weinstein, M. 1983. Nonlinear Schrödinger equations and sharp interpolation estimates. *Comm. Math. Phys.*, **87**, 567–576.

Whitham, G.B. 1965. Non-linear dispersive waves. *Proc. R. Soc. Lond. A*, **6**, 238–261.

Whitham, G.B. 1974. *Linear and Nonlinear Waves*. New York: John Wiley.

Wu, M., Kalinikos, B.A., and Patton, C.E. 2004. Generation of dark and bright spin wave envelope soliton trains through self-modulational instability in magnetic films. *Phys. Rev. Lett.*, **93**, 157207.

Yang, T., and Kath, W.L. 1997. Analysis of enhanced-power solitons in dispersion-managed optical fibers. *Opt. Lett.*, **22**, 985–987.

Zabusky, N.J., and Kruskal, M.D. 1965. Interactions of "solitons" in a collisionless plasma and the recurrence of initial states. *Phys. Rev. Lett.*, **15**, 240–243.

Zakharov, V.E. 1968. Stability of periodic waves of finite amplitude on the surface of a deep fluid. *Sov. Phys. J. Appl. Mech. Tech. Phys.*, **4**, 190–194.

Zakharov, V.E., and Rubenchik, A.M. 1974. Instability of waveguides and solitons in nonlinear media. *Sov. Phys. JETP*, **38**, 494–500.

Zakharov, V.E., and Shabat, A. 1972. Exact theory of two-dimensional self-focusing and one-dimensional self-modulation of waves in a non-linear media. *Sov. Phys. JETP*, **34**, 62–69.

Zakharov, V.E., and Shabat, A.B. 1973. Interaction between solitons in a stable medium. *Sov. Phys. JETP*, **37**, 823–828.

Zharnitsky, V., Grenier, E., Turitsyn, S.K., Jones, C., and Hesthaven, J.S. 2000. Ground states of dispersion-managed nonlinear Schrödinger equation. *Phys. Rev. E*, **62**, 7358–7364.

Zhou, X. 1989. Direct and inverse scattering transforms with arbitrary spectral singularities. *Commun. Pure Appl. Math.*, **42**, 95–938.

Zvezdin, A.K., and Popkov, A.F. 1983. Contribution to the nonlinear theory of magnetostatic spin waves. *Sov. Phys. JETP.*, **51**, 350–354.

# Index

Printed in the United States
by Baker & Taylor Publisher Services